SOUND TRACKS

Sound Tracks is the first comprehensive book on the new geography of popular music, examining the complex links between places, music and cultural identities. It provides an interdisciplinary perspective on local, national and global scenes, from the 'Mersey' and 'Icelandic' sounds to 'world music', and explores the diverse meanings of music in a range of regional contexts.

Sound Tracks traces the ways in which music has informed complex globalisations, the role of companies and technology in diffusion, innovation and commercialism and the wider significance of cultural industries. It links migration and mobility to new musical practices, whether in 'developing' countries or metropolitan centres, and traces the recent rise of 'music tourism'. It examines issues of authenticity and credibility, and the quest for roots within different musical genres, from buskers to brass bands, and from rap to rai. *Sound Tracks* emphasises music's contributions to the contradictions, illusions and celebrations of contemporary life. It situates music and the music industry within spatial theories of globalisation and local change: fixity and fluidity entangled.

In a world of intensified globalisation, links between space, music and identity are increasingly tenuous, yet places give credibility to music, not least in the 'country', and music is commonly linked to place, through claims to tradition, 'authenticity' and originality, and as a marketing device. This book develops new perspectives on these relationships and how they are situated within cultural and geographical thought.

John Connell is Professor of Geography and Head of the School of Geosciences at the University of Sydney and **Chris Gibson** is Lecturer in Economic Geography at the University of New South Wales, Sydney.

CRITICAL GEOGRAPHIES

Edited by Tracey Skelton
Lecturer in Geography, Loughborough University
and Gill Valentine
Professor of Geography, The University of Sheffield

This series offers cutting-edge research organised into three themes of concepts, scale and transformations. It is aimed at upper-level undergraduates, research students and academics and will facilitate inter-disciplinary engagement between geography and other social sciences. It provides a forum for the innovative and vibrant debates which span the broad spectrum of this discipline.

1. MIND AND BODY SPACES
Geographies of illness, impairment and disability
Edited by Ruth Butler and Hester Parr

2. EMBODIED GEOGRAPHIES
Spaces, bodies and rites of passage
Edited by Elizabeth Kenworthy Teather

3. LEISURE/TOURISM GEOGRAPHIES
Practices and geographical knowledge
Edited by David Crouch

4. CLUBBING
Dancing, ecstasy, vitality
Ben Malbon

5. ENTANGLEMENTS OF POWER
Geographies of domination/resistance
Edited by Joanne Sharp, Paul Routledge, Chris Philo and Ronan Paddison

6. DE-CENTRING SEXUALITIES
Politics and representations beyond the metropolis
Edited by Richard Phillips, Diane Watt and David Shuttleton

7. GEOPOLITICAL TRADITIONS
A century of geo-political thought
Edited by Klaus Dodds and David Atkinson

8. CHILDREN'S GEOGRAPHIES
Playing, living, learning
Edited by Sarah L. Holloway and Gill Valentine

9. THINKING SPACE
Edited by Mike Crang and Nigel Thrift

10. ANIMAL SPACES, BEASTLY PLACES
New geographies of human–animal relations
Edited by Chris Philo and Chris Wilbert

11. BODIES
Exploring fluid boundaries
Robyn Longhurst

12. CLOSET SPACE
Geographies of metaphor from the body to the globe
Michael P. Brown

13. TIMESPACE
Geographies of temporality
Edited by Jon May and Nigel Thrift

14. GEOGRAPHIES OF YOUNG PEOPLE
The morally contested spaces of identity
Stuart C. Aitken

15. CELTIC GEOGRAPHIES
Old culture, new times
Edited by David C. Harvey, Rhys Jones, Neil McInroy and Christine Milligan

16. CULTURE/PLACE/HEALTH
Wilbert M. Gesler and Robin A. Kearns

17. SOUND TRACKS
Popular music, identity and place
John Connell and Chris Gibson

SOUND TRACKS

Popular music, identity and place

John Connell and Chris Gibson

LONDON AND NEW YORK

First published 2003
by Routledge
11 New Fetter Lane, London EC4P 4EE

Simultaneously published in the USA and Canada
by Routledge
29 West 35th Street, New York, NY 10001

Reprinted 2003

Routledge is an imprint of the Taylor & Francis Group

Typeset in Perpetua by Taylor & Francis Books Ltd
Printed and bound in Great Britain by St Edmundsbury Press Ltd,
Bury St Edmunds, Suffolk

British Library Cataloguing in Publication Data
A catalogue record for this book is available from the British Library

Library of Congress Cataloging in Publication Data
A catalog record for this book has been requested

ISBN 0-415-17027-3 (hbk)
ISBN 0-415-17028-1 (pbk)

This book is dedicated to some of those who died during its long gestation period and in varying ways helped to make it what it is:

Rob Buck (of 10,000 Maniacs)
Steve Connolly (of the Messengers)
John Denver
John Lee Hooker
David McComb (of the Triffids)
Curtis Mayfield
Rob Pilatus (of Milli Vanilli)
Dusty Springfield OBE
Tammy Wynette

And, to our fathers,

Leo Connell
who would never have believed that popular music ('you call that music?') and scholarship might ever be combined, and

Ian Gibson
for knowing that music has no boundaries of age or class, and believing that music was a sound track to life.

CONTENTS

ILLUSTRATIONS

Tables

Figures

PREFACE

In some respects this book emerged from an article in an Australian newspaper, the *Sun Herald*, which covered the arrival of a Canadian band Junkhouse (who were never to be heard of again in Australia) and stated:

> In the Canadian steel town of Hamilton there are limited choices for teenage boys in their final years at high school – unemployment, a life in the steel mills or a career as a rock 'n' roller, bluesman or jazz performer. . . . They knew that if they were going to work in Hamilton the music had to be rootsy. 'That sort of music has always been what the town is about' . . . [the lead singer] said. 'A blue collar town, and most of the people from Hamilton who decide to play wind up with a pretty direct human type of music. When you're learning how to play music . . . you have to be able to relate to your audience, to communicate. We learn to communicate in the simplest terms in my home town, because it is an industrial town.'
>
> (28 August 1994)

In the popular music world similar statements enshrining environmental determinism are legion. Indeed, a week or so later, the same newspaper, in an article about a rock group from the small Victorian country town of Ballarat, stated simply 'rock bands don't come from Ballarat'; out there presumably was country and western land. Here were two fundamental themes: that music is somehow linked to place – the idea of a 'Hamilton sound' – and that music is also about mobility, diffusing sounds to the world while enabling the social mobility of musicians (in the same way that jazz, boxing and basketball were supposed to provide a route out of the ghetto for African Americans). In that single short paragraph there were basic geographical suppositions about place, identity and movement. By contrast, Steve Kilbey, lead singer of Australian group the Church, once claimed that 'music is magic. It's got nothing to do with geography. It's got nothing to do with industry or standard; it's magic' (quoted in Howlett 1990:

33). In this book we seek to show that he is both wrong about geography and right about the magic.

When David Harvey concluded the Preface to his distinguished *Explanation in Geography* (1969) with an acknowledgement to, among others, 'Miles Davis, John Coltrane, Dionne Warwick, the Beatles and Shostakovich' it suggested a whiff of scandal: low culture intruding into the high culture of intellectual enterprise. Surely he could not be serious? Intellectuals were expected to subscribe to classical music. Many such cultural prejudices were nothing more than undisguised elitism, about the spirit and context of performance, yet music is not without merit or meaning because it is popular, as will become evident here.

Perhaps significantly, as the first draft of the book took shape, Sydney was the Olympic City and at every venue popular music accompanied athletic endeavour. No sport was too ascetic for rock music, and escape was impossible as the music was sponsored by the International Olympic Committee. Taxi Ride sang 'Get Set' and Bruce Springsteen did 'Born to Run' at the athletics, Midnight Oil's 'King of the Mountain' accompanied the mountain biking, and Vanessa Amorosi's 'Heroes Live Forever' and Yothu Yindi's 'Calling Every Nation' were everywhere. National anthems capped all these efforts. In New Zealand National Party leader and former Prime Minister, Jenny Shipley, was still recovering from a decision by organisers of a National Party conference to play the Troggs' 'Wild Thing' as her entrance song. On the other side of the world, American presidential candidates were using popular music on the campaign trail: George W. Bush with Hispanic music to woo the 'migrant' vote and Al Gore preferring Sting's 'Brand New Day'. A year later, as the book went to press, over a hundred popular songs had been banned from the air in New York, with lyrics too close to the reality of terrorist destruction. There was little doubt that popular music had to be taken seriously.

This book then seeks to redress the rather neglected place of geography in any analysis of popular music, by tracing the links between music, place and identity at different scales, from inner-city 'scenes' to the music of nations. It examines the influence of culture, economics, politics and technology on the changing structure and geographies of music at local and global levels. Initially we trace the role of music from an expression of local culture in indigenous societies, through dispersals and expansions towards a still incomplete globalised industry, where 'local' sounds remain vibrant. Migration and ethnic diversity have contributed to hybrid, diasporic sounds, at the same time as new technologies of production and distribution have moved 'local' sounds, whether rock, 'world music' or trance techno, to global audiences. Innovative recording and distribution systems and legal challenges have unsettled old certainties. The local has not however disappeared and has even become formalised in the depiction of particular scenes, as in Seattle, Detroit and Liverpool. Such local spaces are very different across genres, from the invented tradition of some folk music to rapid cycles of fashion in club cultures. Music tourism has emerged from affluence and nostalgia, to emphasise contemporary and continuous reconstructions of space through music. We have sought to provide a distinctive and critical spatial perspective, which embraces globalisation, local acquiescence and resistance, drawing on diverse musical texts and practices to develop a geography of popular music.

We must declare our interest and biases, and thank some popular muses, for one author the Incredible String Band, Leonardo's Bride and 10,000 Maniacs, and for the other the Church, New Order and, of course, Barry White. For both of us, Spinal Tap added a necessary extra dimension. We would also like to thank Chloë Flutter for her assistance with Chapter 9, Wendy Shaw for advice, Krste Trajanovski for technical assistance and Linly Goh and Jessica Carroll who led the way. In Berlin we must thank Jutta Albert; in Montreal, Will Straw; in the UK, the Price household, Frank and Kath Robinson, Gonnie Rietveld, Kurt Iveson and Arun Saldanha; in Sydney Janet Witmer, Ali Wright, Kate Lloyd, Anthony Hutchings and Andrew McGregor kept us sane. We must also thank Ann Michael, the last in a line of Routledge editors who somehow kept faith with us, and Emma Hardman for her proofreading. Above all, we would like to thank Robert Aldrich, who carefully read every chapter and reminded us that popular music was a global phenomenon, and David Bell, David Keeling and Tracey Skelton who provided detailed comments on the first draft.

John Connell, Chris Gibson
September 2001

ACKNOWLEDGEMENTS

The author and publisher would like to thank the following for granting permission to reproduce material in this work:

Margaret Antaki, for figure 10.4a, an advertisement of the '2001 African Drum and Dance Tour Senegal'.

Berkeley Breathed, for figure 12.1, a Bloom County cartoon.

Karen Buchan, for figure 10.4b, an advertisement of the '2002 Percussion and Dance Cultural Study Tour to Guinea'.

Michael Colton, co-editor, Modern Humorist, for figure 11.1, a cartoon of 'Pro-MP3 Propaganda'. (www.modernhumorist.com)

The McGraw-Hill Companies for figure 2.1, after figure 7.19: 'Folk song regions of the US', from page 243 of *Human Geography*, by Fellman *et al.*

The Royal Geographical Society, London, for figure 3.1, 'Mt Hagan Natives Listening to a Gramophone'.

STA Travel, Advertising and Promotions, Australia, for figure 10.3, the full image of the advertisement 'Rave the World'.

Tandoori Space, Leeds, for figure 8.1b, their 'Sub Dub presents Iration Steppas etc.' flyer.

Think Tank, Leeds, for figure 8.1a, their 'East Village' flyer.

The University of Chicago Press, for figure 7.2, after figure 2: 'Musical Influences on Zouk', from page 49 of *Zouk*, by Guilbault.

Viking Sevenseas NZ Ltd, for figure 7.1b, the full cover image of the EP 'Polynesian Playmates', VE 292.

Aejaz Zahid for figure 1.1, his 'Disoriental' flyer.

Every effort has been made to contact copyright holders for their permission to reprint material in this book. The authors and publishers would be grateful to hear from any copyright holder who is not here acknowledged and will undertake to rectify any errors or omissions in future editions of this book.

1

INTO THE MUSIC

This book explores the many ways in which popular music is spatial – linked to particular geographical sites, bound up in our everyday perceptions of place, and a part of movements of people, products and cultures across space. It seeks to develop an innovative perspective on the relationship between music and mobility, the way in which music is linked to cultural, ethnic and geographical elements of identity, and how all this, in turn, is bound up with new, increasingly global, technological, cultural and economic shifts.

The cover image for this book suggests one starting point for exploring these themes: a South American panpipe band in Times Square, New York, in 2001, providing a seemingly authentic Andean musical experience in a different hemisphere. The band, playing panpipe ballads over pre-recorded keyboards, was accompanied by a colleague selling home-made CDs of the group. In one sense, it is an unsurprising image: South American panpipe busking groups became common in cities around the world in the 1980s, especially after the international success of Inti Illimani and the rise of New Age music. Indigenous knowledge of musical traditions provided quick resources for migrants keen to earn an income, as with, in other contexts, Cantonese violinists, Caribbean steel drummers and flamenco guitarists. Yet the image reflects much more than just an incidental part of a city streetscape. It is a busy scene in a unique city, created and constantly transformed by migration. It suggests links between music, tradition and authenticity, reinvented in the public spaces of the city; it demonstrates how technological changes (notably the digitisation of music) have informed local music production, generated new home recording cultures and small-scale entrepreneurialism. Music is caught up in multiple layers of networks that come together in that one street scene. In contrast to the low-key musical economies of the foreground, the background confirms the corporate domination of music: the Virgin megastore, and the global headquarters of BMG, one of the world's largest entertainment companies (in Times Square, perhaps the archetypal global entertainment space). Passing cars boom with a wealth of sound – R&B, techno, rock classics and current hits; MTV's American television studios, behind the photographer, advertise exclusive interviews with Janet Jackson; souvenir stalls replay stage and show classics out into the street. Music fills the scene and affronts – or soothes – the senses.

While Times Square may be atypical, music surrounds us, in shopping malls, cinemas, lounge rooms; as a soundtrack to fashion, TV channels and video games. Yet, despite the presence of music in most people's lives, this area of popular culture has been largely neglected as a 'serious' academic pursuit. This is perhaps due to music's relative invisibility, and the apparent lack of tangible ethnographic material to be analysed and explained in ways that other material aspects of culture have been studied. As Smith has argued, it is as if a claim for 'the non-social, implicitly metaphysical qualities of music has almost succeeded, making music perhaps the last of the arts to be looked at from a critical cultural perspective' (1994: 235). Here, we trace the various links between music, place and spatial identity, and introduce a plethora of examples – from artists and their output to global distribution, from local 'scenes' to national music traditions – which map out diverse geographies of music. From its origins as sound experienced only in 'live' circumstances to sound waves captured in a computer chip – music in its varying forms has become almost inescapable. Similarly, popular music transcends geographical scales, from live performances in corner pubs to global tours; equally, we can shut ourselves off from the outside world in the private space of a Walkman, while governments create anthems and cultural policies aimed at representing a sense of nation. This book brings an explicitly geographical approach to popular music – thinking of music in terms of place and movement, of proud heritages and dynamic, fluid soundscapes.

In many respects, the academic world has shifted in ways that make it possible to take music seriously. The move away from sometimes rigid theories of society and economy, towards studies of social diversity and heterogeneity, has illuminated the complexities of how members of communities create and sustain meanings and identities for themselves. Consequently geographers have given greater attention to the ways in which our understandings of space and place are mediated by popular cultural forms such as television, print media, film and music, hence some geographers have called for more intense examinations of musical texts in studies of society, polity and culture (Gill 1993; Smith 1994; Leyshon *et al.* 1995, 1998; Kong 1995a; Nash and Carney 1996; Carney 1998; Romagnan 2000). This has marked a conceptual shift within cultural geography, from its historical concern with producing 'objective' studies of cultural landscapes, to interpretations (or 'readings') of human-made spaces as a form of 'text' – discursively constructed arenas that are shaped by wider social relations and representative of divisions and tensions in society. As Leyshon *et al.* have argued, 'space and place are...not simply...sites where or about which music happens to be made, or over which music has diffused, but rather different spatialities are...formative of the sounding and resounding of music' (1995: 424–5). The centrality of music to youth sub-cultures (particularly since the 1950s), the links between music and social movements in the 1960s (such as soul music and the American civil rights movement), the role of music in mediating stories of place (from urban decay in punk to the rural utopias of country music), the widespread array of venues and sites in which music is now encountered (from concert halls to airport lounges) and the more globally integrated nature of music distribution are just some examples that have amplified the necessity for critical analysis.

Popular music has appeared in some university settings, gradually filtering through the curricula of music, sociology, media and communications departments, yet has been accompanied by considerable scepticism. In some circles, this scepticism stems from notions of popular culture as fanciful or irrelevant at 'serious' universities; at times popular music is subject to more extreme attacks, written off as a legitimate area of study by those with conservative views of music, who see it as inconsequential. While literature, film and art have been graced by an abundance of work from a cultural geographical perspective, popular culture, and particularly popular music, has remained enigmatic territory. Kong has traced the lack of popular music studies in geography to a tradition of cultural elitism – researchers privileging those 'serious' and enduring cultural artefacts over popular cultural forms, which have 'been regarded with disdain as "mere entertainment", trivial and ephemeral' (1995a: 184). This is part of the wider priority attached to vision (Smith 1994; Ingham *et al.* 1994), reflected in both the empirical underpinnings of 'science', and in post-modernism's origins in architecture and art. The omission of music from mainstream geographical inquiry can also be attributed both to the belief that it is not 'geographical' and to the complexity of its expression, engaging sight and sound simultaneously. As long as the written word remains the dominant academic medium, visually experienced cultural forms are likely to remain the most widely studied texts (Smith 1994), ensuring music's relative neglect.

What is 'popular' music? Is it simply that which sells the most? Is 'classical' music distinct from 'popular' music (see Box 1.1)? Music also implies much more than just texts (whether lyrics or musical scores). Musical practices include whole constellations of social uses and meanings, with complex rituals and rules, hierarchies and systems of credibility that can be interpreted at many levels. Music can represent a highly participatory art form or a passive consumption experience, from karaoke or busking on streetcorners to hearing easy-listening 'muzak' in shopping aisles, or dancing in a club – hence geographies of music are inextricable from the various contexts of performance, listening and interaction in space. As a sometimes-living exhibition and art form, as fluid, invisible sound, popular music refuses to provide a uniform or static text to manipulate or deconstruct.

Box 1.1 What is 'popular music'?

Any attempt to distinguish popular music reveals basic disagreements: criteria to differentiate 'classical', 'folk' and 'popular' music are artificial and at best localised. All music that is heard and enjoyed can be interpreted as 'popular' in some sense. Whether talking about 'traditional' music styles that remain important in the social practices of indigenous communities or migrant groups, the mass-produced output of major record labels, or the categorisation of music in record shops, music involves the broadcast of sound by individual performers or groups beyond the performance context (stage, radio station, recording studio), to audiences in a variety of places that understand and recognise the noises as 'music'. This marks a spatial trajectory away from highly localised and contained origins, to absorption into the musical styles and consumption patterns of a wider community (to varying degrees). Yet boundaries of meaning are consistently erected between what is deemed 'popular' or otherwise. Adorno regarded popular music as a mass-produced, commodified and standardised product, involving minimal creativity. Consequently 'serious' music, 'art' music and 'experimental' music were portrayed as structurally distinct from 'popular' music (see Adorno and Horkheimer 1977), yet, considered in their own social and historical contexts (however narrow they were), these too were popular. In many societies, these divisions had no meaning. In Italy and elsewhere, opera was genuinely 'popular' across social classes, yet in many contemporary contexts it has been associated with refinement, and an educated 'cultured' elite. The notion that some musics are of 'objectively' higher status or quality remains common. (As an example, in 1996 court action was brought against Italian group FCB concerning their dance remix of Carl Orff's *O Fortuna* (entitled 'Excalibur'), in which the companies who owned the rights over the original piece claimed the techno remix 'debased' the original. FCB won the case.) Similar perspectives exist over what constitutes folk music or world music. Others have consulted more quantitative, seemingly democratic techniques to define genres – 'popular' music is simply that of the masses, that which sells the most copies, or draws the largest crowds. Manuel (1988) argued that popular music could be distinguished from other types of music, since it was largely disseminated by the mass media, and this substantially influenced its form. Then there is presumably some sales number (or level of media exposure) below which music of all sorts is 'unpopular' or merely the domain of cult enthusiasts. Yet highly influential releases by artists are sometimes distributed within very narrow parameters (such as dance tracks, popular in clubs but the sales of which are limited to a smaller number of DJs). Aesthetics and economics are not easily unravelled. Grossberg provides some clarity to this ambiguous notion of the 'popular' by warning against quantitative or aesthetic judgements about particular artists, recordings or 'scenes' (such as arbitrary record sales criteria or distinctions based on personal taste):

> [the 'popular'] cannot be defined by appealing to either an objective aesthetic standard (as if it were inherently different from art) nor an objective social standard (as if it were inherently determined by who makes it or for whom it is made). Rather it has to be seen as a sphere in which people struggle over reality and their place in it, a sphere in which people are continuously working with and within already existing relations of power, to make sense of and improve their lives.
>
> (1997: 2)

This approach is central to this book, and further suggests that notions of authenticity, and credibility, however these are defined, are key elements of much popular culture. Given the fluidity of meaning that surrounds the term 'popular music', we have sought throughout this book to avoid narrow definitions and boundaries (although it would be impossible not to implicitly reflect our own tastes and perspectives). We have referred to a number of examples and case studies of popular music in very different spatial contexts, and to musicians of varying commercial 'popularity'. Much can be made of these comparisons, yet it remains impossible to cover all contexts, regions and interpretations of musical spaces, across the multitude of genres, performers and time-periods in which musicians have created sound. Moreover, the 'popular' not only involves cultural products (CDs, music videos, concert performances) that are numerically or financially successful in different countries, but constitutes the whole realm within which tastes come and go, the social contexts in which 'fans' emerge with distinct cultural attachments to a sound or artist, and the human spaces that are created for the enjoyment of music. There can be no formal definition of popular music.

See: Adorno 1988; Brackett 1995; Kassabian 1999; Middleton 1990; Shuker 1994.

Ironically, the allure of popular music as a site of research inquiry is intensified because it is so tangled up in the activities of everyday life. Ward has even argued, 'music-making is, more than anything else you can think of quickly, the cement of society' (1992: 120). While this might be exaggerating the social role of music somewhat, many everyday understandings about places (whether particular sites such as concert or festival venues, regions with music traditions, or national institutions) are mediated through engagements with popular music. Everyday associations with places may come to be defined by musical expressions, on a number of levels. Just as Hollywood has become a mythological site through its proximity to the global film industry, so too Nashville, Seattle or Memphis in the United States, Liverpool and Manchester in England or Tamworth in Australia have come to be known as key sites of musical production, dissemination or festivals for particular audiences. Myths of place are often reinforced in music itself; examples cut across genres and eras: the many music texts dedicated to the cities of New York or Los Angeles (from Frank Sinatra and Billy Joel to Public Enemy and 2Pac); the numerous country and western artists, such as Willie Nelson, who are nostalgic about home and the land; or the often heavily geographical discourses of hip hop or reggae. Analyses of popular music therefore demand diversity, considering the cultural forms, ideologies, identities and practices in place that provide the individual a 'plausible social context and believable personal world' (Eyles 1989: 103), both in material trends and matters of popular discourse.

Music as culture and commodity

In order to understand music's spatial dimension, we explore a series of dialectical relationships that define how music operates in places and across geographical distances. One such dialectic stems from a tension between music as a commodified product of an industry with high levels of corporate interest, and simultaneously as an arena of cultural meaning. The initial stages of the production process for music (whether a work ends up as a recording or a performance) involve small-scale creativity – bands and songwriters creating music in garages, recording studios or local pubs, illuminating 'the ways in which music is used and the important role that it plays in everyday life and in society generally' (Cohen 1994: 127). Beyond its importance as a cultural pursuit, music is captured, transformed and broadcast in a range of ways, involving complicated trajectories of production, distribution and consumption. These reveal tensions about how sounds circulate as both economic and cultural value. Sheet music captured melodies, words and arrangements, allowed rapid dissemination of songs and also established publishing companies as pivotal players in the emerging music industry. Later, recording technologies were established by music companies as attempts to capture clearer sounds and sell more copies of albums, yet in many contexts they also allowed more complex and more numerous grassroots musical cultures, and new informal networks of small-scale production (Chapter 3). However, academic study of music has largely evaded complex connections between cultural and commercial trends, assuming either that music, as an immediately cultural expression, 'belongs' in cultural studies or cultural geography, or that questions of culture and identity are frivolous diversions, compared to the 'real' tasks of examining music's function as a nucleus of economic growth or a possible means of job creation (Sadler 1997: 1919). There are many potential reasons for this. Aspects of popular culture such as film, music and television lend themselves to cultural analyses; much of the rich mythology surrounding these activities is related to their distinctive consumption and interpretation by audiences in cultural milieux. These include sub-cultural settings, where heterogeneous visual markers confront researchers most quickly (for example the safety pin and mohawk fashions of punk, or the pan-African imagery of reggae); geographical settings (the festival, the cinema, the theme park); depictions of cultural encounters in television serials and films (such as *Hi-Fidelity* or *Brassed Off*); and in academic arenas themselves, where certain styles, movements, artists or cultural products are afforded authenticity in intellectual circles. Academic celebrations (and alternatively, deconstructions) of musical texts can return to old debates, variously affirming key expressions, such as 'art', against the tainted stuff of 'commerce'. Notions of academic credibility, attached to selected musical styles, artists or pieces, without self-reflexive critique, stimulate acts of snobbery associated with everyday music consumption (the more 'credible' inner-city clubs, more 'genuinely alternative' bands, a more 'refined' classical music). Academics actively add critical currency to cultural products, as do music critics, retailers, recruiting agents and sub-cultural elites. As Breen has put it, 'we have been too eager to be culturalists – promoters of our musical obsessions – rather than analysts and critics' (1995: 490), writing and reflecting on music's affective qualities as consumers with particular tastes, ideological predilections and preferred readings of musical texts. Such distinctions cannot be sustained: all music, of 'high' or 'low' culture, is commercial in some way, while all cultural materials, even those of mass consumption, provide openings and alternatives for audiences.

Recognition of the commercial dimensions of music has informed a limited number of studies examining music as a cultural economy (Caves 2001; Brown, *et al.* 2000). Music has been bound up in local development strategies, alongside other 'new media', entertainment and 'content' industries (such as film, multimedia, publishing). Music is an industry (or more accurately, a series of economic clusters and networks), like any other, geared towards commodity production, and can be situated within frameworks more familiar to economics. Various studies have examined the patterns of employment in music, locational factors underpinning the music industry, agglomerations of cultural industries (including music) in particular cities and the regulatory and institutional settings within which music development is promoted. Yet such approaches sometimes lack the resonance, emotive importance, politics and cultural meaning bound up in music, although business concerns heavily influence who hears what music. A focus on particular firms, or clusters of businesses, has tended to disassociate the 'economic' from the 'social'; 'it is almost as if the form stood outside of society; interactions with a whole range of public and private bodies and individuals are lost' (Pratt 1994: 1–2). Moreover, music often allows for capitalist and non-capitalist economic activities to take place, linked as it is to big business, but also to an informal sector: a 'black economy' with its own unpredictable multipliers, and networks of economic exchange. Examining an 'economy of culture' also requires recognition of the need for ethnographies of the cultural economy; music is channelled through gatekeepers, individuals in a range of settings who manage and promote certain flows of music. Such gatekeepers may exist in sub-cultures, in record and publishing companies, at radio stations and record stores, while state policy makers have attempted to more forcibly influence international flows, to varying degrees of success. Particular sets of gatekeepers have become a 'critical infrastructure' governing the cultural economy (Zukin 1991; see also Mitchell 2000: 83), involving workers in cultural industries (such as music, book and film critics, designers, television producers) who flesh out ideas of 'culture', and create, consume and reroute music through particular channels.

Popular music remains an industry permeated by gendered norms and expectations at all levels; some of the most unequal labour relations can be found there. (In some countries male employment in the music industry outnumbers that of women by 5 to 1.) In part due to the persistence of male domination of gate-keeping positions, gender assumptions, biases and exploitation permeate other aspects of the production and social consumption of music (a point returned to at various points in this book). Some authors (e.g. Pratt 2000; Leyshon 2001) have acknowledged the socially constructed linkages between individuals, institutions and agencies within cultural economies – networks of influence, trade and knowledge transfer that are always being generated, negotiated, renewed (or severed). Such approaches attempt to rework the 'cultural' into economic analyses, emphasising the embeddedness of economic activities in social relations, production spaces and consumption districts, and in discrete cultural milieux. The economics of music cannot be divorced from the networks of people who make and promote it – music is an inherently risky and often vulnerable industry, hence the importance of cultural knowledge and contacts for generating a 'buzz' surrounding an artist or release. Thus, Straw located the Canadian music industry 'within and between a wide range of institutional and social spaces', distinguishing it from other cultural industries such as film and television that had narrower professional boundaries within which most production took place. Yet music activities 'unfold within artistic communities which resist definition, constituted as they are in the overlap between the education system, sites of entrepreneurial activity (such as bars or recording studios) and the more elusive spaces of urban bohemia' (Straw 1993: 52). Musical cultures are commodified, but music never leaves the sphere of the 'cultural' or 'social' even when it is being manufactured, bought or sold.

Complex interactions between economy and culture exist across all geographical scales. Local musical cultures are bound up in questions of economy (how much musicians are getting paid for gigs, the companies involved in producing and selling musical instruments, the commercialisation of local sounds, the changing economy of music retailing); meanwhile economic aspects of musical activities are always socially and culturally embedded, relying on aesthetic judgements, and particular networks of actors that are not always obviously economic. Music conjures up representations of place, identity and culture, but the 'economic' is evident in how companies create images of products, brand names and concepts, and in reactions to the economics of music in musical texts themselves (hence the Sex Pistols' aversion to EMI). The interconnection of the cultural and the economic in music will become apparent throughout the book: early chapters explore the role of technology and capital investment in the growth of globally integrated markets for popular music. Subsequently connections are mapped between, on the one hand, musical infrastructures and production networks in certain places, and, on the other, the mythologies of musical heritage that help shape popular impressions of those same places. Music endows social status, it moves with migration, attracts tourists, fosters a sense of belonging, generates jobs, all of which are reflected in perceptions of the cultural and economic role of music.

Music as fixity and fluidity

If music is simultaneously a commodity and cultural expression, it is also quite uniquely both the most fluid of cultural forms (quite literally, as sound waves moving through air) and a vibrant expression of cultures and traditions, at times held onto vehemently in the face of change. This tension, between music as itinerant and fleeting, and music as something static, fixed and immobile, underpins much of the discussion in this book. Both 'fixity' and 'fluidity' operate as umbrella terms that reflect a range of spatial practices, tendencies, decisions and physical objects (see Table 1.1). 'Fluidity' or 'spatial mobility' indicates flows of music, people, capital, commodities and money across space. This emerges in a number of ways in music. Music is, at its most basic, sound transmitted from the microlevel (in a bedroom, pub, car, between headphones) to the macroscale (through various means, including the global media). Music is also an artefact moving with people, whether as indigenous knowledge, oral traditions or recordings. Mobility also maps out musical economies, in the desire of entertainment companies to capture dispersed markets and seek new sources of

Table 1.1 Interpretations of 'fluidity' and 'fixity'

	'Fluidity'	*'Fixity'*
Material processes	Distribution of products	Agglomeration tendencies at an industry level
	Migration	Insularity
	Capital flows	Fixed production infrastructure
	Mass markets	Domestic markets/tradition
	The act of tourism	Sites of tourism
	Broadcasts and transmission	Consumption of products
	Technological diffusion	Sources of musical product
Discursive processes	Cultural flows and stylistic influences	Territorial assertions
	Discourses of styles and symbols	Tradition/heritage/authenticity
	Trans-continental and cross-cultural alliances	Cultural 'resistance'
	Hybridity	Appeal to 'roots'

music. Music in all eras is characterised by particular sets of networks, technologies and institutions that map out cultural connections at different geographical scales.

Music has been transformed through spatial mobility. Simultaneously, music has influenced the manner in which wider global economic and cultural change has occurred. Music has been a feature of campaigns to both open up global trade and protect the intellectual property of entertainment corporations, and a part of debates about global cultural change and the erosion of local differences. Corporate interests in modernisation and standardisation have always been aligned with spatial mobility – a keenness to overcome spatial distance and differences – and rhetorical constructions of a 'borderless world' (Ohmae 1990). As the operations of entertainment companies increased and became international, distribution networks expanded in reach: musical expressions from diverse geographical locations could be purchased, broadcast and heard in many locations. Music may well reflect the increasingly global reach of popular cultural technologies, but it does so only because of particular investment decisions in the entertainment industries, within parameters defined by the state, and because sub-cultures and audiences have accepted outside sounds. Moreover, some have suggested that popular music around the world has been caught up in a process of convergence where, economically, different forms of media such as film, Internet, print and sound are owned by the same few corporations and, culturally, texts are received via the same channels, leading to a 'standardised' repertoire of sounds, styles and images of place, as more locations are incorporated into a global popular cultural 'matrix'. Similarly, some argue that a process of cultural imperialism has utterly changed global geographies of music: localised music traditions are slowly being erased by sounds determined and distributed by global corporate entities. Yet, not surprisingly, musical influences have become increasingly hybrid, with numerous indigenous, local music traditions being fused with those from production centres such as in the United Kingdom and the United States, both questioning and transforming cultural identities: mobility and new forms of internal diversity are not inconsistent.

Music in place

Concurrent with the emergence of global 'mediascapes' (Appadurai 1990), processes of musical fragmentation and diversification have occurred within countries. As media corporations distribute popular sounds beyond their origins, niches and more subtle markers of musical difference have sprung up in unexpected places, from amateur local scenes, community musics and sub-cultures to new transcontinental sounds. Reactions to globalisation have differed – musicians have sought out new sounds or returned to 'roots'; sub-cultures have emerged that, even if momentarily, evade the products and commercial logic of media corporations while audiences – fans, critics, occasional listeners – receive and interpret music in the diverse contexts of their own lives. Accordingly, many geographies of music have tended to locate analyses in more detailed local circumstances, generating place-bound theories and regional ethnographies of music scenes, audience cultures and experiences of place, ranging from work on buskers and amateur scenes (Tanenbaum 1995; Cohen 1991a, 1991b, 1994, 1995; Finnegan 1989), sub-cultures and 'indie' scenes in various world centres (Maxwell 1997a; Mitchell 1996; Goh 1996) and 'traditional' music in remote parts of Brazil or Papua New Guinea (Seeger 1987; Feld 1982). At various scales, and in a vast range of locations, music has been linked to place – whether in the subways of New York, the rainforests of New Guinea or the clubs of Manchester – as cultural geographers and others traced links between music styles, sub-cultures and place.

As early as the 1920s, music, and sound more generally, were considered part of an early 'landscape tradition' within regional geography. Regional distinctions could be drawn not only through those characteristics experienced visually (in terms of physical landscapes and the built environment), but also through encountering sounds, noise and tunes. Sounds were a central part of landscapes, and thus were crucial to the maintenance of a sense of 'balance' within those landscapes, suggesting an aesthetic 'order' that linked certain sounds with particular settings, constructing 'harmonies of scenery'. Geographers such as Cornish (1928, 1934) and Abercrombie (1933) attempted to align sounds with scenes of appropriateness and inappropriateness; the honk of a car horn in a country lane, or the sound of a gramophone in the open air, were considered aesthetically dissonant and unwelcome: 'If this quietude of the senses be broken...the mind begins to pay attention to imperfections instead of dwelling upon an ideal, and we no longer live in Arcadia' (Cornish 1928: 277).

More substantive approaches underpinned later efforts; cartographies of production and diffusion of musical forms such as country and western, jazz and folk music attempted to 'capture' cultural processes in maps. Popular music, like other aspects of culture, could be represented spatially, explained and described in terms of the location and origins of musical scenes, styles and pieces; the movement or diffusion of musical genres and styles across space, or the networks of musical tours, patterns of trade of musical product, or the locations of supposed 'hearths' of musical cultures (e.g. Carney 1974, 1978). Music thus provided cartographies of cultural production, reception and consumption. This research, originating from North America during the 1970s and 1980s, dealt with the spatial location and distribution of specifically North American music styles (Carney 1978), the delineation of musical centres and the diffusion and communication of musical expressions further afield (Ford 1971; Horsley 1978). Since the advent of more sophisticated, global communications technology and more rapid networks of production and distribution, 'local' music is now transmitted and received far beyond regional and national boundaries. This fusion is particularly evident in 'world music', to the extent of it being part of promotional club flyers, but in a manner that mythologises the extent of global reach (Figure 1.1). The recent trend for musicians to record their own music and distribute releases via Internet technology, and the practice of reproducing and disseminating pre-recorded music in various computer formats through email (also known to the music industry as a form of digital 'piracy'), suggest quite different networks and cartographies of music in the digital age (Chapter 11).

Variants of the cartographic tradition attempted to evaluate the cultural 'distinctiveness' of musical expressions at a variety of spatial scales. Aldskogius (1993) and Waterman (1998) explored music festivals in Sweden and Israel, assessing the presence or absence of particular regional music traditions, while Lomax and Erickson (1971) and Nash (1968) attempted to map world musical styles. Other research with this descriptive, 'regional' flavour has included articles on the images of place that are evoked through music lyrics (Henderson 1974; Lehr 1983), depictions of sacred and profane places in country music (Woods and Gritzner 1990), images of cities such as New York (Henderson 1974), descriptions of regional 'sounds' (Bell 1998; Curtis and Rose 1987; Gill 1993) and even an atlas of the life and performances of Elvis Presley (Gray and Osborne 1996). Such cartographic depictions of music styles and scenes provided valuable detail on particular styles of music, the ways in which these were disseminated and the lines of migration undertaken by performers. Such cartographies of music hinted at the need for greater emphasis on ethnographies of music, and developing a musical 'sense of place' as a starting point for more developed analysis. Yet, ultimately, such studies were limited through their 'failure to engage with the social and political contexts in which music is produced' and the socially constructed nature of human understandings of place and space (Kong 1995b: 186). Cartographies – just like printed maps – need to be situated in networks of economic, social and political relationships.

Studies of local 'scenes' in Europe and North America have shown how musical forms and practices 'originate within, interact with, and are inevitably affected by, the physical, social, political and economic factors which surround them' (Cohen 1991a: 342), resulting in the construction of diverse representations, or identities, for those regions (see also Kruse 1993; Cohen 1994, 1995, 1999). Music is

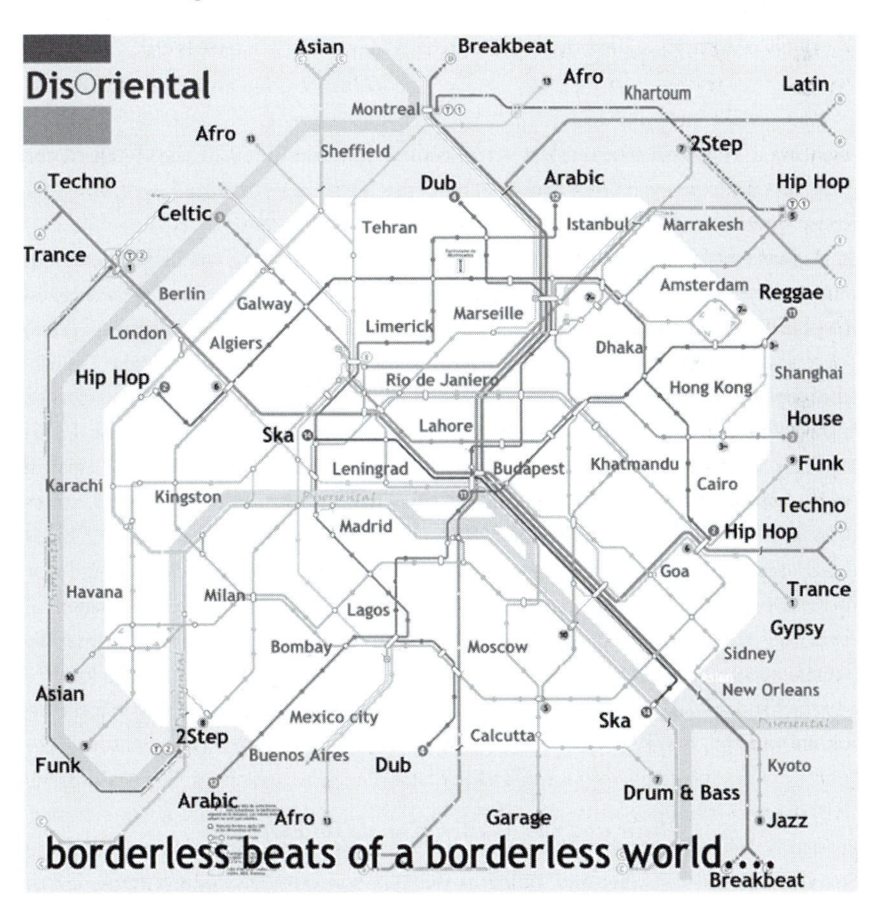

Figure 1.1 Disoriental club flyer, Bradford

bound up in places as 'articulated moments in networks of social relations and understandings' (Massey 1994: 154), illustrated in a range of examples, from sub-cultural formations (in clubs, pubs or churches), to particular events. Such networks are layered and differentiated in various ways – new cultural alliances do not form on a 'blank slate'; rather they inherit the particular set of circumstances, traditions and social relations of older generations of cultural producers and consumers. The consumption of music opens up a mass of possibilities for enjoyment, pleasure, engagement or for developing wider cultural traditions. Yet, in emphasising ethnographic methods, Cohen recognised the limits to such approaches and the dilemmas of relativism:

> Ethnography in the anthropological sense has its limitations. It is small-scale and face-to-face, and this raises the problem of typicality – whether the small part studied can represent the whole – and the problem of incorporating detailed description which may seem banal or tedious.
>
> (1993: 125; see also Chan 1998)

Associated with this is a tendency to become too enmeshed in the detail of the local at the expense of recognising how the local is constituted within wider flows, networks and actions.

A sense of place thus transects with activities operating on larger scales in many ways: local and regional 'sounds' are captured, marketed and transmitted through the worldwide distribution networks of music multinationals; musicians, sub-cultures and audiences in a multitude of localities receive and interpret music from other places, while local narratives of experience and identity can be sustained by dispersed populations across national boundaries (Mitchell 1996; Stokes 1994). Many sites, or wider geographical regions in which musical production and consumption occur, become linked with particular sounds, styles or musical approaches (such as the 'Motown' sound, New Orleans jazz). This is often due to concentrations of infrastructure for music production and for musical cultures in particular areas (for example, cities with an abundance of recording facilities, live venues or access to inexpensive technology), but may be equally attributable to a process of mythologising place in which unique, locally-experienced social, economic and political circumstances are somehow 'captured' within music.

Music and the politics of space and identity

Social conflicts, tensions and political debates sometimes become struggles to control a process of representation in various media. From tabloid reporting of 'gang violence' and 'ghetto life' to literary writing about 'exotic' cultures, from travel brochure images of island paradises to protest songs, it has become evident that visible and audible media are spheres in which narratives of place are generated and articulated. These may include associations between music and ethnicity, nationalism, class and gender – creating 'identities' for individuals, social groups and even for whole regions (Stokes 1994). National anthems, music traditions such as the blues, funk and disco (and the ways these styles are mimicked and absorbed in other places), and the work of artists as varied as Elvis Presley, kd lang, Moby and the Afro-Cuban All Stars, all suggest different ways in which spaces of expression are created, and how places (the inner city, the concert, the community hall) are experienced. As with cultural sites and identities more generally, musical spaces remain contested. Conflicts and contests are both discursive (in the varying representations of identity), and concrete (in struggles over public and private spaces).

The processes of identity formation in and through music occur in uneven ways, mediated by relations of power: the racial, gendered and socio-economic filters that can act to polarise sections of society, marginalise groups of people from mainstream economic and political power, and silence oppositional or 'alternative' cultural voices. For example, all music, whether considered 'independent' or mass consumed and commercially oriented, is racialised in complex ways. Moreover, the majority of music is in English, while musicians from other backgrounds are not always afforded the airplay or opportunities to participate in Anglophone markets. Despite such barriers, numerous non-Anglo-American musical traditions – from salsa to rai – have continued to feed into the commercial music industry. 'Peripheral' sounds can and do alter trends, inspiring musicians at the 'cores' of English-speaking music production in Britain and the United States, and voicing experiences of place that negotiate (and sometimes oppose) the global processes commodifying musical heritage (Rose 1994; Lipsitz 1994). Musical identities can challenge accepted social norms, configuring reactions to 'mainstream' cultural practices, and asserting new styles. Popular music, alongside other media such as art and literature, operates at many levels, providing a platform for the expression of marginalised voices, while illuminating global alliances and cultural flows. The performance and reception of popular music in particular local circumstances may be 'an effective form of resistance to the homogenising forces of the culture industry, not necessarily by producing an alternative sound, but by enabling people to experience music in distinctive localised ways' (Smith 1994: 237). Popular music can then be seen as an integral part of the process by which spaces are created for social interaction, entertainment and enjoyment, including the plethora of sites designated exclusively for the production and experience of music, such as small live music venues, nightclubs and discos, record bars and concert halls; and even spaces not normally associated with music, but where the broadcast or infiltration of music serves various political or commercial intents. Music is one way through which ordinary acts of consumption and movement throughout daily life could constitute 'tactics' of subtle opposition (de Certeau 1984) that emerge from within the cultural spaces governed and controlled by others, occurring as they often do in the private spaces of home, in the corners of the night-time economy, beyond the panoptic gaze of the state.

Aspects of 'sub-cultural' and 'oppositional' style frequently become valuable in themselves, in particular through musical genres, fashion styles and attitudes. There are always struggles and points of exchange between musical communities and the wider music and fashion industries, as musical space is defined as cutting edge, authentic or 'underground', in opposition to the commodified or 'simulated' products of the mainstream. Music can be a means of accruing 'sub-cultural capital' (Thornton 1995; McRobbie 1994), authenticated in cultural studies itself through examinations of the role of music in everyday life – ethnography thus cannot escape the contradiction that it authenticates as it describes. In turn, many theorists have retreated from celebrating such expressive musical identities, stressing qualities of heterogeneity, difference and hybridity.

A question of scale?

Many examples of musical identities are spatial, as they relate to physical sites and the movements of culture, commodities and people across territory. Yet music is also audible, and the public spaces where musical identities are constructed (and new evasions of control figured out) are both physical (as in the performance spaces of music) and virtual (as in the spaces of public broadcast). New hybrid identities are created by the spread of musical sounds (as with the global distribution of reggae or country and western), while musical sub-cultures and sites can be politicised and subject to scrutiny as part of struggles for local spaces.

Authors writing on the politics of popular culture have grappled with this complexity and how it relates to geographical scale. McLeay (1997) and Shuker (1998) both questioned the terms 'global' and 'local', through the tendency to reify certain actions as belonging to one scale or another, or to posit cultures, institutions and individuals exclusively within particular geographical limits. Thus, the 'global' and 'local', oversimplified concepts for what are complex and multi-scaled actions, are replaced by new terms such as 'glocalisation' that attempt to indicate the simultaneous 'global' and 'local' elements of economic processes and cultural identifications:

> The hybrid term 'glocalisation' has emerged as a more useful concept, emphasising the complex and dynamic interrelationship of local music scenes and industries and the international marketplace.
>
> (Shuker 1998: 132)

Mitchell (1996, 1999) has similarly appealed to the 'glocal' in a more sub-cultural approach, looking at ways in which global 'templates' have enabled local mobilisations of hip hop musical languages and diasporic identities (see also Bennett 2000). We intend to side-step such discussions rather than dismiss them, suggesting that the terms 'fixity' and 'fluidity' reflect more dynamic ways of describing and understanding processes that move across, while becoming embedded in, the materiality of localities and social relations. The 'global' and the 'local' happen simultaneously, not as a mere coincidence, but often as part of a formal contradiction – they are constructed in part because of endogenous actions, productions and expressions, but are also defined against what they are not. Consequently more active terms are needed than 'global' and 'local' or variants such as 'glocalisation', which reify the status of geometric space over the dynamic conditions under which space is actively constructed and consumed by companies, institutions of governance and by individuals (Lefebvre 1991). Dialectical explanations of inter-scalar relations have appeared from work in cultural geography and political economy, examining the complex dynamic of scale in global economic and cultural change. In many respects, a tension between movements of objects, people and money across space, while also seeking to establish permanence in place, has always characterised spatial dimensions of human activity: a dynamic of 'globalisation/reterritorialisation' (Brenner 1999: 436). There is a tension:

> between preserving the values of past commitment made at a particular place and time, or devaluing them to open up fresh room for accumulation...between fixity and motion, between the rising power to overcome space and the immobile structures required for such a purpose.
>
> (Harvey 1985: 150)

But, here, our use of a similar dialectic involves discussions across economic and cultural domains – fixity and fluidity operate at multiple levels, from the formal and institutional to the personal, embracing much more than just circuits of capital. Music illustrates that such dynamics are difficult to tease out in any ordered manner; music cannot be contained within a single explanatory theory – it is dynamic and unpredictable, involving movements of sound and people, expressing mobility in certain periods, stability in others.

The following chapters of this book explore multiple ways in which music can be understood as geographical, and the various debates that cut across this: Chapter 2 examines music and connections to place, establishing the logics of 'authenticity' that ground performers, styles and songs in geographical locations. Subsequently, Chapter 3 explores the converse, considering music as mobility and change. Further chapters then expand on a number of themes relating to this dialectic, both discursive (in lyrics, 'sounds', in the politics of identity and world music) and also material (in the built spaces of music, in tourism and in the digital realm). Throughout these chapters various tensions are revealed that shape music as fixity and fluidity: between innovation and continuity, in nostalgia and return, between authentic and fake. Places, and their specific socio-historical, economic and political circumstances, shape musical expression; musical recordings act as catalysts for the construction of spatial identities that sometimes last (for example in the associations between the Beatles and Liverpool) or fade. Music traditions can alter places (whether through generating employment or through influxes of tourism), while music flows across space (from oral traditions to Internet distribution) in directions and along pathways that are sometimes directed, often random, but always mediated by (or constituted as reactions against) flows of capital, new technologies and styles. Music creates places and networks of cultural flow, but does not do so beyond the worlds of politics, commerce and social life.

2

MUSIC AND PLACE

'Fixing authenticity'

This chapter examines the idea of spatial fixity, where continuity is valued over change, stability preferred to cycles of fashion, and which links music to particular places and establishes those links as traditions and genuine aspects of local cultures. It examines how authenticity is constructed for particular styles, genres, artists and releases. Such authenticity, as we discuss here, is in part constructed by attempts to embed music in place. This occurs in a number of ways: through ethnomusicological practice, in various mobilisations of tradition, in discursive constructions of place by songwriters and in the way that audiences receive music. Fixity is thus complex – no one theory could examine all permutations of the links between music and place. A sense of fixity is usually implied whenever music is discussed for pre-capitalist societies, both in relation to the cultural and geographical origins of music, but also through the link to nostalgia – related to yearnings for past glories, lost youth and claims for styles of music that evade the 'corrupting' influences of contemporary society and economy. The early part of the chapter considers the role of ethnomusicology in examining remaining fragments of 'traditional' music, which imply discussion of ancient and unchanging times; yet ethnomusicology is primarily part of a modern era, to some extent a response to cultural change under colonial and capitalist expansion. The moment of commodification – as music is transformed from cultural expression to product, as traditions are usurped by change – is crucial. Binary relations established in considering such commodification constantly appear – between 'tradition' and 'contemporary' in folk revivals, in 'regional' traditions and in various expressions of 'roots' in music. Unpacking fixity implies that we begin by examining more closely notions of 'traditional' music and commodification, as they establish the character of a particular form of authenticity.

Ethnomusicology and fixity

It is a truism that no society is without a musical tradition, and it has also been stated, less frequently, that music and dance are often defining elements of particular cultures. In many 'traditional' societies people argue along such lines as 'Music is of intrinsic importance in our lives. As such, it is a social issue.... Music is a social event, a way we celebrate our culture together' (a Melanesian islander from the Torres Strait, quoted in York 1995: 28; Seeger 1994). Scholars have taken similar views both of the social context and the musical content; thus the songs of the Dyirbal Aborigines of northern Queensland provide 'a kaleidoscopic view of their environment and life-style, an expression of likes and hates and fears, and a reflection of customs and laws' (Dixon and Koch 1996: 16). Every human culture appears to include some kind of music, even if this relies on a very broad definition of music. All cultures seem to embrace some means of producing extra-verbal sounds, listening to these and deriving pleasure, meaning and utility from them; the presence of musical instruments, excavated from archaeological digs from at least 7,700 years ago (Zhang *et al.* 1999), suggests that this may have been true throughout history.

Ethnomusicology has concerned itself with the distinctiveness and dynamics of music in its socio-cultural contexts throughout human history. For many it has come to represent the study of music in 'traditional' societies. Central to the ethnomusicological tradition is a sense of endogeny – of musical expressions emanating from within relatively unique social landscapes, rather than interacting with outside flows, consuming and reproducing the products of others, or mimicking international sounds. At its most basic, ethnomusicology's concern with the endogenous relies on the 'traditional' as the subject of study, and, by inference, on constructions of 'modernity', 'contemporary' and 'non-local' in music as opposites of this. When Adorno and others railed against the 'culture industry' in the industrialised world (with its devaluation of music as an art form, its blatant dumbing down of music), ethnomusicologists offered recordings of distant peoples, far removed from the recording industry, radio, corruption or commercialisation. Given the pace of technological change, the increasing range of trade and commerce, and the force of colonial impositions, musical traditions, as with many other facets of indigenous life, were under pressure. Ethnomusicologists not only documented surviving traditions, but also brought music's emotive force to bear on arguments about the destructive potential of industrial capitalism and European colonialism. As 'gatekeepers' of traditional music – returning from remote contexts with field recordings that sometimes made their way to Western specialist record labels – they opened up an avenue between indigenous production and distant consumption that would later reappear as 'world music'. Such recordings were reified as 'traditional', in contrast to commercial styles from Europe and the United States, establishing that dichotomy, while subtly grounding music in particular social, cultural and political spaces. The music was preserved, not because of its commercial appeal, or inherent artistry, but because it had, despite the odds, been transported, even rescued, from distant and vulnerable places.

Recent ethnomusicological work has broadened to recognise and celebrate the diversity of musical production in both 'traditional' and 'modern' contexts, examining migrant musics in urban settings, examining and acknowledging the ways that all cultures have been implicated in, and affected by, colonialism and capitalist expansion (for example, Roseman 2000; Turino 2000; Wade 2000). Yet, important binaries established in earlier efforts persist in the current era, and have created a framework that most readings of contemporary music work within, or react against. Distinctions between 'traditional' and 'modern' have informed vast numbers of studies, yet are hard to sustain upon closer examination. Traditions associated with 'ancient' peoples remain in the contemporary world despite colonialism and commoditisation; these have been variously retained, reinvented and sold for particular purposes; what appears to outsiders as 'traditional' may be entirely contemporary in authorship. Australian Aboriginal composers, for example, have been adept at mobilising contemporary themes within traditional musical expressions, with songs about modern consumer lifestyles, romance and so on, often performed to unaware tourists, who perceive the sounds as redolent of 'serious' encounters. Recognising the constructed nature of what is called 'tradition', and problematising the ways in which these constructions might cloud perceptions of 'other' cultures, remains crucial in an era of rapid communications and insatiable commercialisation. However, while 'tradition' is fluid and constructed, some cultural expressions have persisted despite, or in reaction to, cultural change. Music can enable the maintenance of indigenous knowledge. Retained fragments of music hint at matrices of musical diversity now mostly lost, but also reveal much about how music has remained adaptable, malleable and vital. It is thus important to consider music in 'traditional' contexts as part of a trajectory that links particular discourses of cultural production and reception to other Western forms. 'Traditional' music, as a subject of study (rather than as a chronological starting point), sets in train particular ways of thinking about 'fixity' that appear in a plethora of other contexts.

Locality and social organisation

No cultures were ever wholly isolated from external contact, and from material transactions and culture change, but until recent phases of world history many cultures were for the most part separate and distinct from others only tens of kilometres away. Even so, communities exchanged goods with their neighbours, whose languages and cultures bore some resemblance to theirs. Autarchy was non-existent. In all such localised, small-scale societies music existed. For the Siwai of Bougainville Island in Melanesia in the 1930s:

> singing is highly popular and accompanies many kinds of occasions. Parents sing lullabies to their children and use songs to teach them new words. Lovers compose boastful songs about their adventures, and relatives express their grief in moving laments. Most dramatic of all however is the harmonised singing of large numbers of men and boys, which takes place at some feasts.
>
> (Oliver 1955: 36; cf. Clay 1986: 204–18)

Similar statements have been made of most small-scale societies in every world region, emphasising the role of music in both everyday and ritual life, and its link to social organisation. Among the Suya, of Brazil's Amazonia, 'Suya society was an orchestra, its village was a concert hall, and its year a song. Their singing created a certain kind of settlement in which sounds revealed what vision could not penetrate' (Seeger 1987: 140). In such societies, not only was music inescapable, but since music was integral to social structure, all participated; in upland Papua New Guinea:

> It is broadly assumed that every Kaluli must become a competent maker, recognisor, user and interpreter of natural and cultural sound patterns....Soundmaking is highly valued and considered necessary for survival, expression, and social interaction for all Kaluli men and women.
>
> (Feld 1984: 389, 397; Box 2.1)

Such comprehensive significance of music may have been most evident in relatively egalitarian societies where specialists and leaders of any kind were few or non-existent.

Some emphasis on music and song as elements of everyday life and ritual is ubiquitous. While music invariably provided pleasure, it was rarely performed merely for entertainment; rather it was linked to economics, politics, other arts and, above all, to social and ritual organisation and language (but in most societies differentiating between these categories made little sense). In certain contexts music – and ritual life – created, defined and maintained community:

> To consider song and ceremonial life to be mechanical products of other aspects of social life is to miss the essential nature of musical and ceremonial performances. Suya ceremonies created euphoria out of silence, a village community out of a collection of residences, a socialised adult out of physical matter.
>
> (Seeger 1987: 86; see also Lomax 1959: 929; cf. Kaemmer 1993: 23–4)

Consequently, the most memorable and aesthetic Suya musical performances were those that brought the most people together, 'made a village beautiful, and a person euphoric' (Seeger 1987: 132, 131). Music and songs provided links to the past, ties to the natural and spiritual worlds and expressions of the contemporary.

Attempts have been made to demonstrate a relationship between complex social structures, and complex musical systems, evident in Alan Lomax's cantometric project (Lomax 1962, 1976). Lomax's neo-evolutionary perspective argued that song styles shifted with productive range, political development, the extent of social stratification and class formation, 'severity of sexual mores', the nature of the relationship between males and females and the degree of social cohesiveness – in short, the gamut of economics, politics and society. However, although the Kaluli, for example, exhibited particular relationships between socio-musical practice and social organisation, through an aesthetic relationship with the forest ecology, and with an emphasis on autonomy and valuing self in relation to others, the shape of the Kaluli musical system could not simply be predicted from their mode of production and techno-economic complexity. Moreover, similar musical forms are found in societies of widely different social complexity (Feld 1984: 400–5), although there was usually some crude relationship between socio-economic structure and musical form.

Musical forms were usually linked to conceptualisations about the organisation of society. In Hindu society the idea of reincarnation caused life to be considered as an endless series of cycles; the form of classical Indian music, based on cycles, reinforced this basic element of Indian life (Kaemmer 1993: 120). Where social organisation was based on divisions within society, music may have represented these divisions. Suya communities, like many Amazonian societies, were divided into two social groups (moieties), replicated in the structure of ritual and spatial organisation, and in the divisions of songs into halves (sung by the different moieties) and the structure of the songs themselves (Seeger 1987). In Malaysian Temiar society musical participation in curative medical practice was often collective, bringing together divisions in society, specifically to overcome social and political factionalism, and thus sources of ill-health; some forms of illness were perceived to result from the soul being lost or mislaid, hence treatment involved singing a 'way' to bring the soul back home; metaphorically music linked domains of travelling, knowledge, singing and healing (Roseman 1991: 9). Temiar healers were 'singers of the landscape' and the songs themselves were referred to as 'paths' (Roseman 1998: 110; cf. Friedson 1996: xvi). Nature and society were indistinguishable, music linked rituals and everyday life, and aesthetic configurations were part of a comprehensive pattern of reality.

Relatively simple divisions and relationships are more intricate in complex societies. Amongst the Aymara of the Peruvian Andes mountains, groups of panpipe players accompany community festivals. Panpipe players are divided into three groups – broadly high, medium and low voices – and in contrast to European musical forms where the high and low voices are most important, the Aymara consider the middle group the most important. Giving importance to the centre, with both peripheries of less significance, also occurs in Andean weaving, and parallels social organisation, with primary village communities in the centre, and peripheral communities at higher and lower elevations in the mountains (Turino 1989; Solomon 2000). Music commonly replicates and emphasises gender divisions, alongside authority, age and class (Feld and Fox 1994). Songs could be, however, both forms of social control – reflecting the ethos of a particular culture, in terms of values, sanctions and problems – and a means of challenging norms, through the expression of feelings that could not be spoken in other contexts (Merriam 1964: 190–7). Music and song were more flexible than other elements of social structure.

Complex divisions in society were sometimes reflected in song ownership. In some cultures songs belonged to particular individuals, usually their composers, or, more frequently, to lineages or clans (Clay 1986: 203–4; Merriam 1964: 82–3, 248; Moyle 1986). Where daily life was more individualistic and even lonely, as among the Sami herders of northern Scandinavia, music could be individualistic, with songs recounting personal genealogies and depicting particular herding environments (Stockmann 1994: 10). Individuals had rights to specific musical compositions in many parts of the world. On Tanga Island (New Ireland, Papua New Guinea), during the 1930s:

> A dance master...has the inalienable right to his own compositions, and to such songs and dance arrangements as he has seen elsewhere and introduced to the island. Although a dance is a public affair...no other village would dare to plagiarise an original composition.
>
> (Bell 1935: 108, quoted by Merriam 1964: 82)

Individual or sub-group ownership of music and song was most evident in stratified societies, where specialists existed (such as the Tanga 'dance master') in musical and ritual fields. In some societies, musicians were a distinct component in social stratification, as in West Africa where *griots*, troubadours who praised ancestors and important men, held hereditary positions and were paid for their services. Few could genuinely be described as professionals, in the sense that they could sustain a livelihood from music, but most societies had specialists or experts (Merriam 1964: 123–30) who were occasionally paid for their services, even in pre-capitalist societies, such as Siwai (Oliver 1955: 306). Yet in other societies musicians were outcasts or deviants, not bound to the norms of society, and at the bottom of the social order; Barongye musicians

(Congo/Zaire) were regarded as 'lazy, heavy drinkers, impotents, hemp smokers, physical weaklings [and] adulterers' (Merriam 1964: 136). Patronage, and the presence of paid composers or performers, tended to result in greater musical complexity.

The same performance or piece of music may simultaneously carry different meanings and offer different interpretations, even within small-scale societies. This ambiguity enables people to manipulate symbolic meanings for their own purposes (Kaemmer 1993: 110), emphasises the lack of homogeneity within societies and points to the role of individual agency (evident, directly, in the presence of composers). Music is composed and performed by people who 'are creating something that is at once a re-creation and a new creation under unique circumstances' (Seeger 1987: 85, 86). Just as music is apparently universal, so too are distinct skills, in composing and playing, which take music far beyond the mundane (Merriam 1964: 67–70, 114–16) and thus also beyond the simply functional. Music is invariably held to have aesthetic qualities, which extend its role beyond utilitarian goals (although achieving high performance standards may be regarded as essential to particular ritual ends), to become a means of entertainment and a legitimate art form.

Such analyses of music relied on the relative isolation imposed by distance, and the intensity of local distinctiveness. The characteristic subsistence orientation of small-scale societies gave primacy to the role of the environment in everyday life; natural and social worlds were intertwined and drawn together through social organisation and song. For the Temiar, and for many others, songs 'mark the natural and social landscape of the people, naming it, locating it in time and place, in history' (Roseman 1991: 175), though the relationship between songs and natural landscapes, historical origins and cosmological systems is complex, and both technical and metaphoric. Quite literally, Temiar 'sing their maps: theoretically, in their epistemology of song composition and performance; melodically, in contours of pitch and phrasing; textually, in place names weighted with memory' (Roseman 1998: 106). This complexity has been particularly well elaborated in Chayantaka communities in highland Bolivia, where places 'were discursively imagined in song texts' (Solomon 2000: 272), and for the Kaluli, whose musical repertoire was closely and intricately linked to the tropical forest environment (Box 2.1). Music and songs were sometimes linked into periodic cycles, as among Australian Aboriginal societies where such song cycles (song lines) recounted ancient journeys, with each song representing a stop on the way (Kaemmer 1993: 112; Weiner 1991: 196–9). Natural environments provided a particular range of sounds and distinct local options for instrument production, and both were relatively limited. As small-scale societies became involved and incorporated in a wider world their musical characteristics began to change.

Box 2.1 Music and the Kaluli environment

The approximately 1,200 Kaluli people occupied tropical rainforests on the southern fringes of the New Guinea highlands, and were mostly swidden agriculturalists and sago growers. Their environment 'is like a tuning fork, providing well-known signals that mark and coordinate daily life. Space, time and seasons are marked and interpreted according to sounds' (Feld 1984: 394). Seasons and cycles were conceived in terms of sounds – of birds, waterfalls and other natural phenomena – as much as by counting moons. Kaluli songs were consciously modelled on bird sounds, the flowing of water and other auditory elements of the forest environment, and involve interlocking, overlapping and alternation, and the layering of parts and sounds, rather than any performances in unison. When Kaluli die their spirits are said to live in birds. Bird songs were considered to be the voices of the ancestors speaking to the living. Several species of doves have calls with a descending pitch pattern, and this pattern was the basis for much Kaluli song (Feld 1982). Because birds sing, whistle and also weep and speak, they also provided 'a simultaneous index of the environment as well as a deeper symbolic understanding about self, place and time' (Feld 1984: 395). Kaluli perceived the land and the forest as mediators of identity, enjoyed listening to the sounds of the forest and improvised duets with birds, cicadas and other forest sounds, in the course of everyday life. At Kaluli ceremonies, songs were sung by visitors to the hosts: 'the hosts find the song texts in particular to be sad and evocative because they concentrate on maps and images of the places in the immediate, surrounding forest, places to which the hosts have a sentimental attachment' (Feld 1984: 393). The Kaluli relationship with the forest was neither antagonistic nor destructive, production of material needs was relatively easy and the symbolic and pleasurable dimensions of the forest reinforced this materialist basis, so the forest became a 'mirror for social relationships, particularly as mediated through the poetic imagery of songs that concern maps, lands and identities, as well as through formal structure and singing style' (Feld 1984: 395). Hence 'Kaluli songs map the sound world as a space–time continuum of place, of connection, of exchange, of travel, of memory, of fear, of longing and of possibility' (Feld 2000b: 199). More than in most other cultures the environment contributed to the creation of a musical ethos.

See: Feld 1982, 1984.

Unravelling authenticity: roots and regional traditions

Traditional music is imbricated within local culture, and inseparable from it: an expression of the ethos of small-scale society. In the discourse of both indigenous peoples and external observers, contrasts have constantly been made between tradition and transformed contexts where culture as a way of everyday life, with every facet derived from the ideology and practice of distant ancestors, has become, at best, culture as a reified symbol of a way of life (sometimes the manipulative rhetoric of those who seek to preserve the old) rather than lived experience. Yet, such a dichotomy is invalid: 'traditional' societies were never wholly unitary essences, and indigenous peoples were not 'peoples without history' until the West brought 'social change', enlightenment and economic development (Wolf 1982). Indigenous peoples were aware of differences between cultures, drew in alien elements of culture and technology where appropriate and were rarely wholly sedentary.

There is, therefore, no particular moment at which any culture somehow becomes inauthentic, in its incorporation of external elements, since society was and is always in flux. However, prior to the advent of colonialism (in itself an imprecise and fluctuating phenomenon, and not restricted to solely 'Western' transformations) cultures were more likely to have had a shared suite of values. Since colonialism (and particularly that of Europe) changes in values, and music, have been much more rapid. Sometimes the transition was straightforward; the Temiar musical historiography was highly adaptive enabling them 'to call upon the spirits of old mountains and new wristwatches, rainforest birds and parachutes, connecting them with the past while moving them into the future' (Roseman 1991: 175). Yet in many societies notions of authenticity were zealously guarded. In Mandak, when women's dances at mortuary feasts incorporated songs from Lihir (an island 50 km away), men objected to this 'impurity' of tradition (Clay 1986: 208). Among older Dyirbal Aborigines in Queensland, twentieth-century songs were regarded as not 'the real thing', though most songs in the existing repertoire dated from the nineteenth century, a period of exceptionally rapid colonial change, which was reflected in the songs (Dixon and Koch 1996: 17). Such apparent resistance to change has been widely documented (e.g. Kaemmer 1993: 183) yet even in relatively isolated societies musical forms changed, and notions of authenticity – in all spheres of life – were constantly contested, and challenged by innovation and the delights of difference.

Authenticity remains an intangible concept. Essentialist perspectives construct authenticity in relation to concepts such as 'spontaneity' ('live'), 'grassroots' and 'of the people', in opposition to their antithesis: 'manipulation', 'standardisation', 'mass' and 'commercial', yet no genre of music could 'walk on to the historical stage in uncontaminated form' (Middleton 1990: 6). Authenticity in its strictest sense applied to museum objects – a process where historical artefacts were verified: scientifically 'proven' to have originated from particular places and be genuinely what they were claimed to be. However, in terms of more fluid and ephemeral aspects of culture, such as music, it is impossible to measure authenticity against any given scientific criteria. Discussions of musical authenticity imply a different use of the term, constituting interpretations of the validity of music from particular contexts and in certain modes of consumption. What is 'authentic' is socially constructed in various ways.

One way to understand how authenticity in music has been constructed starts with assumptions about the moment of commodification, when indigenous or folk musical traditions came into contact with wider musical economies. More specifically, notions of authenticity derive from how music is valued, and the shifts in value that occur as music is perceived to have been disembedded from its social and cultural origins. In more abstract terms, this involves the relationship between music's use and exchange values as sounds are commodified. At the heart of the social relations of the process of commodification is a distancing of producers of goods from audiences:

> Commodified labour produces commodities, things that are produced for sale and therefore for consumption by someone other than the person whose labour produced it. Instead of being organically and transparently linked within praxis, the relation between production and consumption is indirect and mediated through markets, money, prices, competition and profit – the whole apparatus of commodity exchange.
>
> (Slater, 1997: 107; see Jackson, 1999: 96; Fine and Leopold 1993: 259)

This process is not only metaphorical – it is geographical in a material sense – distancing the factory from home, productive activities from social activities. Indeed, the sphere of commodification, and the corresponding distances between productive consumption (of musical styles, scenes, of musicians' labour itself) and audience consumption, have expanded enormously, especially given music's inherent fluidity and mobility (the focus of the next chapter). The ways in which this has occurred are complex. For most commodities, visible linkages between a product's marketed image and the realities of its production method (and the labour, environmental and cultural politics that go with this) are discouraged, lest consumers discover any unethical aspect of the commodity's creation. Conversely, for music products, as cultural objects, value is reliant on a sense of connectedness between consumers and producers – in this case emotionally, with the singer/songwriter/band. Bloomfield suggests that the successful popular song creates an 'imaginary identification' between consumer and artist, where the perceived use value of the song is its 'emotional conversation':

> In the mass consumption of music, pop songs at their most effective provide the listener with the illusion of entering into a direct and immediate (unmediated) relation with a human producer that is capable of gratifying the listener's individual need, that speaks directly from one subjectivity to another.
>
> (1993: 16)

These ideas are critically related to 'ideologies of authenticity' underpinning popular music consumption, where the performer reflects on personal experience that resonates with emotion, embodies the results of that reflection in a musical-narrative form, delivers a performance that serves to bring out fully its (inner) meaning and where listeners read this emotional meaning by bringing their personal experience to bear on the performance (Bloomfield 1993: 17). 'Traditional' music exemplifies ideas of such unmediated communication, at least within the direct context of consumption in indigenous societies. It establishes a 'benchmark' for constructions of authenticity, deeply embedded in the daily life of pre-capitalist societies.

In this 'ideology of authenticity' 'live' music is paramount: it is the context in which communication between musicians and audiences is perceived to be the most direct and where musicians may struggle for their art despite poor incomes and labour relations. More frequently because of distance or time, live music was impossible hence notions of authenticity were derived from a context where directness remained important, in the face of the encroachment of commercialism. For Grossberg authenticity had three dimensions: aesthetics (the skill and creativity of the artists); the construction of a rhythmic and sexual body (often linked to dance and black music); and the ability to articulate private but common desires, feelings and experiences, through a common language that constructs or expresses the notion of community (Grossberg 1992, 1993). These ideas consequently reappeared in 'authentic' scenes, in music that appealed to emotional triggers, such as the blues, folk and country musics where directness of emotions remained crucial. More generally, artists and promoters have always sought a musical quality of 'rawness' and 'honesty' that is seen to go beyond 'the trappings of showbiz' (Frith and Horne 1987: 88) through proclaiming some expression of identity (Roberson 2001), as with the reception, at various times, of jazz, punk, grunge, hip hop, world music and 'indie' rock.

Migration and diffusion: reconstituting the folk region

Authenticity was particularly sought in folk and country music, and recognised in terms of continuity and oral transmission, usually from a rural, working-class, community tradition (Frith and Horne 1987: 128) though none of these latter concepts could stand up to sustained scrutiny as polar types. Authenticity throughout popular music invoked a genealogical inheritance, implying 'forms of artistic expression that were the genuine expressions of total forms of life' (Coyle and Dolan 1999: 26), which had descended from folk music variously described as:

> Music that is collectively owned, of ancient and anonymous authorship and transmitted across generations by word of mouth; a canon celebrating life in the past and urging change for tomorrow, the performance being on simple instruments in natural settings...the joyful performance by specially gifted but not 'professional' artists.
>
> (Ennis 1992: 88)

and with claims to being embedded in particular places, and earlier times. The quest for music that derived from such apparent tradition was central to folk and country revivals, even stimulating the transformation of the visual images and names of country performers to meet appropriate rural images. Thus, the Possum Hunters, whose name was redolent of rural life, shifted in the 1920s from matching suits to 'the agrarian garb of nature' (Tichi 1994: 152–3; Peterson 1997: 146–50); somewhat later Willie Nelson, Dolly Parton and others underwent similar career transitions, all emphasising roots and locality (Yates 1975; Malone 1985; Lewis 1997). Revival and transformation were the principal elements of 'invented tradition', the outcome of the processes of inventing, constructing and formally establishing behaviour and images that automatically implied continuity with a suitable, historic past (Hobsbawm 1983: 1). In the case of folk and country music, particularly, but not exclusively, that was often paralleled in an 'invented geography' that sought to tie particular genres to particular regions. The conjunction of history and geography was evident in the emergence of an apparently distinctive bluegrass sound and region (Box 2.2). Though there is a fine line between invented traditions and geographies, and practices that have continued over long periods in particular places (e.g. Rosenberg 1993a: 20; Peterson 1997: 217), the conjunction points to the idealisation of a culture, and its musical heritage, seemingly in danger of disintegration, out of ideological (and later commercial) concerns.

As long as tunes and songs were transmitted orally, and transmission by word of mouth has been one definition of folk music (Kaemmer 1993: 66), they were inevitably transformed in that process, sometimes deliberately so to reflect local circumstances, ensuring distinctiveness and difference between regions. Largely in response to this apparent diversity, various attempts have been made to demarcate music areas, and 'fix' music in place, such as that which produced three 'comprehensive musical culture areas' (defined as 'Europe and Negro Africa', 'Orient and Oceania' and 'USSR, Mongoloids and Amerindians') based on tonal

Box 2.2 Creating authenticity: bluegrass

Bluegrass music (characterised by rapid ensemble playing of non-electrified string instruments, usually dominated by the banjo, and with the infrequent vocal accompaniment typified by high-pitched 'high lonesome' sounds) is often equated with Appalachian mountain music, and seen by some as the only true, natural, authentic form of country music (Cantwell 1984) or, at the very least, 'a refuge from the "progressive" and pop styles that have inundated mainstream country music' (Malone 1985: 323–4). Bluegrass music however emerged only in the 1940s, as the sound created primarily by Bill Monroe, Lester Platt and Earl Scruggs. Monroe, eventually a mandolin player, grew up in western Kentucky, far from the bluegrass district, while Scruggs and Platt, both guitarists, were from North Carolina. Bluegrass had a brief boom before it was ignored by radio stations because of its connotation of 'old-time hayseeds', only to benefit from the folk-song revival of the 1950s and 1960s when it was seen as 'the living link back to the older acoustic forms of country and folk music' (Peterson 1997: 213) or 'the latest expression of poor, rural, working-class pioneer America' (Rosenberg 1985: 13). It eventually achieved some mainstream success through the role of 'Foggy Mountain Breakdown' as the theme of the 1967 film *Bonnie and Clyde*, the 'Ballad of Jed Clampett' for the television series *The Beverly Hillbillies* and further exposure in the film *Deliverance* (1972), all of which emphasised both 'hillbilly' and 'red neck' themes. By contrast, those who valued the apparent authenticity of bluegrass emphasised the role of the banjo, the lack of self-consciousness of musicians who created the best music when they were 'starvin' to death' (which ignored the commercial success of Bill Monroe) and playing as a group, rather than celebrating individual performance; both the past and obscurity were romanticised (Rosenberg 1993b). Just as bluegrass music was a recent creation, it had no real link to the Kentucky bluegrass region, but mainly originated in western North Carolina, based around a network of local musicians and social institutions, including fiddle contests, music festivals, radio stations and recording studios (Carney 1996). However the label 'bluegrass' (whose origins come from Bill Monroe's first band name) gave it rural authenticity through an apparent base in Kentucky: 'a ready made organic metaphor that helped to verify its traditionality' (Rosenberg 1993b: 198), and an elegant combination of history and geography.

organisation and reduced 'the otherwise endless variety of styles into some semblance of order and simplicity' (Nettl 1956: 141–2). Another version established four global musical zones, linked to various forms of polyphony (Nash

1978: 12, 41), and yet another nine world regions, based on 'cantometrics', a holistic description involving performance style, phonemic patterning and overall conceptual patterns (Lomax 1976), which primarily sought to link musical structures with social and cultural structures. Such classifications, though not without validity as crude indicators of regional similarities, are however both drearily functionalist (Keil 1994) and timeless, with limited relevance in relation to the dynamic nature of contemporary cultures.

Long before such regions were identified, they were changing, sometimes substantially so, through the diffusion of new musical styles and instruments. Moreover within such global regions there were major regional and local variations. (Lomax actually referred to six large regions: North America, South America, Insular Pacific, Africa, Europe and 'Old High Culture' – an Afro-Eurasian region – with three 'troublesome but interesting' isolates – Arctic Asia, Tribal India and Australia. Other than the isolates, each of these regions were subdivided to make a total of fifity-seven sub-regions, which also displayed some degree of homogeneity.) Consequently, any music region could be differentiated into sub-regions, at various scales, according to the intent of particular approaches. But diffusion and migration constantly influenced all regions, with the introduction of new musical forms (and new peoples), in parallel with other social and economic changes, hence most regional divisions have been challenged (see, for example, Pickering and Green 1987: 21–2) as far too simplistic and of little contemporary relevance.

Because of migration, cultural diffusion and acculturation the notion that there are authentic folk musics – and hence regions – has always been open to doubt. Distinctive folk musical traditions were easier to recognise where cultural change was less evident, written music and professionalisation absent and unity of style more common. However, other folk music regions, notably those of North America, were largely products of migration and diffusion. Until the twentieth century the folk music of the United States 'was based primarily on a complex of borrowed, adapted and improvised musical forms that still largely reflected the musical culture of England and Germany' (Carney 1978: 287). Some of the best-known folk music regions in the world had little or nothing to do with the early history of those regions. Folk-song regions found their way into introductory cultural geography textbooks, where it was argued that 'folk music provides an excellent basis for delimiting formal regions of nonmaterial items of culture' (Fellman *et al.* 1996: 233), and reproduced in map form (e.g. Figure 2.1). Divisions were made between the Northern region ('based largely on British ballads') and the Southern region ('where the solo is high pitched and nasal' and ballads 'more guilt ridden and violent than those of the North'), adding a 'French-Canadian' region and a 'Mexican-American' (Spanish) region to those of Lomax, but failing to cope with much of Florida. None of the regions had any relationship

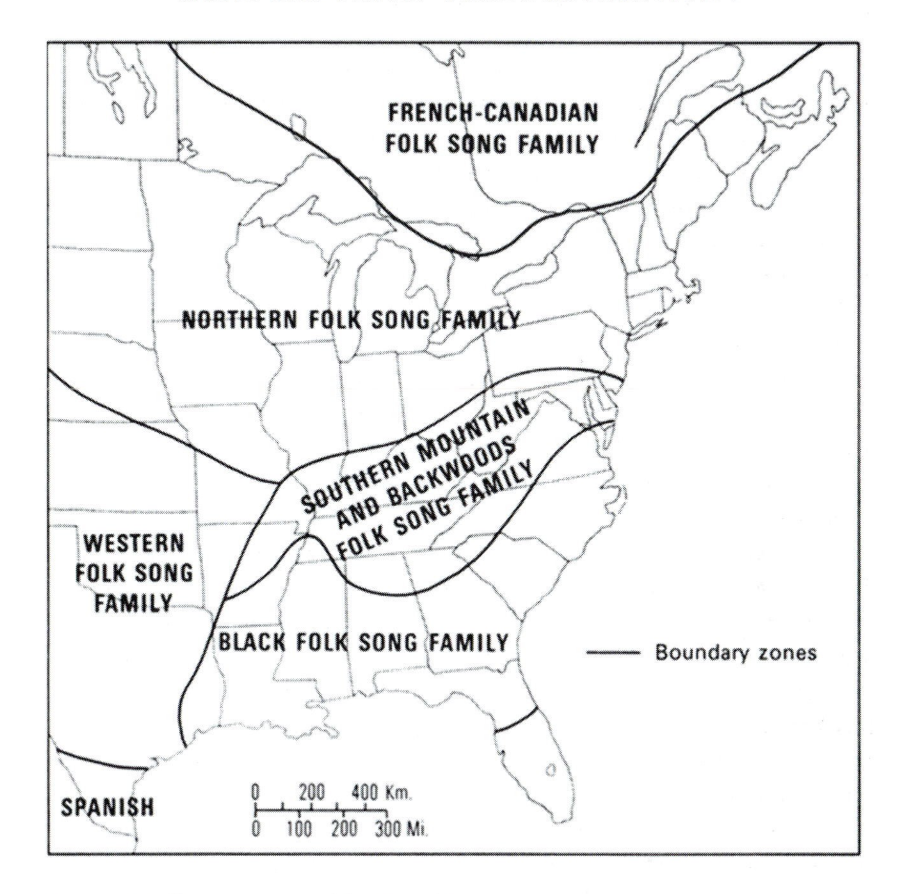

Figure 2.1 Folk song regions of the United States
Source: Fellman, Getis and Getis 1996

to earlier Native American music regions. Such divisions thus arbitrarily emphasised exogenous influences, combining content, style and geographical origin. The folk regions were again functionalist and timeless; fixity, through mapping culture, rendered a dynamic landscape immobile. Other than indicating the role of migration and diffusion in the creation of musical style, regions merely froze contemporary culture into a mythical but indeterminate past and sought to create and stabilise authenticity.

Regional differences were ubiquitous, just as the influence of colonialism on indigenous music was not homogeneous. In Britain the north-east of England had its own forms of music and speech. Ireland, Scotland (differentiated into, at the least, the highlands and lowlands) and Wales had separate traditions and instruments (the harp in Wales and Ireland and bagpipes in both Scotland and Ireland) that had not existed, or had largely died out, elsewhere in the British Isles. Instruments varied between and within regions. The pipes of Scotland were different from those of Ireland. In north-east England, the Northumbrian pipes – one of the only instruments indigenous to England – were a local favourite. In County Clare (Ireland) the concertina, invented in England around 1829, gradually came to dominate folk music – a cheap alternative to bagpipes. Like so many local 'traditions', the reasons for its popularity are unknown, but have been attributed either to it being introduced by sailors along the banks of the River Shannon, or as a result of Clare having more musicians per capita than any other part of the country, and thus being particularly receptive to all musical influences (O'Róchàin 1975). The Industrial Revolution placed pressure on folk music and particularly on regional variations, as populations became more mobile (assisted by railway construction) and more urban, and church hostility opposed the music, musical instruments and dancing of several Celtic areas. In many urban areas new musical traditions emerged, through the creation of songs that protested against harsh industrial conditions (Palmer 1974) alongside increased numbers of street musicians, organ grinders and, later, brass bands, which though associated with England had often come from Germany (Russell 1997). By the 1870s newspapers had displaced ballads as a news form, but singers had moved into pubs and eventually became some of the first music hall performers.

By the twentieth century folk songs faced competition from other forms of popular music, especially in urban areas; while regions retained some degree of vitality, that vitality was emphasised by the perception on the part of practitioners, audiences and, increasingly, promoters, that folk music required and was sustained by local roots and traditions (Middleton 1990: 127–8). Such ideas began to feature more prominently in advertising, and were stressed on record sleeve notes, as in the description of the period immediately after the First World War when one of the last great Northumbrian pipers, Billy Pigg, learned his skills from Tom Clough:

> Billy used to sit in Tom Clough's kitchen, while the master took a class of aspiring young pipers. At the end of the evening Billy would cycle the six miles back....furiously humming the tunes that Tom had played, memorising them so he could practise them as soon as he got home.
>
> (Charlton 1971)

A decade later Pigg moved to work on a new farm in northern Northumberland, no more than 50 km away:

> When he deserted south Northumberland for the northern hills, Billy took with him a store of tunes gathered from Tom Clough and the other pipers of the district and was warmly welcomed by the traditional musicians of his new environment...a mutual exchange of tunes was

> inevitable. There thus became available to him many old tunes passed down by generations of country dance musicians.
>
> (Charlton 1971; see Feintuch 1993: 189–90)

Similar processes went on in Ireland, at different scales; in winter in Donegal:

> Relatives, who had spent the summer working in the South or in Scotland, would, on their return, sing every night for a week or so, teaching fresh songs or lilting new tunes for the fiddlers to play.
>
> (Yates 1975)

In this way, and through widespread references to locality and region, folk music was wholly imbued with a sense of place, history and timelessness – pre-literary and pre-recording technology – though by then local and regional processes of personal diffusion were being supplanted by sheet music, broadsheets, the gramophone and later the radio.

Outside the developed regions of North America and Europe, regional traditions were similarly breaking down under the various diffusions and cultural change associated with an acceleration of global linkages. Latin American folk music had acquired what was increasingly seen as a Hispanic heritage, though in many areas – typified in the Caribbean – there were significant African influences. Africa itself was influenced by Islamic and other musical traditions – via coastal and continental trade – and Asia also had absorbed significant European and other influences long before the twentieth century. Paradoxically music regions, at any scale, could only become recognised as they were in the process of transformation; as regions were defined, so they faded away. Transformations, initially limited by the slow oral transmission of music and diffusion of instruments, accelerated over time, with the rise of literacy and the technological transformations that followed the Industrial Revolution. While most cultures welcomed innovations – musical and otherwise – the extent of change depended on local circumstances. By the end of the nineteenth century what was perceived as a longstanding nexus between the local region and musical traditions could no longer be universally upheld, and the practice of folk music increasingly constituted a quest to claim authenticity by recognising the virtues of the local and traditional.

The survival and revival of local authenticity?

Although local musical traditions remained in place even in the most industrialised parts of the world – diminished but nonetheless significant as indicators of local, regional and national identities – concern had mounted by the end of the nineteenth century that the demise of some traditional forms was imminent. Though many musical cultures had disappeared into oblivion long before then, the impact of the Industrial Revolution and the rise of global capitalism (including large-scale and long-distance migration) also contributed to a rise in middle-class nostalgia for a world that had been lost. Migration of people (and of instruments) was perceived to threaten purity and integrity, as music, instruments and songs were detached from their cultural context – especially as industrialisation and urbanisation ended the dominance of rural life. Just as soon as scholars began to take an interest in folk music, its authenticity was argued to be at risk; even at the social and geographical margins of European life, some folk songs and dances had ceased to be performed by the middle of the nineteenth century (Boyes 1993: 1). By the end of that century, on both sides of the North Atlantic, collectors were already seeking to preserve what was 'authentic' in the face of rapid change. In the late nineteenth century, for example, Francis Child produced the five-volume *English and Scottish Popular Ballads*, in which he sought to differentiate 'traditional' ballads from recent compositions. Folk song was being avidly recorded and preserved, especially in rural areas, where agricultural life was in decline, and there was an assumption that folk song was dying out; active preservation intensified after the founding of the English Folk Dance and Song Society in 1898. At about the same time morris dancing was revived in England, 'probably because the English thought they invented it', though its origins were in African fertility dances that had long before spread to Europe (Clarke 1995: 69). Uncertainty in the face of industrialisation triggered cultural conservatism.

In North America and Europe folk music had gradually become disengaged from the small-scale, sometimes domestic, context in which it originated. In the United States twentieth-century fiddle contests emphasised prizes and public display, as opposed to the older tradition of playing at house parties and barn dances, a simultaneous process of commodification and the convergence of different regional styles. Many genres of music were not at risk, but declined and were transformed in the twentieth century. Songs that offered both information and entertainment increasingly became obsolete, as the radio replaced them. Work songs accompanying or celebrating particular activities, especially in rural areas, declined as newer capitalist modes of production displaced them. Transport, radio, records and the cinema threatened some local musical forms, by providing various alternatives, where few were previously available. In India many folk genres only survived by incorporating film songs into the repertoire (Manuel 1993: 56). Similar syncretic forms emerged in most contexts, with cross-over between folk traditions and more homogenised popular music. Folk revivals and preservation celebrated an older order: local music, community and amateurism, untainted by urbanisation, transformation, modernity and reliance on a centralised recording industry.

The recording of music and song – even on paper – began to alter it, formalising it rather than allowing gradual evolution and transformation (Attali 1985). Electrical recording was even more effective in standardisation. From the very first recordings of Indian music at the start of the twentieth century, for example, performance existed in two worlds: the extended live performance and the two- or three-minute duration of the record. The short duration of records immediately posed problems. From the earliest days many great performers of Indian music refused to be recorded 'because they found recording to be contrary to the spirit of the art' and audience feedback, an important ingredient of performance, was no longer possible (Farrell 1998: 75). Preservation changed, restructured – and even reduced – the content, format and quality of what was being recorded. As the revival of morris dancing indicated, what exactly was authentic about 'traditional' or 'folk' music and performance was impossible to define and discern. Even when Child was collecting his ballads many songs had already become commercial – and had re-entered the oral tradition as performers learned them from broadsheets (and later recordings) – while other initially commercial recordings had entered the oral tradition.

Folk revival symbolised attempts to preserve early musical traditions and, to some extent, the cultures of which they were part: working-class, rural areas where elements of community life remained, suggesting continuity with earlier times. Revivalism favoured rural areas, as popular music became more urban, rural songs seemed to epitomise dying lifestyles and 'folk music' was something played in an appropriate setting, beyond the reach of urban transformation. Some perceived 'genuine folk music as that where authorship had been lost or forgotten over the years' (Carney 1996: 74), denying the possibility of urban folk song, the creativity of known authors and performers or the need for change. Authentic music, whether folk music ('country music' rather than Native American music in North America), or the music of less developed countries, was also often seen as that performed live, hence no electronically transmitted performances could be authentic (Peterson 1997: 239). Music and songs created and performed by urban workers – including miners and railway workers – were unlikely to be seen as authentic, and worthy of preservation. The English folk revival had parallels in the promotion of English church bell ringing, which entailed evoking a sense of continuity with a medieval past. Church bells, and the art of bell-ringing, were seen as symbols of the continuity of English church life: 'bell-ringing has been described as the last authentic sound of medieval England' (Johnston *et al*. 1990: 5) though it is now an invented tradition with contemporary sequences of changes in bell-ringing bearing little resemblance to those of two centuries ago. Their symbolic significance was revived and enhanced in the twentieth century through the growth of bell-ringing, as a musical form, and in literature through the work of writers such as the Poet Laureate, Sir John Betjeman, who wrote of bell-ringing as a 'classless folk art', with the bells, a testimony to the expertise of English craftsmen, ringing 'through our literature as they do over our meadows and roofs and few remaining elms' (quoted in Johnston *et al.* 1990: 20, 33; see Johnston 1986; Train 1999.) In Scotland revivalism emphasised Gaelic language songs that preserved Scottish national identity (Munro 1996). These pre-empted efforts in the United States, mainly in the southern mountain regions of Appalachia, to collect and preserve what was 'native and fine' (Peterson 1997: 239). Similar attempts were made in continental Europe and, rather later, globally, to preserve the music and culture of modernising communities in developing countries.

Increasingly attempts to seek authenticity took on commercial connotations, as impresarios of commercial popular music sought to find new market niches. When country music suddenly became popular in the United States in the 1920s the music of the earliest entertainers:

> relied on untrained, high-pitched nasal voices and simple musical accompaniments, evoking images of farm, family and old-fashioned mores along with more than a dash of sexual double entendre. Entertainment industry impresarios [consequently] sought out old men steeped in tradition, playing old songs in traditional ways. The performances of these old-timers, even if historically and aesthetically accurate, were...taken by the radio audiences and record buyers as bemusing novelties....To the consumer, authenticity was not synonymous with historical accuracy.
>
> (Peterson 1997: 5)

For three more decades the industry tried numerous permutations on the theme of 'rustic authenticity' (Peterson 1997: 5). Folk-song revivals elsewhere has less to do with preserving the past (which in the sense of rural community life was ever more distant) than with invoking a particular historic image, such as the supposed solidarity and class-consciousness of agrarian workers, and hence the fabrication, rather than preservation, of authenticity.

Fabrication of authenticity, and challenges to invoking continuity, became more evident over time. Simultaneously nostalgia and contemporary songs played a growing role. This was evident in the second English folk-song revival, which coincided with the 1930s Depression years, and in the third revival in the 1960s, which was linked to the radicalism of the Campaign for Nuclear Disarmament (Harker 1985). During both periods, in Britain and in the United States, new songs of struggle emerged, famously 'The Ballad of Joe Hill' (a tribute to the leader of the International Workers of the World), and there were deliberate attempts to link music to social protest. In the second half of the twentieth century professional folk singers, such as Pete Seeger, Joan Baez and Ewan McColl, revived old songs and composed new ones as 'folk music' became one element in commercial popular music. Folk clubs boomed in the 1960s in most industrialised countries – to some extent the successor of student jazz clubs, primarily supported by the middle class – in an era when folk music was firmly entrenched in the electronic media that, according to one observer, simultaneously 'made it available to a larger public and castrated it' (Watson 1983: 40). Yet, despite commercialism, folk song never replaced other forms of popular music in variety theatres and folk dance failed to influence the crowds in urban ballrooms (Boyes 1993: 113). Many performers, such as the Copper Family of the Sussex Downs, retained a repertoire of agricultural songs, but performed in new circumstances (Box 2.3), and stressed the significance of history, rurality and place (either English village life, for the Copper Family, or rural Ireland and Celtic traditions, for such groups as the Chieftains and Boys of the Lough). In each case, as in the linking of bluegrass music to Appalachia, and the nineteenth-century folk revival, the invented geography that was produced saw each place as 'pristine remnant of a bygone natural environment...unspoiled by the modernist thrust of urbanisation and industrialisation' (Peterson 1997: 215). While brass bands too had faded, for a variety of reasons, their powerful links to local community always hinted at revivalism. Yet forms of music were constantly evolving, and attempts to capture and preserve an essential, authentic past were doomed to failure. Local traditions remained, partly by default but partly re-created.

Box 2.3 A song for every season

During the 1960s and 1970s folk revival in Britain, one of the more popular and acclaimed performing groups in the Sussex area was the Copper Family, an extended family group who, in Sussex accents, performed unaccompanied songs of English rural life in earlier centuries. The Copper Family descended from agricultural workers long present in Rottingdean, a coastal village east of Brighton (and now effectively a suburb of it). The songs they sang, passed on orally between generations, were dying in the 1920s, though in the village pub the old men 'clung on to the singing of the old songs as one of the few things that remained constant in a rapidly changing world' (Copper 1971: 183). Such places and songs reeked of authentic English rural life.

> In that jam-packed room where the atmosphere was near curdled with strong shag tobacco and even stronger language...some old fellow, bowed nearly double with years of work and weather, got up to sing. He would take up a sooted-up clay pipe from his mouth with

> a hand like a clump of blackthorn, clear his throat, spit approximately in the direction of the spittoon, then, raising his eyes to heaven as if listening for a cue from some celestial prompter, would sing naturally and tenderly a song like 'Sweet Lemeney'.
>
> (Copper 1971: 184)

Such men were perhaps 'the last of a dying race of old-time countrymen' whose songs generally met 'a good deal of antipathy. They were old-fashioned and out of date and not nearly so exciting as the American importations that came through the horn loudspeakers in the corner of almost everyone's living room' (Copper 1971: 184–5). In the 1930s Jim Copper recorded the words of over sixty such songs in his *Song Book*, to ensure they would be remembered and 'recapture the moods and feelings of the mid-eighteen hundreds' (Copper 1971: 3) at a time when song collectors could never be sure that even the oldest singers 'would produce a real old tune, or one of the music hall songs which had been popular in the 1914 war' (Wightman 1971: xii–xiii). The first revival of the Copper Family songs was on a BBC radio programme, *Country Magazine*, in 1950 when two of the Coppers were broadcast live, singing 'Claudy Banks' in the open air looking down over the South Downs 'in the heart of that part of Sussex to which we all – both songs and singers – belonged' (Copper 1971: 185). The BBC then initiated a national search 'for the remnants of traditional singing to...preserve them for posterity', and the Copper Family sang at the Royal Albert Hall and eventually made a number of radio and television broadcasts. The Family:

> did our damnedest to bridge the gap between the tap-room and the concert platform [where] the pallid ghost of the old singing tradition was living on only in the artificial world of folk song recitals where...elegantly-gowned sopranos...or moustachioed baritones in evening dress made brave attempts at resuscitation.
>
> (Copper 1971: 189–90)

They sought to return folk music to its roots. Bob Copper's account of the Copper Family and rural life in the South Downs provides a sentimental view of bygone rural life, where, at the turn of the century:

> Village folk were insular to a degree that is difficult for us even to imagine. They remained on the whole much the same as their fore-

> bears, uninfluenced by outside forces in their dress, their speech, their method of work and ways of spending their leisure.
>
> (Copper 1971: 52)

In agriculture

> many of the methods employed were still precisely the same as they had been for a thousand years. They...threshed out grain with flails as their Saxon ancestors had. Their lives were hard but uncomplicated and through living and working close to nature they had a clearly-defined and well-balanced sense of values. In the main they were content...in the words of one of their old songs 'Peace and Plenty Fill the Year'.
>
> (Copper 1971: 53–4)

It was seen as an idyllic, self-contained world, replete with folk wisdom, dialect ('A shepherd, 'e don't say nothin', only swear at ye') and a simplistic naivety about music; at the first BBC broadcast, its musical director commented 'Just listen to those diminishing fifths', to which Jim Copper is recorded as retorting 'What the 'ell d'you mean – "diminishing fifths" – are we gone wrong or summat? That's the way we always 'ave sung 'er' (Copper 1971: 186). Despite Bob Copper's book being published in 1971 there is no record in it of the manner in which it was the folk-song revival that brought some degree of fame to the Copper Family and enabled its publication.

Local linkages were increasingly traced within more mainstream popular music, some of which, from Bob Dylan to Bruce Springsteen, made deliberate obeisance to folk-music traditions of amateurism, performance and the notion of community (see Box 2.4). Rock music was generally perceived as being more credible, as it appealed to both 'folk' and 'art' sensibilities, to the music of the youth community and to creative sensibility (Frith 1987). Rock music thus articulated post-modernism and youth alienation. It combined entertainment with social identity and values such as resistance, refusal, alienation and marginality. Where this continued to occur popular music could claim authenticity; where it was influenced by the economic interests of 'establishment' it lost political edge and hence its authenticity. New forms of apparent working-class rebellion, and expression, such as punk, metal and rap music, were thus imbued with authenticity, despite their novelty, through their links to amateurism, performance and even 'ordinariness' (Leach 2001). Popular music

Box 2.4 Bruce Springsteen, place and authenticity

More than most other performers, Bruce Springsteen sought to emphasise the relationship between place, community and identity. His songs are primarily about working-class issues, evoking notions of community and local identity: 'the effects of poverty and uncertainty, the consequences of weakness and crime...the murky reality of the American dream' (Frith 1988: 98). They honour yet transcend ordinariness, though they have been criticised for sexism (Moss 1992; cf. Smith 1992). Songs are constantly set in place, in the small-town United States, notably Asbury Park, New Jersey, a postcard of which was on the cover of his first album *Greetings from Asbury Park, NJ* (1973); subsequent albums such as *Born to Run* (1975), *Darkness on the Edge of Town* (1978) and *The River* (1980), and tracks such as 'My Hometown' (1984), were full of references to the places where Springsteen had lived in the 1960s. He has also provided cameos of other American landscapes as in 'Badlands' (1978), the various tracks on *Nebraska* (1982), a solo acoustic album that offers ravaged tales of despair, defeat and defiance, in the context of an American state rarely celebrated in song lyrics, and *The Ghost of Tom Joad* (1995) (inspired by John Steinbeck's vision of rural America). Numerous songs, such as 'Born to Run' (1975) are American 'road movies' in miniature, while 'Streets of Philadelphia' (1993) above all (assisted by the video and soundtrack in the film) pointed to the anguish of the inner-city United States. Springsteen provided tributes to diverse working-class landscapes: songs that in many respects were populist, nationalist invocations of the United States. The most celebrated of all, 'Born in the USA' (1984) was both interpreted, as Springsteen intended, as opposition to Ronald Reagan's 'colonisation of the American dream' (Frith 1988: 101), but, with the Stars and Stripes flag fluttering aloft, as wholesome patriotism (see Chapter 9). However interpreted, they represented and articulated the United States. His regular live performances (with a stage show and band), including frequent free performances at benefit concerts (and willingness to experiment with acoustic albums), suggest spontaneity and commitment, emphasised by the 'work clothes' of blue jeans, singlet and headband. The images of class, place, nation (and himself) that Springsteen created, and adhered to, granted him a credibility shared by few other established performers.

See: Frith 1987; Sandford 1999; Palmer 1997.

that was placeless, created electronically and highly commercial, was regarded as a major disjuncture between producers and consumers, and a denial of authenticity. What distinguished authenticity in rock music was its emphasis on spontaneity and improvisation (Middleton 1990: 53) rather than any long-term historical continuity of performance as invoked in folk and country music. Yet even this was a source of apparent authenticity in a different kind of way, as some rock music, even that of the most commercial performers such as Bruce Springsteen, cherished continuity with earlier popular music and evoked a sense of place. Performers whose success was national, such as Midnight Oil (who raised issues concerning Australian national consciousness) or primarily local, such as the Whitlams (who sang of gentrification in inner-city Sydney, and demanded audience participation) had little difficulty in being attributed integrity and authenticity, in the face of commodification, commercialism and mass communication. Bands – such as Ladysmith Black Mambazo, Lindisfarne or Cypress Hill – chose local names, which were evocative of place, and often of history and rurality (and thus of community). More generally, where any form of popular music has provided some link with place and community (including the fans), displayed some sense of history, or claimed some heritage (in instruments, local performers or ethnicity) and evoked lived experience there have been claims to authenticity.

Music in place

Music functions as a form of entertainment and aesthetic satisfaction, a sphere of communication and symbolic representation, and both a means of validating social institutions and ritual practices, and a challenge to them. Music may comment upon and reinforce, invert, negate or diffuse social relations of power (Roseman 1991: 175). As societies changed, and music became commodified in different ways, direct links between local cultures and musical practices competed with the more rapid diffusion of musical instruments, tunes and songs. Early incorporation of cultures in a wider context, such as through the partial globalisation of the Islamic world, brought more substantial change than hitherto, where diffusion was slow and over short distances, though even then there was a proto-capitalist trade in music and song (evident, for example, in several Melanesian societies), and some evidence of both syncretism and displacement.

The notion of roots and origins has remained important to many styles and releases, and mobilisations of tradition at various times have constituted imaginative affronts to disempowering economic and cultural change. Yet while there have always been searches for genuinely local music, correspondingly there have always been problems in 'freezing' the music and cultures deemed to be authentic. Searches for roots and celebrations of tradition became more complicated and contradictory. Authenticity could provide the raw materials for depictions of exotica and otherness (see Chapter 7), which only trapped, rather than liberated, subaltern identities. Many were moved enough to subsequently denounce the quest for authenticity altogether, abandoning stories of commodification and cultural loss, and celebrating the contemporary, arguing that

> newly emergent hybrid forms, and the middle-class cosmopolitan cultural world to which they belong, do not necessarily constitute a degenerate and kitschy commercial world, to be sharply contrasted with a folk world we have forever lost. In fact, it may be the idea of a folk world in need of conservation that must be rejected, so that there can be a vigorous engagement with the hybrid forms of the world we live in now.
>
> (Appadurai 1991: 474)

Various slippages always occur, as 'authenticity' starts to blend in with what might (for want of a better word) be described as 'credibility' (which all subscribe to in one way or another through personal tastes). Both authenticity and credibility are constructed in relation to how continuity and change are perceived. Folk music has been 'authentic' because it endeavoured to maintain an oral tradition, yet credibility accrues to the innovative (who themselves gain 'authenticity' over time as they are recognised as the innovators of an important style), and to the skilled (even though those with little skill have been seen as credible, as in 'do-it-yourself' punk). This also happens with the places that music has come from. Regions of dynamism and creativity, places perceived to be the origins of novel sounds, become credible as sites of innovation, and subsequently become authentic, as they are increasingly depicted in media and imaginations in relation to music. Yet places to which music has been linked in this kind of way can become dated, where 'scenes' have come and gone (see Chapter 5). Authenticity was constantly sought by embedding music in place, yet such efforts never guaranteed commercial or critical longevity. In contrast to attempts to concretise music in place and in social practice, popular culture reflected the fluxes and fluidity of contemporary life, unsettling binary oppositions established in earlier phases of modernity (tradition/contemporary; authentic/inauthentic; local/global) by refusing to be pinned down. Music had been, and would continue to be, mobile.

3

MUSIC AND MOVEMENT

Overcoming space

This chapter introduces the different forms of mobility linked to music. Music, like all forms of sound, is inherently mobile. Mobility also involves movements of people, and the music they bring with them (Chapter 8). Recordings themselves move as objects – as treasured artefacts of lives and places, traded through migrant links across countries, underpinning musical economies or sustaining diasporic connections. Mobility is also responsible for performance types and musical spaces rebuilt in new circumstances as the result of flows of music, rekindling traditions or inspiring unexpected borrowings and appropriations. The movement of sonic objects – CDs, instruments, recording technologies – highlights the centrality of technology in discussions of mobility and music, a theme that reverberates throughout this chapter. Music is rarely, if ever, made beyond technology (even a cappella vocal performances often take place in sites chosen because of their particular harmonic qualities), and technology invariably prefigures mobility. Technological developments in music have been intimately connected to industrialisation, colonialism, urbanisation and associated social change. Technologies transform spaces of consumption, as different generations navigate the trends and styles of contemporary popular culture.

Mobility engenders a sense of dynamism in time and space, a heightened sense of transformation most evident in the contemporary era, what Harvey (1989) called 'time-space compression'. Mobility is both metaphorical and material in the modern world in various guises, as 'lines of flight' and 'nomadism' that create new 'spaces of flows', in a series of cultural 'scapes', such as media-scapes and ethno-scapes (Appadurai 1990). Chambers emphasised this state of flux, when characterising new spaces of consumption, even 'hyperspaces' (Kearney 1995: 553), opened up by contemporary media:

> There are no eternal values, no pure states: everything, including cultures of resistance and the oppositional arts, is destined to emerge, develop and die...we all become nomads, migrating across a system that

> is too vast to be our own, but in which we are fully involved, translating and transforming bits and elements into local instances of sense.
>
> (Chambers 1993: 192–3)

Complex, and often indeterminate, hybridity is paralleled in histories of music that are told through metaphors of evolution and journey, revealed in how 'progress' and 'modernity' – both as technological advances and as global diffusions of knowledge and style – were measured through music. Yet mobility only makes sense as a counterpoint to the trends discussed in the previous chapter. Metaphors of hybridity, and of fluid, virtual spaces explain only part of the story; mobility triggers new attempts at fixity – holding on to traditions despite losses of popular appeal, constructing spaces for local expressions (both material and discursive), developing cultural industries, marketing music through place and marketing place through music. As discussed in Chapter 2, some music has been constructed as authentic because it suggested a directness of communication between artists and audiences – an emotive 'honesty' in words and sounds. Apparently authentic music, often the subject of ethnomusicological study, became so in part because of an assumption about being embedded in place, the extent to which music was a product of fixed social and spatial co-ordinates. However, such interest in fixity only emerged as a reaction to the sense of disjuncture apparent in commodification, and those ideas too have always been mediated by various technologies and through a range of cultural gatekeepers.

When music became commodified as sheet music and as recordings, the distancing of those who made, as opposed to those who owned the rights to music, expanded immensely. Socially embedded musical practices were transformed into exchange value (the rights to profit from selling material musical objects), defining a moment where music became a commodity. This moment for many might have begun when a recording desk entered the picture, but commodification was as much when the rights to 'capture' and distribute sounds exchanged hands and became owned by others (those with the means of production), in what was to eventually become known as copyright. Copyright remains the conceptual and legal mechanism of ownership and rights of music, irretrievably connected to technology, and thus a crucial concept underpinning fluidity (see Chapter 11). Applying regimes of copyright ownership to music implies diffusion. Owners of rights in music, themselves cultural gatekeepers, earn profits from selling music by physically removing music from its sites of production and initial consumption. The sets of cultural alliances and networks that were required for movement were always uneven. Music follows particular tracks linking producers and consumers in different places. Sometimes such tracks are a product of migration, while others are part of promoters' efforts to 'penetrate' new markets. Many factors, alongside copyright, including similarities in tastes, linguistic and other cultural barriers (such as the persistence of local traditions) shaped the structure of mobility. This chapter thus explores the flows of music that oppose fixity (Chapter 2), examining the ways in which musical movements have shaped particular geographies of music across eras. Later chapters expand on thematic aspects of tensions between fixity and fluidity, in tourism, in various musical texts, in sub-cultural formations and music 'scenes'.

Early diffusions

In contrast to the kinds of assumptions made about 'traditional' music, diffusion of musical forms was often extremely important prior to European colonialism and commercial expansion. By the nineteenth century the Chopi of Mozambique possessed semi-professional xylophone orchestras, but the entire Chopi system of orchestras, techniques of composition and use of heterophony was probably imported from Indonesia, via Madagascar, while xylophones had arrived in Madagascar around AD 500 (Nettl 1956: 11–12). At similar time periods as many as three separate waves of musical style reached North America from Asia, Arabic musical forms reached East and West Africa and Indonesia, and various long-distance transfers occurred elsewhere (Nettl 1956: 118). Numerous other examples of musical change and mobility prior to exchanges with Europeans (Kaemmer 1993: 173–4, 191–6) were vividly apparent through the expansion of Islam and the resulting spread of musical instruments:

> The Kurdish spike fiddle then was found in the Balinese gamelan as well as in the hands of the Egyptian bard; the old Semitic frame drum was struck by the Arabian as by the Spanish girl; the Persian oboe was blown by the Dayak in Borneo as in Morocco.
>
> (Sachs 1940, quoted by Kaemmer 1993: 192)

Diffusion involved movements of people and ideas, introductions from both neighbouring and distant places.

In New Guinea new songs were 'purchased' by Arapesh musicians (exchanged for shell rings and other goods) to replace 'songs which are remnants of long forgotten dances', remaining in use for a few years until new imports replaced them (Mead 1935: 9). For Suya too, even in communities without shops where money was largely absent from village life, visitors from beyond 'usually brought cassette tapes recorded in their villages, which they would trade for those recorded by the Suya of their ceremonies' (Seeger 1987: 58). Despite capitalism being in its local infancy, its ramifications were sometimes quickly felt in indigenous societies. In Siwai it was regarded as an accomplishment to be able to sing the songs of neighbouring communities, even

Figure 3.1 Music in culture contact, Papua New Guinea
Note: In parts of the New Guinea highlands a gramophone accompanied the first moments of contact in the 1930s.

if the words could not be understood (Oliver 1955: 369; see Clay 1986: 207–8). Here and elsewhere any innovation was readily incorporated when perceived to be of benefit.

The pace of musical change accelerated with the more pervasive influence of colonialism in various guises, including administration, capitalism and missionisation. Migration (to plantations and mines) and marriage crossed old borders and furthered mobility, education introduced new languages (and diminished old ones) and missions brought hymns in the new languages, often opposing traditional rituals, ceremonies, dress and decoration. The merits of European music (especially hymns) were imposed and local music was condemned as 'primitive', 'backward' and 'immoral'; administrations sometimes introduced military bands and settlers brought familiar songs and instruments from the 'old country'. In some places, as among the Kaluli and nearby groups, traditional music (and related activities) were actively suppressed (Schieffelin 1978, Stockmann 1994). Missions contributed to rapid changes in musical style; in Japan, polyphony was new, European music was not tied to guilds nor did it entail other social restrictions, and church music became a considerable force in the conversion of many Japanese to Christianity (Kaemmer 1993: 181). The pace of change accelerated further where prestige was linked to innovation, as in rural India where weddings were increasingly accompanied by hired musicians with amplified equipment.

Where change was particularly rapid, and there was external opposition to elements of local culture, the impact on musical culture was devastating. Among the Munduruku (Mundurucu) Indians of Amazonia, who had grown rich from panning gold and were able to buy batteries, record players and records:

> The arrow has no meaning to men who stalk the jaguar, the tapir, the peccary, the deer with rifles; myths make no sense to children clustered about the transistor radio. The sacred flutes gather dust, their reeds rotten and their tunes almost wholly forgotten, and pose no challenge to the phonograph.
>
> (Buckhalter 1982: 203, quoted by Seeger 1987: 137)

Such extremes of abandonment were rare (and were not always permanent), and few if any musical traditions were utterly abandoned. Syncretism usually occurred, with elements of different cultures and musical systems being combined together. As in many other facets of society, the introduction of new musical forms, among the Suya and elsewhere, 'incorporated the power of the outside world into their social reproduction', providing new materials for community activities (Seeger 1987: 140). Change both threatened and reinforced identity, hence authenticity was often guarded and proclaimed by advocates for indigenous cultures as central to local musical styles, even when such cultures had incorporated familiar folk songs like 'Clementine' in local repertoires (Kaemmer 1993: 170; Feld 1996a). Claims for embedded tradition were often problematic. While many indigenous communities frequently suffered severe colonial impositions, others clearly chose to adopt new practices within existing social relations. Resulting attempts to bind indigenous communities to notions of 'authenticity' that required static musical expressions could be as disempowering as colonial suppression.

Population mobility, urbanisation, trade and migration – and colonial administration itself – all contributed to parallel changes in musical styles beyond Europe. In Congo local musicians learned French dances and love songs of the 1930s and 1940s from resident European women, most of whom were the musicians' mistresses. Kru sailors from Liberia brought Latin American styles, notably the rumba, from West Africa, while 'the colonial trilogy of religious choirs, scout songs and military parades were fundamental to the formation of Congolese music' and its gendered structure (Gondola 1997: 70–1; Stapleton and May 1987: 8–15). On the other side of the world, half a century later Papua New Guinea went through similar phases, associated with mission music (through churches and schools), European national anthems and patriotic songs, the introduction of Western instruments (such as harmonicas and concertinas), during turn-of-the-century gold rushes, and later the gramophone. The introduction of the guitar effectively defined a new phase of more rapid change. By the 1960s, when electric guitars enabled the emergence of 'power bands', the pattern of musical change paralleled that in other small countries (Wallis and Malm 1984) with original music being composed by groups, 'self-conscious experimentation with the mixing of transnational music style with precontact music forms', the establishment of a local recording industry and greater access to musical technology (Webb 1993: 2–4). Part of this change reflected a 'pecuniary embrace with the local music industry' and, at least in South Africa, a long and 'continuing obsession of black South Africans with things American' (Ballantine 1999: 1), to the extent that, at least by the 1950s, there was anxiety over whether African 'traditional' music could survive 'the onslaught' of Western influences (Euba 1970: 52; Wachsmann 1953). Later, the rapid dissemination of electronic keyboards evoked similar concerns. Though hybridity took different forms in different places, what some took as a threat to local distinctiveness was interpreted elsewhere in a more positive light and, in some cases, fluidity of musical influences provided blueprints for new reclamations of place (Box 3.1).

Box 3.1 Nitmiluk: reclaiming place in song

One particular song, by Blekbala Mujik, an Aboriginal band from Australia's Northern Territory, illustrates the ways in which musical flows gave an indigenous community material for figurative reclamation of place. The Jawoyn community, like others in the Northern Territory, were dispossessed of land after colonial expansion in the 1870s. Jawoyn mobility over their country was restricted, pastoral stock polluted water sources and strangers occupied traditional lands. Jawoyn people were, by the early twentieth century, 'experiencing the violence and cultural shock of colonization' (Jawoyn Association, quoted in Gibson and Dunbar-Hall 2000: 42). After successive government policies of separation, protectionism and assimilation, Jawoyn communities were involved in a long struggle for title rights to their traditional country. Land rights could provide opportunities for the Jawoyn to reclaim some of the ground lost to colonialism, rebuilding communities battered and neglected by Australian governments, and articulating new Aboriginal regional activities. Legal recognition of their territories involved one site, Nitmiluk (known to non-Jawoyn Australians as Katherine Gorge), a stunning landscape and nationally significant tourist attraction, hence the symbolic value of Nitmiluk in colonial imaginings rendered it a highly contested case. The Jawoyn became embroiled in a legal case that took eleven years to resolve, since to prove traditional attachments to territory (and qualify for the return of land), the Jawoyn had to 'prove' their anthropological integrity through a traumatic and drawn out process. The Jawoyn eventually gained ownership of their traditional country and secured a leasehold/management arrangement with Nitmiluk National Park.

Blekbala Mujik, a local Aboriginal band renowned for its blend of country styles, traditional didjeridu-based arrangements and reggae, wrote the song 'Nitmiluk' in 1989 to commemorate the handing back of Nitmiluk to traditional owners, and celebrate Jawoyn strategies of self-determination based on this reclamation. The song was quite literally a direct response to Nitmiluk as landscape, in this case, as songwriter Apaak Jupurrula explained:

> We slept on that earth and it protected us. Our lifestyle revolved around that place and old people came telling us of its importance. One night we were strumming around the camp fire and the song and words came from the wind into our collective mind.
>
> (quoted in Gibson and Dunbar-Hall 2000: 46)

The song mobilised traditional Jawoyn musical motifs, in a didjeridu opening based on a traditional piece, 'White Cockatoo', which then moved into a country-rock style, a mixture of American and Aboriginal regional influences:

> What I do as a songwriter is base new works around what we already have in the way of traditional music...we want to provide some kind of information to the audience that we are strong within our cultural beliefs, that we still maintain our traditional ideology and understanding of a world view, and we would like to share that with the public.
>
> (quoted in Gibson and Dunbar-Hall 2000: 45)

Rock music provided such a platform. Fluidity underpinned a mixing of styles and intents, in the recalling of two regional traditions (one distant, one persistently local), while the lyrics of the song appealed to a new era of post-colonial relations between Jawoyn and non-indigenous users of the Nitmiluk site. A physical site, central to a community's desires for self-determination, was mythologised in song, through expressions that mobilised tradition within more fluid, contemporary soundscapes.

See: Gibson 1998; Gibson and Dunbar-Hall 2000.

Technology: globality?

Music rapidly changed following the technological, political and economic processes that transformed the linkages between parts of the world through the Industrial Revolution. Before that, in sixteenth-century Europe, music went through a period of rapid innovation, leading to the dominance of musical forms marked by organised harmony, metre and tonality, and by its movement from being part of ritual events to secular entertainment. Relationships between musicians, their music and audiences became increasingly commercial, while full-time professional musicians (and buskers) became more visible. The most rapid phase of diffusion and globalisation followed the Industrial Revolution, which began in Britain in the late eighteenth century, and impelled Europeans to seek markets and raw materials throughout the world. That search was accompanied by colonial conquests and settlement of 'new worlds', which often resulted in new transport infrastructure, education and health systems. Literacy enabled songs and music to be printed, published and diffused – especially through churches and schools – and the same songs and tunes were sung and played on different continents in more or less the same way (even if their meanings were transformed or ignored). Literacy and notation eventually stimulated professionalism. The oral transmission of music declined (alongside industrialisation and urbanisation) ushering in what was arguably the 'age of repetition' (Attali 1985) and the widespread expansion and recognition of a popular music industry.

The invention of printing and the rise of literacy, alongside technological change in communications (for example, from sailing ships to steamships), were all central to the earliest phases of diffusion and standardisation. They set in train moral debates that would, in the early 1700s, result in legal protection for artists over their rights in creative work, and later entrench the role of publishing companies in music commerce. Sheet music created early patterns of ownership and control in fledgling national music industries, yet its impact on stylistic diffusion and innovation was slight in contrast to the technical developments that began in the 1870s and continued for more than a century (see Figure 3.2). New recording technology, in the more fluid political, economic and technological context of the twentieth century, encouraged commodification of music, and its growing internationalisation. The invention of recording devices (the first of which was Edison's recording cylinders in 1877) was the catalyst for a second phase of expansion, while the pace of musical change and globalisation accelerated with the advent and diffusion of radio. The possibility of recording music, and hence the ability to diffuse not just songs and sheet music, but the actual sounds, changed the relationship between artists, publishing companies and audiences, spawning new ways of experiencing music, and transforming its status as a commodity.

Recording separated consumption of music from the live context, which in turn influenced live performance; as people became familiar with recordings,

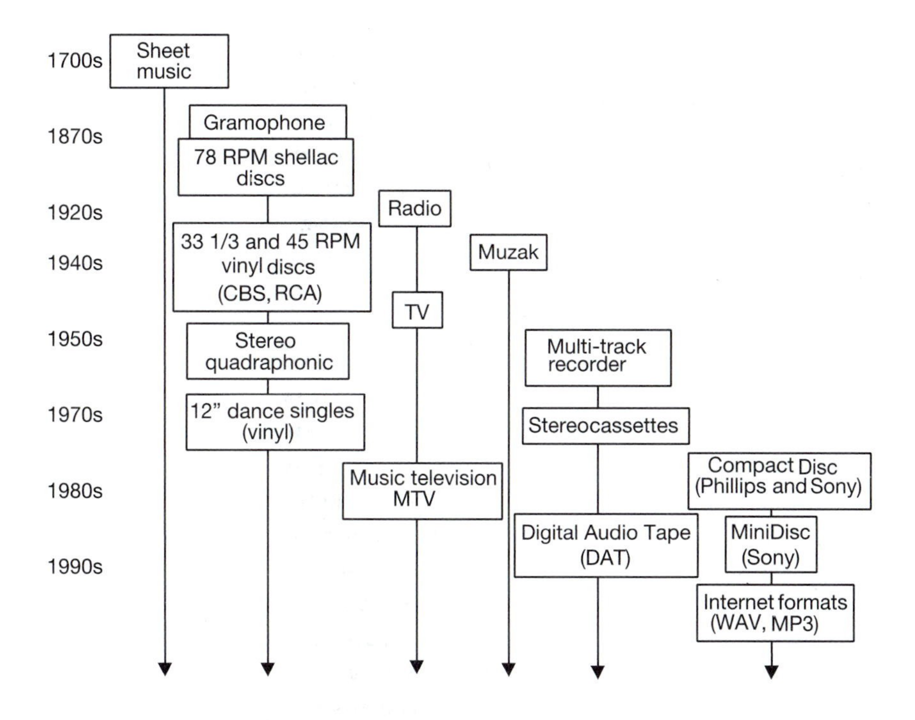

Figure 3.2 Formats influencing the diffusion of music

professional musicians were not expected to deviate far from public expectations, and live performers began to suffer financially (Middleton 1990: 74). By 1914 world record sales had probably already passed the 100 million mark; sheet music sales, which peaked in 1910, experienced a rapid decline. Thereafter the written word played a diminishing role in the spread of music. New recording companies emerged and, from their inception, began to operate internationally; the British firm EMI (Electrical and Musical Industries Ltd) extended its operations to North America, then India in 1907, Latin America in 1910 and New Zealand in 1927. At the turn of the century several United States recording companies went to Hawaii (then simply a United States territory) and began recording what then passed for Hawaiian music, and a Hawaiian music craze swept the United States (Robinson *et al.* 1991: 37–8). After the African craze of the mid-nineteenth century, music companies were searching further afield, for both markets and inspiration. Gramophones and the record industry reached some parts of the developing world little later than developed countries. In India records were being marketed as early as 1902 and the first commercial recordings of Indian music were made in the same year (Farrell 1998: 59). Because sheet music was expensive throughout the nineteenth century (and musical literacy not necessarily widespread) electrification enormously expanded the scope for globalisation.

Such expansion of musical distribution only occurred upon a regulatory platform established by the earlier transformation of music into a commodity via sheet music. Where copyright protection existed, companies were more certain that they would maintain their monopoly over distribution; without copyright they were reluctant to distribute products. Somewhat ironically, copyright, an agent of mobility, relied on rigid legal protection in national legislation. Western intellectual property laws were increasingly writ large upon global landscapes of distribution; the distance between the social context of music and its copyright form (held as a 'bundle of rights' by major recording companies in New York, London and Los Angeles) had been dramatically stretched by the 1930s. That fluidity only existed because of strategic reterritorialisations, as nations conformed to increasingly international legal norms, illustrating the 'dialectical interplay between the endemic drive toward time-space compression under capitalism...and the continual production and reconfiguration of relatively fixed spatial configurations' (Brenner 1999: 435), such as the legal frameworks that enabled expansion to occur.

Radio broadcasting began in earnest in the early 1920s in most developed countries, establishing another infrastructure for the dissemination of sounds. Initially, at least in Britain and the United States, radio was intended to broadcast 'quality' productions, including classical music, but not popular music, which could be obtained on records. Radio and records eventually complemented each other. Records were a cheap form of programming for radio, and radio exposure was the most effective means of advertising and promoting records. The invention of electric recording in 1925, along with broadcasting, signalled the end of the acoustic era, as microphones took over. More recording studios, and the proliferation of radio stations (in the United States at least), encouraged diversity. Radio enhanced the manner in which the recording industry separated the singer from the song, an alienation that paralleled commodification, as the recording took on a social life of its own (Manuel 1993: 7). In different forms music and the music industry were becoming privatised. The combined rise of radio and the record industry allowed songs to be heard far beyond their origins. By the end of the 1960s radio ownership was common even in the most remote parts of the world, and was almost universal in many countries. The number of radio stations grew in parallel, enabling specialisation in particular types of music. Some stations (such as Radio Luxembourg, broadcasting from continental Europe to the United Kingdom and, later, Radio Caroline, broadcasting from a ship moored in the Thames Estuary) were illegal or semi-legal, focused on radical and innovative popular music (or simply songs banned from other stations for their lyrics) and challenged the hegemony, and blandness, of official stations. Radio contributed both to more rapid national and international diffusion and to regionalisation, as specialised broadcasting centres emerged. Chicago, in the 1930s, and then

Nashville, from the 1940s to the 1970s, became key centres for country music, because of the significance of particular radio stations – such as WSM's 'Grand Ole Opry' broadcasts in Nashville – rather than either geographical centrality or commercial development (Peterson 1997: 25). The strategies and market orientations of radio stations contributed to the notion that many regions retained distinctive musical identities.

In Britain and the United States the railways enabled performers to be mobile, and a variety show circuit developed. Initial prejudice against British performers ended around the turn of the century, resulting in the music hall circuit extending across the Atlantic. For the first time individual stars, such as Al Jolson, Marie Lloyd and Maurice Chevalier, acquired considerable followings and ventured further afield. One Cheshire band, which had made a long tour of the United States in 1891, undertook a tour in 1906–7 that included Canada, Hawaii, Fiji, New Zealand and Australia. Other choirs and bands made similar tours, in part to 'cement better relationships' between Britain and Europe, or, in the case of visits to the Empire, to 'generate a sense of imperial consciousness' (Russell 1997: 274). Such ventures, usually reciprocated in some way, emphasised new international links.

Film and television later brought new avenues for promotion and diffusion. Like so many other genres, rock 'n' roll emerged from regional forms (black blues and white country music of the southern United States) but very quickly became a national and international music, as television beamed images of youthful rebellion into previously sanctimonious domestic spaces. Ironically rock 'n' roll had initially 'represented a geographically based revolt by "the provinces" against an old cultural capital (New York City)' (Ford 1978: 213) where the Tin Pan Alley establishment had long regarded country and western and R&B as 'lower class' and unworthy of promotion. After the success of *Rock Around the Clock* in 1956, a variety of films followed, and were a critical influence on the international impact of a new genre. Television complemented films. In the United States *American Bandstand*, begun in 1956, was the model for other American and overseas programmes. The music industry boomed; between 1955 and 1959 record sales in the United States more than doubled – most of this by the indies (small independent recording companies) that had promoted rock 'n' roll. The emergence of rock 'n' roll was seen as part of a 'predictable pattern' in the United States where southern musicians and singers evolved a particular new style, and saw it 'discovered' and developed in the north, given a more 'sophisticated veneer' that disguised its origins and removed its distinctiveness, thus stimulating a new cycle of innovation and diffusion (Gillett 1970: 199–200). However, not only was the southern 'birthplace' of rock 'n' roll as much mythical as real (Hall 1998) but a number of southern musical styles – such as zydeco (cajun), the music of people of French heritage in rural Louisiana, involving accordions and predominantly French lyrics – were also largely ignored in the north. Television, eventually incorporating distinct popular music channels such as MTV (Music Television), brought a new dimension to the diffusion of music that slowly shifted the emphasis of popular music from the aural to the visual.

Beyond the Anglo-American core

Music clearly contributed to the increasingly global nature of cultural and economic linkages, mapping out new networks of commodity flow and entrepreneurial activity. At least at a surface level,

> All countries' popular musics are shaped by international influences and institutions, by multinational capital and technology, by global pop norms and values. Even the most nationalist sounds – carefully cultivated 'folk' song, angry local dialect punk, preserved (for the tourist) traditional dance – are determined by a critique of international entertainment. No country in the world is unaffected by the way in which the twentieth century mass media (the electronic means of musical production, reproduction and transmission) have...created a universal pop aesthetic.
>
> (Frith 1989: 2)

The rise of rock 'n' roll, the success of Elvis Presley and later the Beatles, alongside transitions in other cultural forms, ensured some measure of ubiquity. Jazz had been a precursor; it was argued that, between 1950 and 1960, American jazz had 'created or revitalised viable local music scenes in every major European and Asian capital' (Ennis 1992: 280). By then records were marketed internationally; what was played on radio and television stations in New York, London or Sydney was not radically different – and there was less time lag between continents. The most successful performers (with the exception of Presley) embarked on world tours, but there was only limited evidence of truly widespread international success. Despite statements that the Beatles, even before they were successful in the United States, were successful in 'the rest of the world' or that 'a large proportion of the Western world' recognised their qualities (Gillett 1970: 313), 'the rest of the world' consisted mainly of some European countries and scattered Anglophone outposts. Indeed 'world tours' of even the most successful groups, such as the Supremes (Box 3.2), were anything but global, a point still applicable to many current artists (partly due to poor exchange rates, and hence tour earnings outside the United States or Europe). Most performers confined their tours to the developed world, and primarily to the English-speaking part of it. None crossed the Iron Curtain into Eastern Europe (hence the phenomenal success of the Rolling Stones when they finally played in Russia in 1997, long after the arrival of Coke and McDonalds), and few reached continental Europe. Despite the success of the 'British invasion' in Australia in terms of chart success (in a country where the music industry was less nationalistic and the small market struggled to support a vibrant local repertoire), few performers toured there. When the Beatles visited Australia in 1964, 'to save us, from our isolation, from our inferiority, from our innocence' they were met with crowds larger than ever welcomed them elsewhere, seeking, perhaps, 'a tangible manifestation of the new freedoms which were emerging in Britain and America'. But it proved difficult to attract them to Adelaide – away from the populated east coast – and the tour was seen as of no consequence to the Beatles or their management, and was rarely, if ever, mentioned in biographies or autobiographies (Baker 1995: 54, 56). While radio and television took sounds much further, the major destinations for touring artists remained on either side of the Atlantic, in a very Anglophone market.

Box 3.2 The Supremes on tour

The 'world tours' of the Supremes, typical of the movements of successful groups in the 1960s, were primarily confined to Western Europe (though they frequently travelled in the United States). After a couple of years on the 'chitlin circuit' of black venues in the United States, the success of 'Where Did Our Love Go' and 'Baby Love' in 1964 – respectively top of the charts in the USA and the UK – resulted in their first overseas tour, a two week tour of the UK. The following year they returned to the UK, embarking on another tour, including Bristol, Bournemouth, Cardiff and Newcastle, encountering half-filled houses in the smaller towns where people 'didn't understand what the music was about' (Wilson 1986: 193), and crossed the Channel to France, Germany and the Netherlands. The Supremes first visited Canada in 1966, when they also undertook a Far East tour, which took in Tokyo, Okinawa, Taiwan, Hong Kong and Manila – all places with a substantial American presence – and a brief visit to Barbados. The year after they travelled only to Canada and Mexico (participating, as a trio of nuns, in the filming of a single episode of a weekly American film series, *Tarzan*), but in 1968 they made their most extensive tour of Europe, incorporating England (as all European tours revolved around London, but no longer the provinces), France (Paris and Cannes), Germany (Berlin, Hamburg, Frankfurt and Munich), the Netherlands (Amsterdam), Sweden (Stockholm and Malmo), Denmark (Copenhagen), Switzerland (Geneva),

Belgium (Brussels) and Ireland (Dublin). The following year, the final year in which Diana Ross was in the Supremes, they visited only Canada (Toronto) and had a single brief trip to London, to record a track for the TV series, *Top of the Pops*. In seven years, other than the Far East tour, they had played in just ten countries outside the United States and, after 1965, rarely outside capital cities. They never left the northern hemisphere.

See: Wilson 1986.

After 1979 the record industry entered a period of decline, yet distribution systems were becoming increasingly international, seeking new markets, especially in developing countries, assisted by growing electrification and the rise of the cassette industry. In Europe and North America cassette technology had less impact on the general music industry, since few recordings were issued in cassette format only. However, the popularity of the Walkman ended the dominance of vinyl; in 1980 records had a three-to-one dominance over cassettes, but by 1988 cassettes sold seven times the volume of records (and CDs already sold twice as many). In developing countries the impact of cassettes was even greater, and particularly dramatic in countries like Indonesia, which had very few record players (and records) even as late as the 1960s. Growth was rapid everywhere. In India the cassette boom expanded exponentially in the first half of the 1980s; by the mid-1980s cassettes accounted for 95 per cent of the recorded music market (Manuel 1993: 63). The growth of the cassette industry undermined existing music industries in some countries but, more frequently, established music media for the first time. In Egypt, for example, a vigorous cassette industry was established in less than a decade from the mid-1970s. In 1984 there were almost 400 cassette companies, producing millions of cassettes, though their commercial stability was threatened by hundreds of 'pirate' companies, who were able to market cheap copies within days of the original production. By then three-quarters of the urban population had cassette players and the 'cassette revolution' was in full swing (El-Shawan Castelo-Branco 1987). There and elsewhere, the profusion of cassette companies, the ease and low cost of recording, the cheapness, transportability and durability of cassettes (compared with records) and battery operation, resulted in the much greater availability of a diversity of forms of recorded music at the same time as the more widespread diffusion of 'international' sounds (Box 3.3).

Cassette technology was both an agent of homogeneity and standardisation, but at the same time it was a catalyst for decentralisation, democratisation and the emergence of regional and local musical styles. The lower cost of production enabled small-scale producers to emerge everywhere, recording and marketing music for specialised and local audiences, rather than necessarily catering to a homogeneous mass market (Manuel 1993: xiv). The cassette industry also enabled the growth of new types of urban popular music, as accessible technology resulted in a profusion of local music-recording activities, and the dissemination of the resultant cassettes. Cassettes established niches for regional and local genres, such as the Arab influenced *arabesk* music in Turkey (despite its being held in contempt by the Turkish music bureaucracy) and Jawaiian, a Hawaiian fusion of reggae and local pop music. In some countries cassettes enabled the dissemination of musical genres that were formally discouraged or banned by authoritarian governments, including both sentimental Hong Kong and Taiwanese pop music and the music of dissident rock star, Cui Jian, in China, reggae in Cuba, various protest songs of Palestinians and political movements in Haiti, Chile and elsewhere (Manuel 1993: 33–4; Hansing 2001: 734). Cassettes diffused international superstars and gave local performers an avenue of expression.

Box 3.3 Popular music and cassette culture in North India

A number of criticisms of the culture industry, notably that of the Frankfurt School, associated with Marcuse and particularly Adorno in the context of music, argued that modern capitalism and mass media – and thus particularly such new technology as cassettes, with a global reach – would lead to nothing but standardisation, uniformity, depoliticisation and passivity, emerging from the increased commercialisation of culture. The media would stultify critical consciousness, commodifying oppositional art, and promote consumerism and the myth of a classless society, escapism, commodity fetishism and low levels of community solidarity and participation. The development of 'cassette culture' in North India, however, proved to be simultaneously an agent of standardisation and homogenisation, and a catalyst for decentralisation and the emergence of hundreds of regional and local musical genres and performative styles, despite cassettes emerging in the late 1970s in a country dominated by oligopolistic national media systems. The cassette revolution ended the hegemony of GCI (the Gramophone Company of India), film music and the uniform aesthetic that Bombay music producers had superimposed on hundreds of millions of radio listeners over forty years. Cheap manufacture and easy dissemination of cassettes re-created rather than destroyed regional genres and performances – a flourishing of 'little traditions' that rearticulated in new hybrid forms the relations between high and low culture, rural and urban life, public and private spheres, tradition and modernity, and national identity and the West. Much cassette music was aimed at a 'bewildering variety' of specific target audiences, in terms of gender, ethnicity, region and even occupation (such as truck drivers' songs). The uses and consequences of cassette technology were therefore diverse and ambivalent, contributed to the survival of minority languages, and encouraged the revival, revitalisation (and re-creation) of traditional folk cultures, communities and social identities, being linked to emancipatory socio-political movements (including campaigns against widow-burning and gender inequalities) but also to religious sectarianism, bloodshed and violence. The diffusion of cassettes spread regional music and folklore to migrants elsewhere in India and overseas, who might otherwise have lost touch with their traditions, and stimulated new regional recording studios and markets.

But cassettes also contributed to the decline and impoverishment of local musical performance, especially at weddings; the erosion of traditional Punjabi folk songs and folk theatre; the replacement of Brahmin priests (*pandits*) by cassettes of devotional music and prayers; and the marginalisation of women, since cassettes were mainly produced by and for men. However, rather than being solely a vehicle for secularisation and modernity, cassettes reinforced religious traditions, being the only form of mass media to represent the diversity of India's religions. Yet again, however, cassettes also intensified the trend in which urbanisation and the mass media promoted a shift away from ritual towards popular culture, a transition in which aesthetics, entertainment – and also vulgarity – gained importance at the expense of religious values. Much cassette output syncretised traditional genres with modern elements. The resultant musical hybrids, embodying and responding to the changing social identities of Indian communities in transition, had their own 'authenticity' that made them more expressive and resonant to their particular audiences than folk music. The impact of cassette technology was thus complex, and replete with contradictions and ambiguities, as popular culture became a contested terrain for establishing and negotiating social identity in a rapidly changing society. Far from being inert and passive consumers, Indians repeatedly appropriated and reworked elements of commercial music into their own musical and dramatic discourses. Introduced mass culture became a source of raw materials (divested of dominant meanings) or local community aesthetics and values. It was not simply an arena for the passive consumption of images and products of metropolitan capitalism.

See: Manuel 1993.

While it may be tempting to assume that innovations in musical formats evolved and then were dispersed unhindered, technological diffusions (and their impacts) have always been uneven. This is illustrated in the different trajectories of video-cassettes and CDs, both of which appeared in the 1980s. CDs were more expensive than other forms of recorded music, hence were initially aimed at the classical music market, but by the mid-1980s had extended into popular music and jazz. In the United States CD sales went from 0.8 million units in 1984 to over 200 million in 1989 (Robinson *et al.* 1991: 54). Vinyl production declined rapidly; many popular music retail stores had stopped selling records by the early 1990s, which became a speciality business for small companies, especially for dance music. In developing countries CDs diffused much less quickly, because of their cost and through concerns for the protection of copyright (see Chapter 11). CDs remained largely confined to the middle classes, enabling local cassette industries to continue to flourish. While CDs did not increase the accessibility of music, the spread of television, and video, certainly did. Music videos were soon common currency, initially of live performance, but subsequently became more imaginative and innovative, as late-night and overnight television, and a growing number of TV stations, increased access and available time. Music television (MTV) began in 1981 on American cable television, providing non-stop access to popular music videos (Kaplan 1987); other cable channels followed, not only in the United States, but in other developed countries and then subsequently beyond. The establishment of MTV in Asia in 1992 (with a potential market of over a quarter of the world's population) emphasised the extent of globalisation and standardisation, and the role of image (Box 3.4). But, just as cassette culture had previously done in India, Egypt and elsewhere, MTV stimulated local industries in many countries – effectively creating the video industry in India – and the eventual rise of a number of pan-Asian stars. MTV increased the 'velocity of innovation' (Straw 1993: 8) nationally and internationally.

Box 3.4 MTV Asia

MTV was launched in Asia in 1992 and a year later reached thirty-eight countries from Israel to Taiwan, offering 'a unified vision of a consumerist, pan-Asian youth culture...and the promise of fame for the many Asian singers and musicians keen to grow beyond their home markets' (Wong 1993: 38). It was then argued that MTV Asia's philosophy was 'think global, act local' by creating different playlists and programmes for different parts of the region, introducing local languages for spoken-word programmes, tapping into local cultures (by using Asian VJs and Asian performers) and avoiding frequent Western videos, such as Madonna or the Rolling Stones, which 'do not appear to be made in your backyard'. Its (European) manager argued that 'It doesn't touch your life, and appears to be an American or European phenomenon' (quoted in Wong 1993: 38). The advent of satellite broadcasting throughout Asia also gave rise to new small-scale activities in the domestic film and television industries, especially in Hong Kong and India (though India had no video production until 1982), simultaneously stimulating increased television (and satellite dish) ownership even in remote rural areas, and the growing regional success of Asian performers. Orientation to specific national markets was essential, since, in many countries (such as Taiwan), international acts constituted no more than 20 per cent of music sales (despite the existence of many overseas music-producing Chinese communities). Creating a pan-Asian identity had limited success. The first pan-Asian star was Christina Aguilar (a Thai woman, with Filipino and French parents), described as 'Asia's Madonna', who won the inaugural MTV Award for Asian artists in 1982. She was widely supported in most countries across the region 'even though they obviously didn't understand what she was singing about. They liked the video, they liked her, they liked how she sang' (Wong 1993: 39). The VJs were largely recruited from North America and, despite being of Asian ethnicity, 'all maintain southern Californian drawls'. MTV Asia while providing some stimulus to local industries and performers, only achieved significant local success with performers of great visual appeal, and formulaic presentation modelled on American stars.

A global cash register?

Technological, commercial and creative changes were accompanied by the emergence and consolidation of transnational corporations in the music and communications industries, and the spread of their global reach. By the end of the 1970s the world music industry was dominated by five companies, all based in the core countries of the northern hemisphere. These vertically integrated companies, with their own production, manufacturing, promotion and distribution facilities, became exceptionally large, involving a range of artists and genres, and diversified into commercial enterprises beyond music. The 'majors' exercised a centralising influence on the music industry, contributing to growing homogeneity, at the same time that smaller independent recording companies, the indies – which usually began through recording ethnic or regional music, and which were 'ignored or maligned by the majors' (Robinson *et al.* 1991: 41) – grew to cater for more marginal styles of music, unfashionable, innovative or new performers. Yet, over time, many indies such as Island Records, Def Jam and A and M became linked to the majors, especially for manufacture, promotion and distribution. Periods of innovation in popular music (such as with the rise of rock 'n' roll, punk and techno music) preceded new waves of acquisitions and consolidation of global control by the majors.

Though the majors were oriented to Europe and North America, they were increasingly also the gatekeepers of recorded music for many other areas of the world (Table 3.1). Many parts of the developing world were seen as lucrative new markets for cultural products, as long as copyright laws were changed to protect their interests and prevent domestic 'pirating' industries. The music industry thus emphasised internationally successful performers and styles, aggressively promoting their interests through the control of local media and distribution. Economic development in East and South-East Asia from the 1980s provided a large and affluent market that had scarcely been reached by the international music industry. The Australian band, Girlfriend, five Sydney teenagers, were eventually entirely targeted at Asia, with band members being selected in terms of their visual appeal. Though they had achieved an Australian number one record in 1992, they largely abandoned Australia, learned some Japanese, relocated to Asia and sought to gain access to that market (though the move eventually failed). Endogenous music industries, with limited capital and poor access to broadcasting or sales outlets, remained vulnerable. Despite this, local music had a resurgence in domestic markets around the world by the end of the 1990s. Many recordings were by then

Table 3.1 PolyGram International: selected acquisitions and subsidiaries in the 1990s

Date	*Country*	*New PolyGram subsidiary*	*Previous company*
13/07/1998	Turkey, Turkmenistan, Azerbaijan	PolyGram Plaza	Plaza Müsik
04/03/1996	Thailand	PolyGram Thailand	Far East Bangkok Enterprises Ltd
11/10/1995	Venezuela	PolyGram Venezuela	Rodven Records
16/01/1995	India	PolyGram India	Music India
03/10/1994	Colombia	PolyGram Colombia	–
24/03/1994	Russia	PolyGram A/O	BIZ Enterprises
15/02/1994	Czech and Slovak Republics	PolyGram SRO	–
18/05/1993	Hungary	PolyGram Hungary	Zebra/Multimedia
13/01/1993	Japan	PolyGram KK	Nippon Phonogram/Polydor KK
10/12/1990	Chile	PolyGram Discos Limitada	–

Source: Gibson 2000: 64.

distributed by local arms of the majors, ensuring hegemony of profits for entertainment and media empires, but enabling distinctiveness and local cultural relevance.

The domination of recorded music by a small number of majors created new geographies of music; within most countries at least half and, in some, as much as 90 per cent of the national popular music market (other than pirated recordings) was controlled by five companies (Robinson *et al.* 1991: 43). As in Taiwan (Box 3.4), this did not necessarily mean that the majority of music was imported. In its most extreme form, international marketing resulted in the almost instantaneous transmission of new songs to global markets, vividly evident in the near global response to Elton John's new version of 'Candle in the Wind' (1997), following the death of Princess Diana (with a record number of sales in dozens of countries within days of its release) – an example of the time–space compression that new information and communication technology permitted. A further profusion of technical changes linked into the music industry, including video, cable television, photocopying, personal computers, faxes, satellite communications and the World Wide Web (with individual access to performers' home pages). Various components of flexible specialisation made it possible for certain music to be heard (and seen) almost everywhere (often simultaneously, with global television coverage of major events). Entertainment corporations, such as the Seagram Company and Sony, promoted a global rhetoric. They appealed to mass markets of consumer products across national borders and cultural and economic domains, mobilising the 'globalisation as inevitability' thesis touted by proponents of neo-liberal free trade (Yeung 1998), promoting the circulation of what Ohmae (1995) called 'global products':

> From Osaka to Capetown, New York to Bangkok, London to Beijing, it is coming together. Powerful, inevitable, ever-expanding....A global culture, a global market. People sharing the same experiences, striving for the same goals, entertaining themselves in much the same ways. More and more, it is becoming one world...our world.
>
> (The Seagram Company 1997: n.p.)

These trends are reminiscent of Adorno's predictions of a 'culture industry', where consumption is of a standardised popular cultural 'formula' of products and experiences, commonly under the shroud of 'youth culture'. Somewhat more subtly, and realistically, new transnational cultural trends appeared that, while transgressing previously rigid boundaries, were still bound by continental or linguistic alliances and fractures. Contemporary technology enabled a new sense of pan-Asian music to emerge (see Box 3.5), improved flows of distribution throughout, for instance, the Arabic-speaking world, but also connected tiny dispersed sub-cultures (see Chapter 5). Internationalisation thus contributed simultaneously to greater homogeneity and fragmentation.

Box 3.5 Pan-Asian music

East Asian music industries have historically been perceived as protective and insular, with strong traditional musical styles (as with Japanese enka), language barriers and internal heterogeneity. Hence, Asian markets have been dominated by local music and domestic variations on Western styles, known individually as K-pop (Korean pop), J-pop (Japanese), Canto-pop (Cantonese) or Indo-pop (Indonesian). The Philippines is an exception in that it has consumed more international music and produced more music in English, partly due to its colonial and post-colonial links with the United States. In 1995 the first Asian Song Festival was held, and by 2000 a sense of pan-Asian music was reflected in the participation of twelve nations in the competition (including Vietnam and Mongolia for the first time). Harry Hui, Managing Director of MTV Mandarin (based in Hong Kong), celebrated:

> If pop culture in Asia used to take its cues from Japan and then Hong Kong, now it's almost 'DNA' model, where Korean music is selling in China and Taiwan, mainland Chinese artists are packaged in China like Taiwanese acts, Hong Kong artists are starting to package and position themselves as Japanese-looking acts.…I think what we're really finding now is that all national boundaries for pop culture are meeting in Asia, especially on the music side.
>
> (quoted in McClure 2000a: 49)

Pan-Asian styles and markets, driven by MTV and other cable television programmes, emerged.

The marketing of pan-Asian pop emphasised the crossing of borders and the loosening of historical cultural differences. China's Faye Wang scored success in Japan; Japanese pop artists including Hikaru Utada and Puffy gained audiences in Korea and China; Korean artists sang in Chinese, Japanese and English in order to broaden their market while the Korean hip hop of groups such as Clon was particularly popular. Within pan-Asian pop, sub-regional groupings emerged, reflecting ethnic, religious and linguistic connections. Malaysian releases became increasingly popular in Indonesia (as with Malaysian female pop vocalist Siti Nurhaliza); while mainland Chinese acts were marketed as Taiwanese acts; Chinese-Malaysian acts were popular in Taiwan; and Indian film music was successful in Hindi across the region. These more fluid movements of commodities have been triggered by a range of factors. In the case of J-pop this is through:

> the gradual fading away of wartime memories, Japan's image as an Asian nation with a home-grown pop culture that's an alternative to imports from the West, and a stronger sense of identity among young Asians who want pop stars with whom they can identify.
>
> (McClure 2000b:42)

Only in 1998 were Japanese music products available in Taiwan and it took until 2000 before Japanese performers could visit. Yet even after such transnationalism, reactions occurred, as when Korea and other nations condemned Japanese prime minister Koizumi's controversial visit to war memorials; many countries sought retribution, including bans on imports of Japanese popular culture. Fluidity was not inevitable, nor uncontested.

See: Lockard 1998; McClure 2000a, 2000b.

From sheet music to television, technological change affected the relationship between musicians, audiences and music companies. Companies increasingly became mediators of the flows of sounds between musicians and audiences, seeking to earn profits from the sale of music commodities, or, in the case of higher-quality (and higher-cost) formats, more 'true' recordings, durability and mobility. Radio broadcasts had a dramatic impact on the recording industry during the Depression, providing access to new music without charge, as opposed to the relatively expensive phonograph discs. There were a series of well-documented crises in the recording industry, including slumps in sales of both gramophones and records and national industrial action by the American Federation of Musicians (Clarke 1995). Similar concerns were raised when cassettes and then digital formats (such as CDs, Digital Audio Tapes (DAT) and MP3 files) became common. Yet, while audiences were physically removed from performers in different ways, they also reworked music in their everyday lives, in new social spheres and spaces, through different ways of performing, in informal networks of copying and exchange. The impacts of technological change on local musical practices were easily overestimated and never homogeneous across space.

Not nearly global

The various components of what has become known as globalisation (particularly migration, commerce and technology) have become more accessible and pervasive, and imply a global economy 'that works as a unit in real time on a planetary basis' (Castells 1994: 21). However while relatively 'placeless' Western music often dominates the priority list of the world's major labels, certain 'frictions' are manifest in pervasive local cultures, tastes, musical traditions and scenes. Music, even as a part of the global media and entertainment empires of giant companies, 'cannot escape or transcend the logic of (geographically specific) social institutions, norms and behaviour' (Robins and Gillespie 1992: 154). 'Local' artists have remained popular within countries, prompting the International Federation of Phonographic Industries (IFPI), which represents corporate interests in music, to claim that throughout the 1990s 'figures show an international music industry underpinned by flourishing local repertoire and cultural diversity. All over the world, people are buying significantly more local music than they were at the start of the decade' (1999: 1). Local content varied in different contexts: in the United States local artists made up 85 per cent of sales (IFPI 1999); in Europe, music unique to particular countries averaged around 40 per cent of the total market, with figures usually much lower in markets too small to support extensive domestic activities (as in most developing countries). Hence 'Instead of an overwhelming and irreversible convergence of production, distribution and consumption, a diverse mosaic of geographical patterns in different spheres of economic and social life' (Yeung 1998: 293) is particularly evident for the music industry. Globalisation is far from complete.

Global distribution and cultural links remained selective and often unpredictable. Releases could be hits in one location and failures in others. Australian artists such as Icehouse and the Beasts of Bourbon, largely unknown elsewhere, have generated significant support in Germany, British band Skunk Anansie reached the top in Italy, while the Grateful Dead (and many others), despite the backing of major labels, never found success outside home territory. Cutting across loose notions of 'local–global' interactions are complex networks for distribution and reception – channels of cultural flow – that had to take account of local and regional particularities. While the absorption of 'Clementine' (and similar tunes) into the music of the most disparate societies was evidence of the reach of globalisation, it was also symbolic of the receptivity of particular non-Western audiences to certain Western themes. Country and western music, particularly, found an audience in many developing countries. In the 1970s, the Australian country performer Slim Dusty was even more popular in neighbouring Papua New Guinea and the Solomon Islands, and performed frequently there. Country music has been exceptionally popular in areas as diverse as Jamaica, among the Inuit of Canada and in remote Aboriginal Australia (Walker 2000; Gibson and Dunbar-Hall 2000). In Zimbabwe, country songs have been composed and recorded in at least five local languages, and more people listen to Dolly Parton than Thomas Mapfumo (see Chapter 7). The Dolly Parton Show was broadcast weekly in the early 1990s, and the most popular songs, known locally as 'country classics', were love songs or involved themes of working-class family and values and Christian beliefs, suggesting strength and hope through hard work (rather than any notion of radical change), with simple tunes and comprehensible lyrics. Many listeners, especially women, aspired to Dolly Parton's 'rags-to-riches' story (Zilberg 1995). By the end of the century, 'indigenous Zimbabwean music [was] losing its grip on a country that increasingly wanted to hear the same sounds as the rest of the world (the *Titanic* sound track apparently)' (Ellison 1998: 15). In Southern Africa in the 1980s, globalisation, inherently uneven, both challenged and corresponded to some facets of local identity.

Throughout the world the most popular performers are copied and parodied, and cover bands and imitations are numerous. For example, the prominent north Indian singer, Alisha Chinai, successfully released a Madonna tape, with Hindi-language versions of Madonna's songs, and a cover depicting Alisha dressed in a gaudy bra and no shirt. But a Hindi version of Michael Jackson's hit, entitled *Mithun: the Indian Jackson*, was a commercial failure (Manuel 1993: 135, 190, 273). Copies have not surprisingly been criticised even more than the spread of 'global products'. In India, for example, one music director complained in the mid-1980s:

> As for lifting Western tunes, I think it is bad because rhythmically and melodically we have such a deep musical culture. What makes the whole situation worse is that we're lifting the most superficial branches like disco and ABBA when Western music is actually like a mighty tree.
>
> (quoted in Manuel 1993: 145)

Those who defended such practices argued that borrowed elements were localised in imaginative syntheses, and that 'borrowing' was typical of all cultural forms, was creative and additive, and little different from the situation where classical music was recorded and performed by national orchestras around the world.

In every context 'borrowing', absorption and imitation were selective and often unpredictable, demonstrating that 'many analyses of globalization...fail to take into account experiences of real people' (Taylor 1997: xvi). Much as purists might have preferred Thomas Mapfumo to be more popular that Dolly Parton, Nitmiluk not to have a reggae beat, or songs extolling drug runners not to be popular in Mexican migrant communities in Los Angeles (Simonett 2001), so, in Lhasa, Tibet, 'many Westerners desire to erase karaoke bars from the landscape', perceiving them 'as a sign of the "decline" of Lhasa and the "loss" of an authentic Tibet at the hands of the Chinese' (Adams 1996: 510, 514). Yet 'for many Lhasa Tibetans karaoke is a significant symbol of modernity' where they could indulge in Western dress styles, dancing and drinking, in a context of Chinese entrepreneurialism and music (Adams 1996: 537). The fusion of Western, Chinese and Tibetan style represented 'an accurate representation of the experiences of modernity' and a means of participating in, and benefiting from, rather than being oppressed by it (Adams 1996: 537). Vividly illustrated in the karaoke bars is the complex manner in which particular strands of Western music, here MC Hammer and Queen Latifah, merge with other alien (Chinese) influences, into a syncretic structure quite different from what might have been predicted in corporate headquarters on the other side of the world.

Distinctiveness has not been erased by increasingly global distributions of music. Local cultural differences became evident through another form of mobility, when tourism extended its reach; tourist demands both conserved and dissolved local musical traditions ensuring that some traditional songs, dances (and other cultural forms) were retained, where they might otherwise have been abandoned as the social context in which they originated had disappeared, though performances for tourists were usually transformed and truncated versions of the original form (simplified and shifted into an idiom more appealing to Western tastes or images of the 'exotic'). Despite attempts to create globalised cultural norms, encounters with music are still everywhere situated in local spaces and mediated by local and transnational factors. Culture and music are especially complex and contested terrains as 'global cultures permeate local ones and new configurations emerge that synthesise both poles, providing contradictory forces of neocolonialism and resistance, global homogenization and new local hybrid forms and identities' (Cvetkovich and Kellner 1997: 8). Local themes were sometimes infused with new vitality. In Indonesia, the music and records of Rhoma Irama, based on urban popular music (*dangdut*), were nationally recognised as simultaneous celebrations of Islam, Indonesia and rock music (Ennis 1992: 374). Globalisation provided a vehicle for the celebration and recognition of national and regional identities, at the same time as it enabled the global reach of Western popular music.

Countless studies thus recorded persistent regional variations in folk music performance, as in Italy where in the north there are Galician and Francophone influences and in Sicily there are Muslim connections so that 'at the rural level... here music was mostly circulated orally, Italy was and is by no means a unified country' (Sorce Keller 1994: 45). Equally there are regional variations in radio station choice, in American preferences for country music stations (e.g. White and Day 1997), in places with specialist dance music scenes (such as trance in Belgium, Tel Aviv or North Coast NSW), and in particular cities where hip hop, reggae and dub have unique sites of production, broadcasting and consumption (e.g. Powell 1993: 124–6). Even where government-controlled radio stations ignored or deliberately excluded some forms of local music (as Algerian radio excludes Tuareg music), cassette recordings abounded and were widely sold and exchanged in Tuareg areas (Kaemmer 1993: 201). Values, information and goods were not simply imposed on passive consumers, but constructed and reconstructed by musicians, critics and audiences across geographical and cultural terrain.

There have been no sharp disjunctures between oral, literary and electrical traditions. Oral transmission and composition have never ended and, in many contexts, such as the pub circuit, bands learn from and work with each other (Goh 1996) in a manner akin to that of folk music traditions (see Chapter 5). Greater availability and cheaper access to recording technology enabled local industries to emerge in many countries (even if the products were rarely broadcast or marketed over great distances and in great numbers), retaining local languages and musical styles. And while the ideological construction of tradition and locality has not just persisted but been enhanced in post-war years, such constructions could not have been possible without certain underlying elements of regional distinctiveness – the music, history and geography of Northumbria were different from those of Kentucky (and even more so from those of Taiwan or Papua New Guinea). Local performers dominate many local charts – despite the presence of global superstars on television channels – since, beyond the sounds of music, there are words (local languages), audiences and local industries. Despite fabricated geographies and various elements of globalisation, places continue to give meanings to people's lives and music.

4

THE PLACE OF LYRICS

Nothing should more closely signify the relationship between music, place and identity than the words of songs, especially where performers and audiences have broadly similar interpretations. Yet not all lyrics seek to convey a sense of place or identity, nor is there any simple relationship between lyrics and their reception and interpretation. Throughout history, popular music, like other forms of popular culture, has reflected contemporary issues (though, given its commercial underpinnings, sometimes opaquely). This chapter examines potentially the simplest and apparently the most basic element of popular music: the words and the nature of the images conveyed in songs, where those images relate to place and identity, though it cannot possibly review the diversity of lyrical themes and place references. Most popular music (including much country music) is subtle, ambivalent or vague in its designations and descriptions of place and identity. Moreover the lyrical content is not necessarily central to popular music. Many lyrically powerful compositions have been quite unsuccessful, since the rhythm, melody (or performer) were obscure or simply bad. Not all lyrics were even intended to be comprehensible. When John Lennon wrote 'I am the Walrus' (1967), he was reported to have said: 'Let's see the little fuckers try and work this out' (though some purported to understand the lyrics and it was attacked in a morals campaign).

The archetypal pop song creates an 'imaginary identification' between consumer and performer, where the perceived use value of the song – its emotional 'conversation' – becomes its exchange value and the key to success. Personal experiences, real and imaginary, imbued with emotion, are embodied in narrative form, creating an 'ideology of authenticity' (Bloomfield 1993). Authenticity may be enhanced by a link to place, usually generic and aspatial – 'at the hop', in a car, tenement or cinema – or more evidently spatial – the inner city, 'in the ghetto' or the country – but sometimes linked to particular places and time periods. Even so, the majority of lyrics eschew critical themes and social commentary, hence the classic rock lyric is supposed to describe fast cars, lost, found or elusive romantic love (perhaps with California girls) and drug-induced highs, all of which transport listeners from humdrum suburban and industrial worlds into dream worlds of excitement, recreation and pleasure. Variants of this – and other forms of escapism and enjoyment – have certainly informed many lyrics (Jarvis 1985) in most genres, places and time periods.

Vast numbers of songs (and band names) refer to place, and tempt geographical analysis, yet the sounds and rhythms of names, rather than the 'reality' of place, have often exerted a major role in the choice of location. Carolina and Georgia were more likely to be on the mind than Rhode Island and Connecticut, irrespective of any particular merits, while Tupelo, Pasadena, El Paso, Amarillo (on Route 66) and San Antonio (San Antone) sounded much more glamorous and exciting than Detroit or Pittsburgh: much better to be twenty-four hours from Tulsa than stuck outside Leeds, or a Wichita rather than Luton linesman. Slow boats went to China, while the Red River, the Mississippi (and even the Mersey) sounded better than the Thames or the Seine. In Britain a lassie might come from Lancashire (rather than Berkshire) and in the United States from Tallahassee. The Pretenders, in 'Cap in Hand' (1988), noted that it was almost impossible to pronounce Saskatchewan, though 'I can say Saskatchewan without starting to stutter', hence it had stimulated no popular songs. (The album otherwise managed to include references to Stranraer, Leith and Tupelo.) The resonance and timbre of the words were at least as important as the character and location of the places; onomatopoeia thus contributed to new musical geographies, and states such as Georgia benefited from their disproportionate presence in the lyrics of popular music.

Interpretation also rests in the performance and the conventions of performance. The unfamiliar languages of 'world music' (Chapter 7) do not simply intensify the exoticism but emphasise the music and the performance, hence Peter Gabriel has claimed that 'Senegalese singing…has an intensely spiritual feeling. The voice is such a powerful means of communication…it can transmit a feeling without recourse to words' (quoted in Barrett 1996: 242). Much world music and that of the 'black Atlantic' (funk and rap) is collectively produced and consumed dance music. Consequently 'textual content is foregrounded to the extent that it expresses emotional explosion, but decentered in terms of the specifically linear narratives more common to rock music' (Warne 1997: 145) or the traditions of folk and country music, and it is performed in contexts where comprehension would be difficult. Most audiences have claimed that they rarely attach much importance to lyrics, with one American survey reporting less than 12 per cent of people attaching a 'great deal of attention' to lyrics and 58 per cent stating that their content was irrelevant (Leming 1987). It is quite possible that 'song words matter most, as words, when they are *not* part of an *auteur*ial unity, when they are still open to interpretation – not just by their singers, but by their listeners too' (Frith 1988: 123). Hence particular performances in particular contexts influence comprehension, as may the whole style of the performance.

Listening at home to the radio (perhaps with a presenter providing interpretations), within the headspace of a Walkman, on a car stereo or at a concert, provide quite different contexts for comprehension. Nor can lyrics be separated from performance and style (evident in the intention of the United States Department of Defense to use 'In the Navy' as a recruitment song, until they saw the Village People). For a single song the context may be critical. When 'Go West' was first released by the Village People in 1978, San Francisco had become the American city where gay men had created some notion of community and received a degree of social acceptance. The triumphant rendition of the song was matched by each member of the group dressing as a conventional gay erotic type. By the time that the Pet Shop Boys recorded the same song, their more subdued style suggested, in the wake of the murder of gay San Francisco politician, Harvey Milk, and the rise of AIDS, 'an intense melancholy for a more carefree era of gay male existence' (Textor 1994: 93). Lyrics could thus be recorded, performed and received in quite different ways. Ambivalence and diversity are inevitable; popular music thrives on ambiguity and flexibility.

Inside the city

Themes of place in popular music – nostalgia for lost or distant places, dreams of 'making it' elsewhere, concerns over problem places ('in the ghetto') or simply evocations of idyllic landscapes (from Suya villages, to 'the green green grass of home' or the Mull of Kintyre) – are all part of the ability of music to transport listeners away from their ordinary lives. One dominant theme, the 'gritty' image of the city, and its seeming converse in the country, has been a repeated refrain in popular music (and in much of popular culture, notably in the 'mean streets' of crime fiction). Yet cities too are divided; the 'wrong side of the tracks' defines urban space, from the angry evocations of rap to the less threatening urban angst of Tracy Chapman. From the more classical cartographies of country and western music to the metaphysical spaces of ambient and 'new age' genres, music has represented place in a diversity of ways. History has accounted for the domination of particular places at particular times. In the United States a simple cartography of a century of popular song titles traces the westwards shift of the frontier, from rust-belt to sun-belt, in image and reality, culminating in a fascination for California and idyllic retreats from the city (Ford and Henderson 1974). Nowhere else has there been a similar sense of evolving regional differences, a reflection of its size and distinctive regions, but also of the remarkable twentieth-century shifts in the American population.

Elsewhere capital cities such as Paris, Copenhagen and Rio de Janeiro exerted their charms and attracted songs (and songwriters and bands) but references to small towns and rural areas were rare and as likely to be comedy as 'realism'.

Escape from the mundane and unpleasant was without the spatial connotations of 'California Dreamin'': the Animals urged 'We gotta get out of this place/ Where there's a better place for me and you', but this generic place was more likely to be in the United States than anywhere in Europe. Protest and dissonance invoked some notion of mobility, but that might be transient (perhaps in cars), drug induced (from marijuana to subtle references to cocaine) or, ultimately, to other worlds. Lyrics were more likely to emphasise the merits of the places where bands lived or came from, as in Gerry and the Pacemakers' 'Ferry Cross the Mersey' ('The place where I belong'), Lindisfarne's attraction to 'The Fog on the Tyne' ('Is all mine/ All mine') or even the Kinks' recognition that they were 'Muswell Hillbillies'. Two decades later, inner-city rap groups, such as Compton's Most Wanted and South Central Cartel, linked band names to places. Such place identities, as with so many lyrics, were often ambiguous, vague and inconclusive, and likely to be contradicted. Lyrics were simply ambit claims.

Cities consequently could be places of hope, of new social and economic lives, of racism and unemployment, or simply anonymity. Martha and the Vandellas offered 'Dancing in the Street' and the Loving Spoonful transformed the humid streets of 'Summer in the City' into a night-time party world. Even in difficult circumstances hope sometimes sprang eternal, as in Ben E. King's rose discovered in 'Spanish Harlem'. The city was a promise of things illegal and forbidden elsewhere, of drugs and deals and strange liaisons, a place of excitement and danger, of decay and difference, but not a place of boredom and tranquillity. It offered a walk on the wildside. Uncertainties and challenges were everywhere; in the Doors' 'People are Strange', 'People are strange/ when you're a stranger/ Streets are uneven, when you're down' (1967). Success was sometimes elusive; Otis Redding sat on the 'Dock of the Bay' (1968) regretting: 'I left my home in Georgia/ Headed for the Frisco bay/ I had nothing to live for/ Looks like nothing's gonna come my way'. In the Eagles' 'Hotel California' (1976) material success might be possible but what benefit was that: 'We are all prisoners here/ of our own device/ You can check out any time you like/ but you can never leave'. But the city, one way or another, was the place of action, particularly in the gritty neon strips near the core:

> The urban landscapes of rock are not those of any formal grandeur, the great works of civic design, but the trivial, the ugly and the ephemeral....There is little irony in rock's classic, naive, commercial and traditional celebration of this landscape of consumption and neglect.
>
> (Jarvis 1985:113; see also Gibson and Connell 2000)

If rock celebrated the inner city, suburbia was a place of contempt and comedy, at least in the British music of the 1960s, with the Kinks' 'Well Respected Man' (1966) and Manfred Mann's 'Semi-detached Suburban Mr James' (1966); though the Beatles extolled the 'blue suburban skies' of 'Penny Lane' (1967), it could also be seen as 'subversively hallucinatory...[d]espite its seeming innocence' (MacDonald 1994: 179). Two decades later, Blur reprised 1960s pop, satirising suburbia and again delineating a dreary, repetitive and soulless landscape (Medhurst 1997: 264–7), while Suede celebrated suburban alienation, repressive stability and the sense of lives going nowhere, where 'home' is a place of unease (Frith 1997: 275–7). Suburbia exuded unfashionability and boredom; glamour and excitement necessitated transcending the suburbs. In Australia, the Whitlams championed the virtues of inner-city community life in Newtown, Sydney, and criticised suburban blandness and materialism (Carroll and Connell 2000). Popular music added to wider, more pervasive and negative representations of suburbia.

In many Western societies, but especially the United States, one particular genre – rap – has both expressed and emphasised challenges to the structure of contemporary society, and perhaps especially to racism, and exemplified the critical relationship between popular music and place (Chapter 8). Rap marked the emergence of submerged and repressed voices from the inner post-industrial city; lyrics expressed anger, resentment and resistance, as rappers 'seized and used microphones as if all amplification was a life-giving source' (Rose 1994: 220) and chanted their opposition to marginalisation, racism and capitalism. It is almost a cliché that black struggles to survive in the inner-city 'ghetto', or the (neighbour)'hood', absolutely dominated rap scenes, establishing the Bronx, and putting Compton, in south-central Los Angeles, on the map of urban deprivation (Forman 2000). Its most extreme form, 'gangsta' rap, was linked to crime, which itself could be seen as a form of political and cultural resistance, symbolised in band names such as Public Enemy (Walser 1995; Chude-Sokei 1997; Krims 2000). For some gangsta rap provided a kind of 'redemptive vulgarity' that resisted the 'artificial optimism' of certain African American cultural narratives (Dyson 1996). More commonly it fulminated against oppression, from the state but particularly the police, through rhetorical catch-phrases rather than openly rejected politics, though a number of rappers, such as Ice Cube, challenged nihilism and the absence of self-criticism that dominated rap (Boyd 1994).

The energy, style, anger and resistance of rap, and its validation of depressed urban environments, was attractive to emerging hip hop fans elsewhere in the world, as reggae had been a decade earlier, but with distinctive variations in their appropriation of it. In France rap displayed many similarities with that of the United States, but important differences with respect to racial and national issues (Box 4.1; Chapter 8) and in the outer suburban location of many rap groups. In Germany too rap lyrics reflected social problems, like racism, exclusion, drug abuse and violence, particularly those that were specific to German-Turkish youth, the main proponents of rap, like the diverse issues stemming from 'being between two cultures' (being German in Turkey and Turkish in Germany), blood feuds, intra-Islamic religious animosities and gender issues (Caglar 1998: 247, 252; Box 8.5). Rappers perceived their lyrics as being in opposition to pop, which was claimed to be repetitive, obedient and mainstream with lyrics dealing with love, passive romance and nature, which did not relate to the lifestyle, concerns and sentiments of minorities. Rap thus provided a voice and visibility to marginalised German-Turkish youth, as it did to migrants in France (and elsewhere), but at the same time incorporated them into a specific ethnic space on the margins of German society (ibid.; see Chapter 8). While rap found adherents in most countries, especially where racism was marked, there were exceptions: the 'profane bitterness, antisystemic radicalism and overt sexual warfare would be considered excessive by most Southeast Asians' (Lockard 1998: 263). Regional issues were rather different.

Box 4.1 Rapping in France

In rap themes taken up in France, resistance to racism's impact and discriminatory policies, alongside the problems of the 'ghetto' (here largely transposed to outer suburbs with few facilities), were similar themes to those in the United States. French rap also displayed a vehement anti-nationalism, and an explicit rejection of the integrationist 'republican' model of race relations and of colonialist and imperialist attitudes, especially in Black Africa (rather than North Africa). The social situation of the outer suburbs (especially in Paris) also meant that solutions offered by rappers were not the black nationalist or separatist themes of hip hop's more radical American elements, but were more evidently pluri-ethnic (incorporating black, north African and white elements), transcending divisions within France. Rap has remained confrontational to the capitalist system, denounced in one of MC Solaar's tracks:

> Ce monde est caca, pipi, cacapipitaliste
> Ils augmentent les prix, en tirent les benefices
> (This world is shit, piss, shit-piss-capitalist
> They put up prices and reap the benefits)

French rappers have however tended to refuse to define themselves in terms of traditional political discourses, despite their militancy and anger. The name of one prominent group, Supreme NTM, is short for 'nique ta mère' ('fuck your mother'), incorporating the Arabic slang word for 'fuck', and indicative of parallel misogynistic elements of French rap. If hip hop built some unity across racial lines, it often did so 'by expressing a shared masculine contempt for women' (Lipsitz 1999a: 229). The perceived militancy of rap resulted in the closure in the late 1990s of hip hop clubs in the (right-wing) Front National-controlled town of Vitrolles.

See: Cannon 1997; Gross *et al*. 1994.

Rather like rap, in the blunt anarchic world of punk, a whole genre of music also violently and abusively challenged and condemned existing systems (most notoriously in the Sex Pistols' famous objections to the monarchy in their slanderous version of 'God Save the Queen' (1977)), but rarely offered any solutions or hinted at vehicles for change. The Clash sung 'We want a riot/ A riot of our own' (1977), but disruption itself was enough (Box 4.2). Like rap, punk claimed the inner-city 'wastelands' and opposed the political order, but was more obviously the white music of a displaced middle class with its 'class fantasies' (Lebeau 1997: 285), rejecting suburbia and seeking the credibility of the inner city: dreaming rather than challenging. Punks sought, through inverse social mobility, to tap into a more 'authentic' working-class lifestyle, equivalent to 'real', 'hard' and 'tough' – qualities associated with life on inner-city streets. This was simultaneously a 'hard-edged bohemianism' and an act of deterritorialisation (Trager 2001: 31, 35). In this deliberate self-marginalisation, in response to bourgeois middle-class whiteness, punk was something of a commercial creation in its brief yet vibrant life.

Box 4.2 Punk, posturing and the inner city

Much has been written about the short but ebullient times of punk, reaching its zenith in the second half of the 1970s, which seemed to represent angry inner-city opposition to the global and national economic and political crises that had produced high levels of unemployment in the capitals of Western nations. It was the antithesis of the more gentle rock and folk music that had preceded it. Punk lyrics and style emphasised negation (rather more than absolute nihilism), performers were conscious of class-based politics (seeking working-class credibility and authenticity, especially in Britain) and the necessity to avoid 'selling out' and hence championed spontaneity, and do-it-yourself simplistic musical structures, where skill was equated with glibness. This accompanied a dress style that challenged any notion of expense or fashion, and which pared back the flamboyance of hippies into a minimalism characterised by safety pins, swastikas and hairstyles that seemed to transgress every social standard. The origin of punk is usually attributed to the influence of CBGB's nightclub in New York (whose acronym ironically stood for Country, Blue Grass and Blues) and the influence of bands such as the Ramones and MC5, and the almost parallel emergence of the Sex Pistols in inner London. However, Paris had a punk scene before either of these cities, and similar trends were emerging in Australian and other western cities. Punk also had critical suburban influences, both in the origins of key performers and the manner in which punk sought 'to expose the madness and perversity' of suburban culture (Lebeau 1997: 283), and links, through performers such as Patti Smith, to the bohemian art scene. Punk appropriated and adopted the landscapes of the inner city, a relic landscape of urban crisis that had been bypassed by the economic boom; young people repopulated parts of the inner city, notably in Australian cities where angry music emerged from small, enclosed, grubby pub venues to challenge the conformity and myopia of suburban life. Colleges were also in inner cities, and audiences were as likely to be students and suburbanites as alienated workers. The lyrics of punk emphasised opposition rather than creativity, triumphing cheap drugs and delinquency, celebrating marginality, criticising those in power and contributing anarchy. The Sex Pistols sung 'I am the anti-Christ/ I want to be – anarchy': the epitome of an unsentimental and 'scorched earth, year zero attitude to tradition and the past' (Medhurst 1999: 224) alongside intense pessimism over the future. The Clash offered some possibility of revolutionary change and the Dead Kennedys railed against every element of American consumer society. Yet underlying much of this was a 'camp parody of America's teenage myth' (Osgerby 1999: 166) that situated punk amid suburban fears and impotent rage against boredom as much as about economic collapse and exploitation. Manchester band, the Buzzcocks, simply sang of 'Boredom'. Much later, Blink-182, seeking to convey rebellious punk attitudes, proved to be part of the American suburban school of 'soft punk': 'we're totally suburban punk. None of us grew up in poverty. We're rich suburban white kids, we just rebel against boredom' (quoted in *Sydney Morning Herald* October 27, 2001: 16). Lyrics were as much collages and fractured images as messages for action, or even introspection. Critics were constantly disappointed in punks for not realising their 'true' social potential as a force of liberation, and not even ridding their lyrics and postures from such inhibiting tendencies as sexism, racism, homophobia and violence. Though punk seemed to be inner-city white trash music it never became the counterpoint of rap that both supporters and critics might have liked it to be. Punk mythologised the harshness of the inner city, yet in many places gentrification and economic growth removed the most dreary locales.

See: Laing 1985; Nehring 1993; Marcus 1993; Sabin 1999; Savage 1991; Tillman 1980; Trager 2001.

While punk was ambiguously related to economic depression in the inner city, deindustrialisation was a key context for heavy metal music, at least in the United States. Originating in the declining English West Midlands in the 1970s, metal found its largest American audiences in the declining industrial towns of the Midwest and North-East, and most of that audience were working-class youth (Walser 1993). Metal's lyrics of fantasy and aggression (war, death and madness) provided polar responses to the urban crisis; many songs placed their listeners in the 'subject position of helpless pawn in the game of global politics' (Harrell 1994: 97), but helplessness had a certain romanticism. 'Death metal is neither an example of false consciousness nor a coping mechanism for the stresses of an unequalled world. It is a promise unfulfilled' (Berger 1999: 294). In different ways, rap, punk and metal expressed opposition and resistance to the structures of society, especially as these were exemplified in the inner city or in the bleak wastelands of suburbia. While such urban genres were both transgressive, even shocking and resistant, they eschewed class and were criticised for offering no organisational strategies that might become a cultural politics (Mitchell 2000:161) or would transform urban space and society.

From city to country

Country and western music has always expressed some distaste for urban landscapes and for city life itself; almost by definition it celebrated rural space, though not without doubt and uncertainty or any failure to recognise that the country was without blemish. Nothing epitomises this perspective more than various saccharine lyrics of John Denver and, above all, the best selling 'Country Roads, Take Me Home' ('Country roads, take me home/ To the place I belong/ West Virginia...'). Beyond that there was 'Thank God I'm a Country Boy' and 'Wild Montana Skies', a rather different kind of landscape, and many more. Here and elsewhere, country and western music portrayed a 'nostalgia for a rural paradise, symbolised by a yearning for a simpler way of life, a looking back to an uncomplicated place' (Kong 1995b: 187), and a place where an older social order remained valid, compared with the city where John Denver found merely 'money hungry fools' and 'black limousines', and where others found poverty, violence and unrealised dreams.

Country life was rarely without complications: droughts occurred, dogs and stock died, prices fell and wives or husbands walked out despite invocations to 'Stand by Your Man'. Country music is largely the antithesis of feminist perspectives on gender relations. It is gendered music where women break hearts and destroy relationships or remain the stable source of domesticity (cf. Chapter 9). 'Western' music stressed the individualism, loneliness and self-reliance of the male cowboy world. From at least the nineteenth century, rural life was duplicitous: adultery, divorce, loneliness and tenuous relationships occurred in a bitter-sweet world of bars, back porches, dance halls, betrayal and hope. Simple tunes and equally simple and comprehensible lyrics, which is why much analysis of musical lyrics has focused there (e.g. Blair and Hyatt 1992; Tichi 1994, 1998), evoked working-class and Christian values, and the virtues of hard (usually outdoor) work in opposition to both urban anomie and notions of radical change: localised human relationships amid natural landscapes – even 'sacred places and times' (Woods and Gritzner 1990). Home 'is more than a locale and site of family and friends [but is] the essence of the natural world' so that 'the Tennessee mountain home is nature itself, the Emersonian center of authenticity in America' (Tichi 1994: 25). Here there was no place for hedonism and no challenge to the existing order; it was the music of poor white folk (despite a legacy whereby performers such as Hank Williams and Jimmie Rodgers owed much to Afro-American music), struggling against the odds and the limited environments.

Yet however duplicitous rural life might be, it was superior to the city, though rarely as idyllic as that often portrayed in folk music (see, for example, Box 2.3). Many of Gram Parsons' songs were imbued with country's sense of loss, fatalism and yearning, accentuated by plaintive vocals and pedal steel guitars. In 'Ooh Las Vegas' he, the 'poor boy', is seduced and overwhelmed by the big, anonymous neon city; in 'Juanita' he wakes up ashamed in a drug-littered hotel, hoping to be cleansed by a young woman with a 'conscience so clear'. In 'Sin City', whose title exemplifies the ethos of country music, Los Angeles appears as a contemporary, sunny Sodom, and its deadly charms are irresistible. A sense of regret and guilt, the constant tension between desire and duty, alongside the possibility of redemption, run through these and countless songs by Parsons and others (made more poignant by Parsons' own early death from a drug overdose in a cheap motel room). Many country singers, not only in the United States, have recorded and been recorded in jail (notably Johnny Cash's 'Folsom Prison Blues') where nostalgia is pervasive, and redemption problematic, amid the recognition that urban temptations and sin have been overwhelming.

Alternatives to the city existed in generic rural spaces – fields, mountains, bayous and open ranges, and small home towns – all effectively in southern states that became a 'landscape of nostalgic rural salvation and refuge from the cold impersonal nature of the modern cityscape' (Lewis 1997: 167). In Dwight Yoakam's 'I Sang Dixie' he recalled the warmth of a lost southern home, while cradling a homeless man, dying and alone on a busy Los Angeles street. Similarly the music of Joe Ely, Butch Hancock and others invoked the windy, flat plains and townships of West Texas, tracing a nostalgic landscape of 'straight-line highways, neon-lit bars, cheap motels and Pearl beer' (Gumprecht 1998: 76; see van Elteren 1994). For Jason Ringenberg of Jason and the Scorchers, memories of desolate landscapes were something to actively capture: 'I wanted to get this

American, earthy sort of expansiveness. That emotion you feel when you look out over a plain and see the sun set…when you hear the hog feeders rattling…and a buzzard crowing in the back' (quoted in Webster 1988: 158). Home and homecoming, the essence of yearning, is the charter of country music.

Country music thus glorifies safe home comforts and a place to settle down (Jarvis 1985: 116), though domesticity itself, in much of popular music, is a bland, conservative place from which to escape. Yet country music is full of unresolved tensions about urban life, and about blue-collar work and class divisions:

> A rural–urban tension underlies most country songs. Home is always portrayed as rural – green, welcoming, often with mother or girlfriend waiting. Yet home is inaccessible, because the protagonist has gone to the city and been tainted by it. He can long for home and remember it as he sits in his lonely room, or in a smoke-filled bar, but he can never truly return. Cities are portrayed as unfriendly, dehumanising places, full of temptations, greed and selfishness. The tavern, with its neon lights, music and laughter, is an illusion of solace from the city; in fact it is a disguised version of corrupt city life.…Women are especially susceptible to the lure of urban life. They are either angels (waiting at home, patient and loving) or fallen angels (sitting in honky-tonks with tinted hair and painted lips).
>
> (Jensen 1998: 29–30; cf. Lewis 1991)

The commercialisation of country music brought a greater sense of fun and a willingness to parody the conventions of the genre. Many Australian songs such as 'True Blue' (John Williamson) and 'Good-Bye Lucky Country' (Eric Bogle), are often gently benign, showing affection for rural people and places, even as they sometimes satirised them. In America commercialisation, and the rise of Nashville, turned country music from any residual hillbilly ethos towards a paradigm of the ideal south – rural nostalgia, rose gardens, conservative politics and traditional Christian values – a past that never was.

The simple virtues, certainties and dichotomies of country music have given it a massive global following. Displacement (real and metaphorical) and anguish are universal; so too are its counterparts of sincerity and simplicity. Even in wholly urban settings it offers a counterpoint to urban life; in Singapore many night clubs and local bands specialise in country music and the locally well-known Matthew and the Mandarins plead in their most famous song, 'Singapore Cowboy', 'Singapore cowboy, a long way from home/ Won't you sing a poor cowboy another lonesome song'. Home has been displaced to the other side of the globe and the virtuous Nashville. In Thailand country music ('phleng luktoong': child of the fields) is popular in rural areas, especially in the north-east, and among urban migrants. Considered crude and peasant-like by urban elites, it is based on rural folk music, uses traditional instruments and has lyrics that reflect Isan (north-east Thai) identity, describing both romantic nostalgia and the problems faced by peasants (Lockard 1998: 185–6). In Australia, rural Aboriginal populations quickly adopted country music, both sustaining white performers such as Slim Dusty, whose urban allure had faded, and nourishing many indigenous performers in some part because the lyrics that cherished country lifestyles and rural occupations, such as those of stockmen, had particular resonance, but mainly because the 'sense of loss' of country music could easily be related to by Aboriginal people (Walker 2000: 14). In all these places, Zimbabwe and elsewhere, simple musical structures and instruments made replication straightforward.

Movin' on

Mobility pervades country music, but every genre, above all in the United States, envisages some physical escape from constricting confines. 'It may be that the most interesting struggle is the struggle to set oneself free from the limits one is born to' (Marcus 1975: 19). Music is one part of a celebration in literature, film (the road movie) and art of the American road (and the railroad); a prominent part of American culture and the symbol of mobility in a particularly mobile nation. As the epitaph on the grave of the singing brakeman, Jimmie Rodgers, states: 'His was an America of glistening rails, thundering boxcars, and rainswept prairies, great mountains and a high blue sky'. In country music the mobile were apt to be restless ramblers destined to return eventually to the salvation of home. While men could return, women like Emmylou Harris's 'Queen of the Silver Dollar' (1975) were more likely to be ruined by the experience of migration and their return was shameful (Tichi 1994: 44). In country music the geographical path was also a spiritual one, hence Tichi claimed that 'the idea of the voyage of life is entrusted to country music, as it has been entrusted to other arts in American culture' (1994: 193). In other genres mobility, if less common, was at least as complex.

In blues and country music mobility was often a response to poverty and a necessary escape; travel was accompanied by heartache but rarely by success. Rock 'n' roll transformed that: mobility was about freedom and adventure, born of affluence and access to cars or motorbikes. The Shangri-Las exulted in the dynamic 'Leader of the Pack' and Steppenwolf's 'Born To Be Wild' ('Head out on the highway/ Looking for adventure/ And whatever comes our way') became the theme of the movie *Easy Rider* and a hymn for a generation of escapists. Later Bruce Springsteen's 'Born to Run' (1975) transcended this idiom, acknowledging the futility of escapism even if the city was 'a death trap...a suicide rap'. Truck-driving songs offered a parallel 'fakelore' of endless highways, strong coffee, speed, smoke (Jarvis 1998: 102) and the camaraderie of the truckstops that created an elusive and transient culture and transformed the road into home. Many have recorded 'Will there be any Truckstops in Heaven?'. The travelling life was life itself.

Road songs emphasised the simple pleasures of escape and freedom, restlessness and divorce from dreary conventional lives, escape from commitment and closeness. Rock again traced its legacy to the blues, the folk songs of Woody Guthrie and the travails of country and western. As in 'car songs' images were quintessentially male:

> It is hard to think of the early Dylan or Guthrie or the Delta Bluesmen without seeing a picture in your mind of a lone male figure walking away down a never-ending road with just a rucksack and a guitar for company.
> (Heining 1998: 106)

Highway songs offered the promise of somewhere further down the road – either of escape, adventure or even security; Wayne Hancock on 'Wild Free and Reckless' (1999) was a fiddle-driven 'interstate addict going back to Texas and the only gal I've ever known'. But highways also offer open spaces without borders, flatlands for fostering dreams and illusions that escape is possible. Highway songs, epitomised in the much recorded 'Route 66', are quintessentially American; as Leyshon *et al.* have firmly noted 'the glamour of the transcontinental highway doesn't quite work in Essex' (1995: 430) where clear skies, emptiness and distance are absent.

Wilderness and deserts offered similar escapist prospects, as in U2's 'In God's Country' (1987) ('Desert sky, dream beneath a desert sky/ The rivers run but soon run dry/ We need new dreams tonight') or in 'Heartland' (1988) ('Mississippi and the cotton wool heat/ Sixty-six and the highway speaks/ Of deserts dry and cool green valleys/ Gold and silver veins – shining cities'). The inner heart of the continent, the destination of many highways, was a place for discovering inner truths, as in The Triffids' 'Wide Open Road' (1988) and the Go-betweens' 'Cattle and Cane' (1990), both cult indie hits in the UK and Australia, and Kasey Chambers' 'Nullarbor Song' (2001). It is scarcely surprising, then, that such voyages of escape and discovery were linked to 'frontier' colonial landscapes.

Lyrics, often enhanced by the music, like other cultural products, provide both reflections on past, contemporary and future worlds and escapes to them. These 'other worlds', though alternative imaginary spaces, arise from existing worlds to 'subsequently react in a dialectical process on the world of everyday reality' (Pickering and Green 1987: 3). Despite rap and punk's angry denunciations of inner-city life, though these still pointed to the inner city as a risky, painful yet exciting place in contrast to safe, unfeeling and dull suburbia (Trager 2001: 59), most of the lyrics of popular music have emphasised – indeed over-emphasised – the virtues of places. Thus the Beatles' Liverpool, where the seemingly idyllic Strawberry Fields was actually an orphanage, was 'far more idyllic than the reality of Liverpool life for many' (Halfacree and Kitchin 1996: 51). Popular music, especially country music, glamorised rather than vilified place and actually challenged the need for escape and uncertainty.

See the noise

Popular music is much more than soundtrack alone: images of lifestyles, places and particularly performers are enhanced (or sometimes challenged) by live performance, record and CD sleeves and music video; 'when you listened, you were also seeing the performers and other fans...styles of clothing...images of the sexual body...fantasies and social experiences' (Grossberg 1993: 188). Visual images are at least as diverse as lyrics and, since the 1960s, when television and pictorial record sleeves emerged more or less contemporaneously, have acquired increasing importance. Early record sleeves usually depicted performers, but dress and demeanour were important, maintaining the themes of publicity photographs; country and western performers wore 'country' clothes (usually including Stetson hats and boots) and rock 'n' roll bands wore black leather. Initially even supposedly threatening performers, such as Elvis Presley and the Rolling Stones, wore suits and ties until fame gave them flexibility. In the 1960s, the era of counter-culture, vast numbers of bands cavorted waist deep in cornfields or otherwise indulged in rustic splendours. This legacy of the folk revival celebrated community and vernacular culture: reflections of the discontents of modernity. Two decades later rappers and punk bands struck angry poses in the graffiti- and garbage-ridden streets of the inner city. By then country singers were as likely to be dressed in rhinestones and roses as engaged with nature (Tichi 1994). Heavy metal bands were largely absent from covers, de-emphasising the role of musicians and overt commercialism; the cover of Slayer's *Season in the Abyss* was a hazy, predominantly red and black field of skeletons, coffins and upside-down crosses (Harrell 1994). Such generalisations disguise massive diversity: many covers were abstract; only men were used for rai music; Bob Marley tended to be a 'soft' sexual Bob on covers in contrast to the more revolutionary lyrics (Stephens 1998). Some sleeves went beyond generic landscapes to depict real places: Lindisfarne's Tyneside, Eric Clapton's Los Angeles, Ben E. King's Spanish Harlem, Pink Floyd's London (with suitably absurd blow-up pig floating above Battersea Power Station on *Animals*), U2's Berlin (on *Achtung Baby*), usually in flattering forms that contrasted with those of the rap and punk scene. In certain contexts a sense of place may be deliberately created, as in much Okinawan music, through the frequent portrayal on CD covers of distinctive 'lush tropical greenery...[and] beaches fronting blue seas and blue skies...[and] culturally distinctive and historically important sites, such as houses with traditional Okinawan tile roofs or Shuri Castle, to create a unique Okinawan image and identity' (Roberson 2001: 221).

Videos inevitably enhanced the visual as performers went from studio and concert stage to a range of settings. While television and video tended to emphasise the role of adolescent desire, sexuality and violence, they also depicted generic and specific places. Images partly replaced narrative as the central element (unsurprising when videos were both product and promotion for product) with videos like those of Spandau Ballet, Duran Duran and Wham in the 1980s foregrounding images of travel: exotic settings, beaches, cocktails and models in bikinis, both pastiche and parody of a consumer's paradise (Webster 1988: 160). Almost as frequently 'the rock video world looks like noplace, or like a post-nuclear holocaust place – without boundaries, definition or recognisable location' (Kaplan 1987: 145). Being out there, in the street, defined much of the action especially for male bands; thus when Cyndi Lauper leads a band of girlfriends through New York city streets, pushing through a group of male construction workers, it reflects not just that 'Girls just wanna have fun' but is 'a powerful cry to the privileged realm of male adolescent leisure and fun'. Madonna's 'Borderline' and Tina Turner's 'What's Love Got to Do with it' took similar positions (Lewis 1993: 137). More frequently women were observers, or the objects of attention, rather than participants (Chapter 9). Yet, one way or another, so many videos are tied to place, usually generic but sometimes specific, that as in many rap numbers 'there is a powerful evocation that music of the kinds they perform is originally rooted in some place or authentic locale where an audience of real people, preferably including blacks, are depicted as enjoying it' (Pratt 1990: 27).

The neighbourhood and the ghetto became the focus of funk and then hip hop cultures, both in a discursive sense (through the subjects and sounds of songs themselves) and physically (as the site of 'authentic' performances and cultural roots, and through hip hop sub-cultural experiences). Many of the names of groups, albums and songs, and films concerning hip hop culture, reflected this symbolic reclamation of the American city: the Geto Boys, Cypress Hill and the Watts Prophets; albums such as NWA's *Straight outta Compton* (1989) and Boogie Down Productions' *Ghetto Music: The Blueprint of Hip Hop* (1989); labels such as Sugarhill Records (named after the Harlem, New York, neighbourhood); the films *Beat Street*, *Boyz N the 'Hood* and *New Jack City* ('New Jack Swing' referred to a slower, more soul-influenced hip hop sound); and a plethora of tracks concerning various aspects of ghetto life and culture. Hip hop continued a tradition of representing 'the ghetto' found in soul, jazz and funk.

Numerous record covers depicted critical social issues, which visually exemplified lyrics (Figure 4.1). Film clips were frequently set in ghetto environments and other city spaces – in dilapidated buildings, in cars, late-night trains, alleyways and parking lots. In some instances the depiction of particular cities, neighbourhoods and precincts was crucial, as argued by music video producer, Kevin Bray:

> if you have an artist from Detroit, the reason they want to shoot at least one video on their home turf is to make a connection with, say, an East Coast New York rapper. It's the dialogue...between them about where they're from.
>
> (quoted in T. Rose 1994: 11)

Figure 4.1 Curtis Mayfield, *There's no Place like America Today* (1991)
Illustration: Peter Palombi.

At other times the specificity of the actual locations seemed to be unimportant, where the identification with a generic inner-city locale was sufficient to suggest the location from which the song might have come. Hip hop film clips discursively reconstructed place, as:

> the ghetto comes to be valorized not only as a 'space of death' (and destruction) but also as a space of survival and transcendence. . . . The alternate map of the nation becomes an unmistakable geography of Black America. . . referencing many of the more prominent ghetto crucibles of the country – Compton, South Central LA, the Bronx, Harlem – or more ubiquitous localities like 'the hood', 'the projects', 'the streets'.
>
> (de Genova 1995: 119)

These images could challenge generalised, clichéd representations of place that permeated the mass media:

> Rappers' emphasis on posses and neighbourhoods has brought the ghetto back into the public consciousness. It satisfies poor young black people's profound need to have their territories acknowledged, recognized, and celebrated. These are the street corners and neighbourhoods that usually serve as lurid backdrops for street crimes on the nightly news.
>
> (T. Rose 1994:11)

Not all reception was so insightful. For some, film clips of such 'no-go' districts served a voyeuristic function, as middle-class consumers were allowed into the 'authentic' world of street violence through the distant safety of the television screen; for others, depictions of violence, crime and identity only served to entrench sensationalist representations of 'the ghetto', ironically aiding essentialist portrayals of black American youth and 'gang' culture, what Public Enemy rapper Chuck D called 'niggativity'.

The ghetto became a commodity, enabling the 'marketing of 'ghetto authenticity' (Allinson 1992). The original growth and then renaissance in 'Blaxploitation' films and soundtracks (such as Curtis Mayfield's *Superfly* (1972) and Isaac Hayes's *Shaft* (1971)) capitalised on this urban authenticity, as did much of hip hop in the 1980s and 1990s. Dr Dre reflected on this with regard to their album, *Straight outta Compton* (1989):

> you gotta think about it. . . . We came out with Compton, the NWA thing, so every time somebody sees Compton, they gonna buy that shit

> just 'cause of the name, whether they from there or not. Compton exists in many ways in the music to sell records.
>
> (quoted in Cross 1993: 198; Forman 2000)

Not just individual performers, dancers or characters were bound up in this process of representing and interpreting cultures through music and promotional video clips; territory provided a catalyst for both the celebratory representations of otherwise marginalised neighbourhoods, and stereotypical images of dangerous, chaotic places. Music consumers (from fans to the casual listener and even the accidental audiences of shopping malls and airports) engage with and react to the reputations and associations surrounding musical styles and particular artists, and to the discourses of the places from which they come (or for which 'authentic' origins are inscribed). So, meanings attributed to hip hop in wider public audiences involve both the musical texts and stereotypes of ghetto localities.

Out of the music

Popular music illuminates place, either directly through lyrics and visuals, metaphorically through heightened perceptions, through sounds that are seen as symbolic of place (the 'high lonesome' sounds of country, or southern blues) and in performances that create spaces of sentiment. Thus, dramatically, the steel guitar is said to offer 'the wail of love and loss that men are too stoic to express directly, that they try to drown in beer and whiskey, that they cannot leave behind' (Jensen 1998: 34). Country music is not the only genre whose lyrics can be viewed as 'an expression of humankind's age old existential concern, the search for self and space' (Woods and Gritzner 1990: 238), even if popular music conventionally charts dreams, unfulfilled desires and youthful ideals, hopes and aspirations: 'just about any theme that appeals to young sentiments' (Kong 1996a: 110). Throughout so much contemporary popular music there is a pervasive restlessness, uncertainty and insecurity, affecting love and rebellion, and also sense of place.

Inevitably lyrical contents change over time and space; in dance music they have simply disappeared. Subtlety and introspection have become more evident; irony and realism infuse even country music as it crosses into the mainstream and places may be more threatening than welcoming. In a quite different sense the growing commercialism of the popular music industry brought 'the final triumph of the "new pop"', the eclipse of content by form (Frith 1990: 172), though little was quite so bleak. Popular music has been pervasive enough to create enduring images; thus 'a popular imaginary America in Britain is precisely a meeting of trashy dreams and rockabilly, merging stages of Presley's journey from Tupelo to

Memphis to Hollywood and Vegas, mixing "white trash" with the trash aesthetic' (Webster 1988: 167). Such images are close to indelible.

Music briefly allows escape, an escape that is necessarily virtual, primarily through transient and vicarious participation in particular musical experiences – whether concerts, recordings or videos – where, briefly, individuals and audiences are transported into the music and out of the mundane. Exceptionally, music offers the experience of other realms, imaginary and imagined places, like Jimmy Buffett's 'Margaritaville' (Bowen 1997), or transformed old homes and new destinations. It provides one means, sometimes opaque, of reflecting or challenging social trends, attitudes to place, patterns of mobility and shifts in identity, enabling new identities to be forged through shared experiences in concerts, or simply through the shared lyrics, symbols and a common sense of style. In a more evidently post-modern world a sense of place, however populist and nostalgic, is still part of the lineage of popular music, in various genres that grew out of particular local contexts.

5

SOUNDS AND SCENES

A place for music?

Many places have been known for and through music: from jazz in New Orleans, Soundgarden and the grunge scene in Seattle, Delta blues to Goa techno. Locations where popular musicians have been particularly active, or audiences and subcultures unusually vibrant, have become synonymous (and sometimes eponymous) with specific styles of music. Such local sounds are subsequently heard within national and transnational mediascapes. This chapter seeks to develop a framework for understanding these expressions of 'local cultures' in music, and situating them in contemporary debates about notions of uniqueness within a global context. It examines the emergence of 'scenes' in particular places, and then considers those where local cultures developed to such a degree that a particular city or region became famous for a 'sound' of its own (such as the 'San Francisco sound' or the 'Mersey sound'). Subsequent chapters deal with similar manifestations of 'the local' in music, examining connections between sound and place, from the multilayered tunes of migrant communities in world cities to the relationships between cultural identities, places and nations.

Scenes and sounds

In one sense the uniqueness of local music scenes is straightforward; music is made in specific geographical, socio-economic and political contexts, and lyrics and styles are always likely to reflect the positions of writers and composers within these contexts. Recognition of this embeddedness of musical expressions in particular places was evident in the tradition of mapping regional variations in sounds and styles (Chapter 2), and in the increasing importance placed on ethnographic studies of music 'scenes' (for example Cohen 1991b; Finnegan 1989; Gibson 2000). To understand how musical activities may be shaped by places it is necessary to explore local musical practices, institutions and behaviour. The idea that a deterministic relationship between place and culture exists – as musical styles and sounds emerge from different locations, and as musicians relate to their environment – remains powerful. Links between geography and music are in one sense not surprising, yet representations and understandings of place vary dramatically – as do the musical expressions that emerge from specific local contexts.

Finding places for music, finding places in music

A common element of popular music literature, particularly evident in the discourse that surrounds 'rock' music, is a tendency to search for links between sites and sounds, for inspirations in nature and the built environment, which inadvertently build on early ethnomusicological traditions. Journalists and others have invariably asked musicians what it was like to record in a particular place, what influences their home town may have had on their musical career and their songs, or how they feel about the 'scene' or 'sound' that they are identified with (hence the views of Junkhouse in the Preface). This sense of finding geographical roots for musical sounds and styles, of locating the artist or the scene in physical space, is a dominant theme in the music press, artist biographies and 'rockumentaries'. In 1996 the British Broadcasting Corporation (BBC) documentary on rock music, *Dancing in the Street*, featured artists' testimonies from various stylistic backgrounds and musical eras. Links between place and music were made explicit at many points, both in subject matter and in richly evocative images of place used as backdrops to discussions of popular music scenes (including shots of New York ghettos, London council estates, Mississippi highways, Merseyside docks and Chicago clubs). Artists were asked to explain the influence of their background on the emergence of their particular 'sound'. Such influences were foregrounded in the testimonies of artists such as punk innovator Patti Smith:

> New York is the thing that seduced me; New York is the thing that formed me; New York is the thing that deformed me; New York is the thing that perverted me; New York is the thing that converted me. New York is the thing that I love, too.
>
> (quoted in BBC 1996)

Iggy Pop similarly claimed that his childhood environment had a profound effect on his ability to write and perform, and his attempts to make music that remained locally relevant, but in a quite different way.

> I dropped out of college and went to try to play the blues, but it didn't take me long to figure out that I wasn't middle aged, I wasn't black and I wasn't from Mississippi – but how about if I could take those ideas but do that from the point of view of our life back in Michigan?
>
> (quoted in BBC 1996)

For others such as David Bowie, place had a somewhat different influence on musical expressions, driving him away from his suburban 'wasteland' background (as he put it) in search of exotic, fantasy worlds, as did Boy George a decade later (Chapter 9). Small towns had similar effects:

> Wilco Boumans [*The Keyprocessor*] looked around the small Dutch town of Bergeyk and saw provincial boredom. Having moved to nearby Eindhoven to study, he began listening to a number of synthesiser pioneers, including Kraftwerk and Yello. This music seemed as far as possible from the traditional life he was striving to leave behind.
>
> (Barr 2000: 185)

The presence of music scenes of a substantial size, in larger cities such as London, Los Angeles and Paris, provided the potential for individuals or groups to 'make it rich' and leave behind embattled or uninspiring places of youth.

Such testimonies to the positive and negative effects of place on musical inspiration are legion. For English musician Geoff Barrow, a small place provided a similar impetus to escape yet, ironically, also linked his band Portishead to geographical roots:

> I've lived here since my teens and I wrote music to get out of Portishead as quickly as possible. The main reason we named ourselves Portishead though, was when we became involved in the music industry, we were always known as 'the guys from Portishead'. So I liked the name, but I didn't like the place.
>
> (quoted in Margetts 1994: 31)

Musicians and even fans have migrated to places identified with particular sounds, such as Seattle or Chapel Hill: in a 'spatial trajectory of belonging' where fans may even be 'diasporic pilgrims' (Olson 1998: 282, 283). Large cities usually provided both the economic context (clubs, recording studios, managers) and, perhaps, the inspiration, but less from urban cultural and economic diversity, and more from links with audiences, other musicians and composers. Yet, conversely, small towns have often been associated with 'appropriate' genres of music – notably country (and sometimes folk) music – while denied more 'heavy' sounds; thus rock bands 'don't come from Ballarat', or any other small town (see also Cohen 1991b: 11).

Links are regularly made between artists and their environments throughout interviews, on album covers and in documentaries, yet they are also generated at a collective level with respect to whole communities of musicians, in a process of identifying 'scenes' and 'sounds' for music, whether Seattle's 'musty garages' (Bell

1998: 38) or the Velvet Underground's 'soundtrack to the urban hell of New York' (Gumprecht 1998: 63). Journalists and academics often search for local markers of style – establishing direct relationships between towns, cities, rural spaces, and the styles and influences that might coalesce around recognisable musical forms, instrumentation and lyrics. In this vein, Gordon located the mythological 'birth of rock 'n' roll' in place, emphasising the links between geographical and socio-cultural conditions, and particular styles of music:

> Memphis music is an approach to life, defined by geography, dignified by the bluesmen. This is the big city surrounded by farmland, where snug businessman gamble on the labor of fieldhands, widening the gap between them, testing the uneasy alliance. Memphis has always been a place where cultures came together to have a wreck: black and white, rural and urban, poor and rich. The music in Memphis is more than a soundtrack to these confrontations. It is the document of it.
>
> (1995: 9; see Hall 1998)

The ways in which these processes occur, if indeed they do, are complex, yet the credibility of some musical styles and genres arises from their origins, their sites of production, evident in a number of possible ways: smaller locations, places 'off the beaten track', isolation and remoteness from hearths of industrial production or working-class communities. This is particularly apparent in music with unusual arrangements or vocal sounds, evident in frequent references to a distinctive Icelandic sound (Box 5.1). At a rather different scale generations of musicians from Australia and New Zealand have had varying levels of success in overseas markets, and been identified by record companies, and particularly the English music press, as unique products of an Antipodean 'backwater' industry:

> Australia seemed a strange, Jurassic Park-like environment. Classic 1970s and 1960s influences were somehow isolated and crossbred, away from the decade's predominating synthpop or gothic rock eyeliner trends.
>
> (Lee 1997: 27)

Box 5.1 An Icelandic sound?

Iceland's isolation and the unusual style of its most famous performer, Björk, have suggested to many that there is an exceptionally close relationship between geography and music. Journalists have consistently linked Björk's unusual singing style with geographical origins:

> only a country where engineers defer to elves could have produced such a voice as hers. We used to think she was a freakish one-off, but now a new wave of Icelandic groups is demonstrating otherwise and – as with Björk – once you have heard them, it is hard to imagine them having come from anywhere else.
>
> (Smith 1997: 24)

While the Sugarcubes' 'Birthday' (1987) was 'like music from another planet, the song's reverberating bassline, celestial brass and ethereal production conspired to make this the aural equivalent of a particularly sensual massage', with Björk's 'perversely melodic combination of wide-eyed child and Icelandic banshee' (Strong 1999:75). Similar descriptions have also been consistently directed at Icelandic artists Gus Gus, Mõa and Sigur Rós, and occasionally traced to the linguistic purity of the Icelandic language.

> But every so often, for all sorts of fortuitous reasons, one particular location somewhere on the planet seems to harness its energies around an intense group of individuals to produce a waterspout that is visible from afar.... Amid the general mulch of homogenised and syndicated culture...magical sensations nourish this heartfelt music, this connection with the land that has made its people bond so strongly with it.
>
> (Young 2001: 29, 31)

Sigur Rós have attributed their perceived local distinctiveness to the barren, desolate landscape and 'the presence of mortality' there, and the links to stories, sagas, magic and ritual where 'the majority of the population believes in elves and power spots and stuff like that...the invisible world is always with us' (Young 2001: 33). Yet, Iceland, and especially the capital Reykjavik (with about half the nation's 280,000 people),

> harbours a music scene most European countries would kill for. A mutually supportive network of shop chains, each run independently – often by musicians themselves.... It also has its own music magazines and media, plus apparently countless record labels putting out unconventional sounds...descend to the next circle and you'll find a raft of adventurous music and sound artists working mostly in collaborative ventures
>
> (Young 2001: 32)

That network of co-operation, supported by production and marketing companies and the media, contributes to some elements of common purpose and experiences. However, within Iceland, there is some reaction against Icelandic bands. Commercial radio stations play international music, in turn resulting in opposition to 'American companies that buy up Iceland's radio stations' (Young 2001: 32). Other Icelandic performers play quite different kinds of music. Paul Oscar, who came to (some) international prominence as Iceland's representative in the Eurovision song contest of 1997, later produced his first entirely English language album, the dance-techno *Deep Inside* (2001). Sometimes known as 'the little brother of Björk', and claiming to be inspired by her 'creativity and sounds', he began in musical comedy and moved through disco to his present position (Calmont 2001). Even a small isolated island thus hosts a diversity of musical genres, for which there are equally diverse yet overlapping audiences. The vast majority of the music press has sought to focus on the typicality of its most famous popular musician, Björk, but one who eventually moved away and sang in English.

In some respects these assertions ring true – Robert Forster and Grant McLennan, of the 1980s band the Go-betweens, acknowledged that isolation played a role in the songs they would eventually write:

> we were away from the metropolis. There was no generally held ideal....There was nobody to tell you you were on or off the right track. We didn't meet hip musicians. We weren't culturally aware....London wasn't a 40-minute train ride away. We stayed in and fantasised about records.
>
> (quoted in Lee 1997: 27)

Australian dance music crew, the Avalanches, were similarly celebrated in the UK through ideas of remoteness: one journalist claimed that 'their relative cultural isolation means that they have avoided paying lipservice to American or British trends' (Aston 2001: 10). Yet, ironically, the Avalanches, more than perhaps any other dance group, produced music deeply entwined in global commodity flows, with a cosmopolitan array of influences present in their sample-laden music (Their debut, *Since I Left You* (2000), was reported to contain more samples from diverse places than any record ever released.) Here then was a vastly different sense of scale that nonetheless suggested that even an island continent might be detached from global trends.

In Iceland, remoteness was implied to be associated with levels of independence from market forces and expectations. Ironically, this remoteness and isolation, and an emphasis on pre-capitalist 'craft' production values, became an important marketable option for major labels distributing these artists internationally. As was the case in the 1990s with bands associated with the Flying Nun label from the New Zealand town of Dunedin, success in Europe and America revolved around geographical discourses – identifying music that was innovative because it originated from unique and remote locations. Through this distinctiveness certain places became marketable commodities, where distance lent a degree of enchantment (see Boxes 5.1 and 5.2). However, the desire to gain access to larger markets resulted in performers such as Björk and Paul Oscar recording in English, becoming linked to large companies and moving away.

Box 5.2 The 'Dunedin sound'

Dunedin, the southernmost university town in the world, with a population of about 110,000, became synonymous with a particular 'sound' during the 1990s: guitar-driven 'alternative' music, a 'tendency to sustain or repeat a note or notes, while changing the chords underneath' (Bannister 1999: 71). The origins of the Dunedin scene lay in the intermingling of artists in a closely-knit circuit during the late 1970s and early 1980s. This was an infrastructure of venues, student cafés and housing (for practice rooms), and audiences remote from the larger live music networks of New Zealand's North Island. The work of artists such as Chris Knox, the Bats, the Clean, the Chills and Straightjacket Fits was linked to a 'mythology of a group of musicians working in cold isolation' (McLeay 1994: 39), creating a sound that became known, particularly with American college audiences, as 'pure' – the product of 'craft-style' means of production, remote from the machinations of large record company promotion and production. As King Loser member Chris Heazlewood emphasised, Dunedin's size and intimacy were crucial:

> it's real small. It's *real* small. The music community is very insular: maybe there's about 30 or 50 people, or 100 if you take in the peripheral players in the game.
>
> (quoted in Scatena 1995: 19)

The 'Dunedin sound' was also closely associated with the Flying Nun record label, originally established in Christchurch in 1981 before moving to Dunedin and later Auckland. Other labels, most noticeably Xpressway, were also prominent. A sense of local uniqueness, of remoteness and distance from capital cities and centres of mainstream music production was central to the growth and mythology of a distinct 'sound'. A *New Musical Express* (NME) review of the album *The Venus Trail* by Dunedin group the 3Ds emphasised the importance of the peripheral location of the scene's physical and cultural infrastructure to its sound:

> Pop's least mannered exponents invariably lurk out on the globe's perimeter edges. The propensity for New Zealand's Flying Nun stable to unearth music so drenched with wide-eyed wonder it might as well emanate from another planet is well documented, and Dunedin's 3Ds maintain the noble tradition of the Chills, the Clean, Verlaines et al....it could just as well be an exotic new species of Antipodean fauna.
>
> (Keith Cameron quoted in Mitchell 1996: 223)

Similarly, landscape was seen as having a direct impact on musical expression:

> Their sweet, sticky pop songs are suffused with the tang of something wild and strange...capturing a sense of the South Island landscape – the slow turn of the seasons, and of what it's like to live in that landscape.
>
> (David Eggleton quoted in Mitchell 1996: 227)

Thus notions of a distinct musical 'sound' were related to the contexts in which artists and audiences were able to interact (with appropriate infrastructure, receptive crowds, supportive venues and independent labels prepared to release unknown artists), and to ideas of remoteness from outside influences.

See: McLeay 1994, 2001; Mitchell 1996; Bannister 1999.

Creating places

In most instances places were not merely identified as points of origin of a distinct style but were marketed as the basis of particular commercial enterprises (Table 5.1), even across time periods and sometimes genres. For example, the 'Motown' sound emerging in the 1960s out of Detroit, USA, was a highly popular variation on soul and R&B formats developed by Berry Gordy – a producer and entrepreneur rather than a musician. Known for its combination of 'sweet' vocals (usually delivered in three- or four-part harmonies), dramatic themes of love and loss and exuberant arrangements, the 'Motown' (Mo-tor

Table 5.1 Examples of 'regional sounds'

Sound	*Era*	*Country of Origin*	*Musical Genres*	*Bands/artists/labels*
The Nashville sound	1950–	USA	Smoothly produced country and western without fiddle and banjo of earlier styles	Hank Williams, Jim Reeves, Marty Robbins
The San Francisco sound	1960s	USA	Psychedelic rock, blues influenced	Grateful Dead, Jefferson Airplane, Janis Joplin
The Detroit 'Motown' sound	1960s/ 1970s	USA	Soul groups, harmony vocals, smooth production aesthetics	Motown (label); Diana Ross and the Supremes, Marvin Gaye, the Four Tops
The Liverpool or 'Mersey' sound	1960–	UK	Guitar pop, melodic style, 'Oceanic' music	Beatles, Gerry and the Pacemakers, Echo and the Bunnymen, Lightning Seeds, Orchestral Manoeuvres in the Dark
The California 'West Coast' sound	1970s	USA	Laid-back guitar rock, country influences	The Eagles, The Byrds, Crosby, Stills and Nash, Jackson Browne
The Philadelphia 'Philly' sound	1970s	USA	Soul/funk, with orchestral string arrangements	Philadelphia International Records (label); Kenny Gamble and Leon Huff, O'Jays, Harold Melvin and the Blue Notes, the Delfonics
The Manchester sound	1980s	UK	Dance and rock grooves, Ecstasy influenced	Factory (label), Stone Roses, Happy Mondays, 808 State
Detroit techno	1980s	USA	Electro beats, minimalistic arrangements, synth basslines	Underground Resistance, Derrick May, Juan Atkins, Carl Craig
The Dunedin sound	1980s/ 1990s	New Zealand	Quirky guitar-driven pop, alternative rock, remote 'uniqueness'	Straightjacket Fits, The Chills, Chris Knox, Flying Nun (label)
Hamburg hip hop	1990s	Germany	German lyrics, hip hop, mixtures and fusions of various styles including, reggae, dub	Fettes Brot, Visit Venus, Fischmob, Gautsch, 5 Sterne de Luxe, Thomas D, Yo Mama (label)
The Bristol sound	1990s	UK	Trip hop, drum and bass, 'jungle'	Massive Attack, Portishead, Tricky
Seattle grunge	1990s	USA	Grunge, hard rock and punk influences	Sub Pop (label); Mudhoney, Fastbacks, Melvins, Soundgarden, Pearl Jam, Nirvana
Goa trance	1990s	UK/India, Israel, Australia	Dance music, Indian samples, LSD-influenced 'trance'	Josh Wink, Ollie Olsen, Paul Oakenfold

town) sound became a distinct signifier not only of a particular period in music history, but of a particular place. Ivy Hunter, who co-wrote Martha Reeves and the Vandellas' hit song 'Dancing in the Street' (1964), articulated the mythology of the Motown 'sound' in a lyrical way:

> The sound...is like an old lover. You may be married and have children but if somebody mentions that name, Motown...oh boy. The feeling is not going to go away. Motown is a special place in time that will never come again, but will never be forgotten.
>
> (quoted in Bull 1993: 201)

As rock myth has it, the growth of the 'Motown sound' was explicitly related to the industrial focus of the city; Lamont Dozier, Eddie and Brian Holland, the architects of Motown production techniques, were said to have 'developed a rhythm based on the clattering mechanical beat of Detroit's assembly lines (where many of the Motown staff had worked), and created rudimentary sound effects with chains, hammers, and planks of wood, or by stomping on floorboards' (Perry and Glinert 1996: 191). But it was always a commercial venture: Motown artists were expected to conform to standards of deportment, performance (including rehearsed dance moves) and a common marketable 'sound'. During the 1980s and 1990s Detroit was again attributed its own sound, this time in a more mechanical form: Detroit techno. Championed by 'electro' beat pioneers, including Juan Atkins, Underground Resistance and Mad Mike, the new sound relied on synthesised drums and bass, stripping music of tangible lyrics in favour of a new African American futurism (see Chapter 6). For both genres, the 'mean streets' and industrial ambience of Detroit were seen to stimulate particular – but quite different – local sounds.

Notions of local 'sounds' are always fluid and not always successful: various interests have attempted to promote other places as the 'latest thing' in the music industry, most notably MGM Records' promotion of unheard-of Boston based groups like Ultimate Spinach and Phluph under the banner of a 'Bosstown' sound. This doomed marketing ploy involved MGM attempting to capitalise on the town's large university population and the tide of social changes sweeping the United States in the late 1960s, by signing local bands and investing over $20 million in professional studio production and promotion. (Ironically, Boston eventually emerged as a vibrant 'indie' city in the 1980s, based around its college population and groups such as the Pixies, Throwing Muses and the Lemonheads.) Similarly, the British music press has also tended to search for regional 'sounds' emanating from English contexts. Since the Liverpool or 'Mersey' sound in the 1960s, different centres have been claimed as the sites of unique musical expressions and sounds, including Newcastle, Manchester, Bristol, Oxford and

Coventry. However, as the case of the 'Bristol sound' illustrates, while some within the music press, record companies, retailers and even local authorities may be keen to promote the growth of particular local 'sounds', performers may distance themselves from the kinds of unity, homogeneity and determinism implicit in these representations of local culture (see Box 5.3), and its inherent commercialism.

Box 5.3 A 'Bristol sound'?

In the 1990s a body of electronic music emerged from the city of Bristol, in the West Country of England, labelled variously as trip hop, drum and bass or jungle, from artists including Massive Attack, Portishead, Roni Size and Tricky. Johnson, in *Straight Outa Bristol* (1996), provided an account of how Bristol played host to this scene and 'sound', making connections between the physical infrastructure of the city, its historical legacy as a major English port and site of West Indian migration, and the musical expressions of its artists. The 'Bristol sound' involved music that straddled a number of conventional genres, including reggae, hip hop and electronic dance music, but was characterised by minimalistic arrangements, dub-influenced low-frequency basslines, samples of jazz riffs, keyboard lines or movie soundtracks, and drum loops – 'breakbeats' characteristic of hip hop and rap from the ghettos of American cities during the 1970s and 1980s. Lyrics and vocal lines were often mixed at low volume in tracks, with raps often whispered and melodies resplendent with Jamaican-style ragga tunes. This 'laid-back' sound (as opposed to the rougher, aggressive hip hop of American 'gangsta' rappers) was said to reflect an inherent attitude of Bristolians, and a 'spliff' (marijuana) culture introduced by the city's large West Indian migrant population:

> Bristol's character, particularly its pace, does seem to have an influence on the music produced there. . . . Unlike many musicians in London we've never been rushed, we have time to make music. . . . Bristol is friendly, slow-paced and relaxed.
>
> (quoted in Cohen 1994: 121)

Concurrently, Bristol's West Indian presence was said to have 'opened' up relatively autonomous spaces such as the 'Dug Out' (a club in the inner-city suburb of Clifton), where Jamaican-style sound systems (disc jockeys with portable speakers) were allowed to develop (although the Dug Out was later closed down, giving the club a mythological position of its own within the 'Bristol sound'). Yet serious race riots have marred this harmonious image, while Bristol's working-class council estates gave more negative connotations to this emerging style. As Roni Size argued,

> You can't live somewhere without being affected. If you leave your door and see cows, horses and green fields you'll make different music. We do live on council estates, there are flats around us and dogs and decay – that's our reality. When I look out of my window I see bars. Why should I make music that doesn't reflect that experience?
>
> (quoted in Ferraro 1997: 58)

Thus not all activities of the participants within this local style reflect the mythologised unity suggested by the use of the term 'scene' or 'sound'. As Johnson suggested:

> the notion of a Bristol-sound scene doesn't really exist. Yes, you can see Massive [Attack]'s Robert Del Naja playing football on the downs of a Sunday or drinking in a bar; Daddy G and Mushroom often DJ in the local clubs; Portishead, Massive, and Smith and Mighty have studios here; but it's hardly a trip-hop version of the Bloomsbury group, with the participants living in and out of each other's pockets....It's a case of some local acrimony that the Bristol bands don't actually perform in Bristol.
>
> (1996: 27)

While place may influence musicians' personal expressions, celebrations of a coherent 'sound' remain problematic. As Portishead's Geoff Barrow has succinctly put it 'The Bristol scene exists mostly in people's minds' (quoted in Buchanan 1998: 5). The idea of a 'Bristol sound' was as much a social construction as a reflection of authentic local culture.

See: Johnson 1996.

Communities and sites of production

At the most basic level, before a 'sound' or 'scene' can develop, there should be both a 'critical mass' of active musicians or fans, and a set of physical infrastructures of recording, performance and listening: studios, venues (with sympathetic booking agents) – spaces that allow new musical practices – and even record companies (or alternative labels) and distribution outlets. The most famous scenes (those responsible for distinct 'sounds') have all built upon local popular support (or at least cult status), and featured particularly vibrant combinations of venues, local production and methods of information flow and exchange (such as radio and street press). Infrastructures of musical exchange solidify the presence of scenes, providing concrete spaces and emphasising cultural meaning for participants. Many such spaces have emerged from oppositional or subversive intent: appropriations of urban spaces for subcultural use (as with warehouse parties), new independent media forms (as with pirate radio) and sites of hedonism such as clubs and festivals – heterotopias with more freedom to challenge moral boundaries. Where commercial interests are involved, a sense of the innovative in music scenes is lost, yet despite commodification (but also often because of it) most music scenes are transient, changing and dissipating with the flow of musical and cultural shifts.

Agglomerations of musical infrastructures and audiences are also present in a multitude of places in ways that are not unique. Most capital cities and major urban areas have sufficient population to sustain regional music industries, including venues and professional recording facilities. Domestic output from major cities may be highly significant, as with Paris, Tokyo, Seoul and Sydney, yet these centres have not been attributed a 'sound' in international mediascapes. Local cultures in these locations create a range of sounds, their internal diversity (in some a product of migration) precludes any one given style and international media companies have not marketed their products elsewhere through affiliations with place.

A sense of community, most obvious among audiences, has influenced the parallel growth of scenes in diverse places. The whole notion of a 'scene' involves receptive and enthusiastic audiences and music-listening practices, without which a critical mass of activity cannot develop. Thus, university towns, such as Dunedin and the 'college network' of the United States and Canada, have provided fertile ground for the growth of 'alternative' music styles with disproportionate representations of a particular demographic group – young students more likely to enjoy live performances in their spare time and support idiosyncratic productions. A particularly vibrant music scene developed in Athens, Georgia, during the late 1970s and early 1980s, as bands such as the B-52s and R.E.M. developed their own particular 'sounds' through 'art-student party-circuits', university radio broadcasts and a range of bars, cafés and clubs specifically catering for the concentrated student population (Jipson 1994). While Athens became famous for the 'quirky' sounds of independent bands, other towns hosted similar sets of subcultures – Chapel Hill, North Carolina, became an important scene (with bands such as Ben Folds Five, Seam and the Squirrel Nut Zippers). Austin, home of the University of Texas, dubbed itself the 'live music capital of the world' and

Boston sprouted a plethora of indie artists (Kruse 1993). Similar scenes emerged in such widely scattered university towns as Rouen (where Tahiti 80 formed), Halifax, Nova Scotia (home to Sloan, who met at the School of Art and Design), and Lismore, Australia (where Grinspoon and the Simpletons formed at Southern Cross University).

Port cities such as Liverpool, Hamburg, New Orleans and Oran (Algeria) became sites from where particular 'sounds' were said to have emerged. Ports were important nodes of transport, meeting places for diverse migrants, residents and seamen (thus fostering venues, nightclubs and itinerant musicians), and points of distribution for imported music (rock 'n' roll records were said to have arrived first in Liverpool from the United States during the 1950s). The reputation of San Francisco as a port, meeting place, night-life hub and home for subversive activities, had a significant impact on the growth of psychedelia and the 'San Francisco sound' of the late 1960s. Centred on the Haight-Ashbury precinct near Golden Gate Park, and featuring performers such as the Grateful Dead, Country Joe and the Fish, Jefferson Airplane and Janis Joplin, its sound became entrenched in the counter-culture 'hippy' mythology of the late 1960s. Even more generally, points of confluence of cultural differences were said to stimulate distinctive sounds. Thus in Mali:

> The natural migrations of people throughout the centuries and the trade routes that wend through Timbuktu and the capital, Bamako, have brought a steady influx of foreign cultures, instruments and influences. The spread of Islam along the camel trade routes of the Sahara left one of its strongest traces in Malian singing styles, with vocals that often sound like some muezzin calling the proverbial faithful to prayer. So the adaptation of Western electric instruments onto Malian music is simply one more chapter in a story that's been unfolding for a hundred generations.
>
> (Anon. 1994)

The combination of population diversity, migration and an infrastructure of clubs generated a nocturnal economy of which music was a crucial part. Central to the importance of both university towns and port cities were diverse infrastructures of performance, mobile populations and a vibrant culture of music consumption. In other exceptional circumstances, the combination of an unusual location and a highly successful recording studio could stimulate a particular association between music and place; in Berlin a convoluted cultural and geopolitical past together with Hansa studios have created an unusual heritage (see Box 5.4).

Box 5.4 Berlin and Hansa studios

Many musicians have travelled to Berlin to record and tap into its musical traditions and influences. Berlin has at times become an important site in rock industry mythology, as a place with its own 'sound', and 'feel'. The roots of this reputation date from the 1920s and 1930s, when Berlin was a famous centre for jazz, hedonism, political conflict, cabaret and night life. During the 1970s and 1980s, in a dramatically altered post-war environment, this reputation remained an attraction. For Iggy Pop, it constituted part of a wider mythology of Germany and geographical conflict, and Berlin's status as divided city:

> I'd always been fascinated with the Germans – a lot of guys in rock 'n' roll liked the uniform, you know. I went there with Bowie and we basically dug the whole idea that this was Berlin...this was a war zone, a no-man's land.
>
> (quoted in BBC 1996)

This influence has been translated into musical expression, evident in recordings made by Iggy Pop and David Bowie at the famous Hansa Studios. Hansa, originally built in 1910 as the Meistersaal concert hall, was used by German publishing multinational Bertelsmann (BMG) during the 1950s and 1960s for domestic recording purposes. With the construction of the Berlin Wall in 1961, the studios, once centrally located near Potsdamer Platz, found themselves in a volatile and globally significant geographical position – literally metres from the Berlin Wall and a newly created 'no-man's land'. For the musicians that flocked to play there, that geopolitical significance would 'spill over' into recording sessions:

> *Heroes* [David Bowie, 1977] and *Lust for Life* [Iggy Pop, 1977] were both pure German products – the studio where they were made had a gun turret – a lookout post for the guys who would watch to try to keep people in the East – straight out the window of the control booth. Sometimes they would wave to us – it's a hell of an atmosphere.
>
> (Iggy Pop, quoted in BBC 1996)

For Tony Visconti, the producer responsible for David Bowie's *Heroes* (1977), 'the studio, Hansa by the wall, was a very wise choice. There was no mistaking the sound – we had a feeling on tape immediately' (quoted in

BBC 1996). Other artists have since responded to this sentiment. Lou Reed wrote an album called *Berlin* (1973) without actually going there; U2, Nick Cave, Depeche Mode and Brian Eno have all recorded at Hansa. U2 lead singer Bono attributed much of the 1991 album *Achtung Baby*'s raw, cutting-edge quality to the wider cultural influences of the city, 'There was this sense of Berlin [being] where it was at – you know it's not just being "dedicated followers of fashion", it's more than just a style' (quoted in Alan 1992: 214). The tropes of a 'Berlin' sound and style were reflected in the promotional video clips to singles from the album, in particular the song 'One', filled with images and metaphors of the divided city (including prominent civic monuments and 'trademarks' such as the East German 'trabbi' motor car), urban decay, black and white images of bleak and empty streets, cutting to depictions of bars filled with smoke, band members in transvestite garb – reminiscent of the pre-war years when Berlin was the cultural hub of Europe. Berlin (in particular Hansa Studios) not only provided the physical infrastructure necessary for recording significant albums of this magnitude, but also represented a location, with a political and cultural style, that somehow combined pre-war hedonism with post-war geopolitical tensions.

While some places have justifiable reasons to 'claim' musical heritage as their own, the geographical complexities of musical expression and dissemination are evaded in popularised histories of such places as in the case of jazz, the Memphis blues and rock 'n' roll (Havens 1987; cf. Hall 1998). Almost always in recent centuries, local scenes were the outcome of more fluid, multidirectional spatial flows and influences. Simplistic, place-bound depictions of the genesis of a sound and scene are rarely adequate. Few places, in recent times, have exclusively maintained a particular sound or genre for even a brief period. Moreover regions where one sound or style has apparently dominated, for even a few years, vary enormously in scale alone, from putative musical links with continents (Australia) and nations (Mali), to relatively small towns (Dunedin) or islands. It is implausible that Memphis and Mali have much in common in their ability to stimulate distinctive genres though, like many other places, they are places of mobility. By contrast, other locations, including Dunedin and Iceland, have been credited with the nurturing of distinctive sounds because of their apparent isolation from global trends and movements.

Diversity and repetition

Linking sounds to particular places is especially difficult where similar sets of geographical and cultural contexts predicate the growth of parallel scenes. While

Liverpool is primarily associated with the kind of music played by the Beatles (and their contemporaries), and there are economic and ethnographic reasons why that should be so (despite arguments that the Beatles sound was created in Hamburg), the city also hosts metal and country bands, as does any large city. Each of these genres may be differentiated more from each other than the music of a particular place differs from that of another of comparable size (see below). More challenging questions concern the nature of a genre. What constitutes the 'kind of music played by the Beatles' is open to debate, since they ranged from whimsical ballads to hard rock. Genres are neither static, nor easy to define. Performers themselves often both rejected classification (perhaps seeking cross-over markets) and welcomed a diversity of influences. The vocalist of one aspiring Australian band, Soulscraper, claimed:

> we have something that's been referred to as 'industrial', but even that is just another term. We have material that has a heavy reggae dub kind of feel right through to stuff that's almost grindcore metal. We drift around taking the things that we feel are best for the song. In Sydney everyone thinks we're a techno band, here in Melbourne we're known as alternative grunge. We're not terribly interested in labelling.
>
> (quoted in *Rolling Stone* 483, May 1993: 26)

Combinations of musical sub-genres constantly create new scenes, such as blends of metal and funk (as with Red Hot Chilli Peppers, Living Colour), or hardcore rock and hip hop (as with Korn, Limp Bizkit), creating communities of fans in otherwise widely different settings. Hence a matrix of ska-punk scenes appeared simultaneously in different contexts, from United States groups such as Fastball, Green Day and Blink-182 to Australia's the Living End, alongside a thriving Japanese scene. Links between Japanese ska-punk and similar scenes overseas have increased, including the export of releases by Kemuri to Western markets, the migration of group Pelican Jed to Australia and Pizza of Death Records' licensing of three-piece Hi Standard's products to Fat Wreck Chords in San Francisco. Such flows of product and movements of artists across dispersed consumption spaces illustrate the increasingly transnational character of sub-cultural tastes.

Ideas of community and transnational linkages may be as much symbolic as real, as with the identification of English 'northern soul' scenes (in places such as Manchester, Sheffield and Wigan) with the musical output and geographical imagery of Northern American industrial cities. Northern soul scenes rejected:

> the authority of 'the south' of England, [allying themselves with] the American North, in particular with cities such as Detroit...northern soul is yet another example of the ways in which British working class

> culture has produced an 'imaginary' identification with America as an 'escape' from native cultural traditions.
>
> (Hollows and Milestone 1998: 90–1; see also Nowell 1999)

By the 1990s this transnationalism had spread to other countries with English colonial links or where dance and 'deep funk' styles were popular. 'Northern soul' scenes emerged in Australia, Canada, Japan and throughout Europe, DJs such as Keb Darge replayed old 1960s 'rare grooves' and collated specialist compilation records. Vast numbers of musicians and audiences have similarly claimed affiliation with more ideologically, stylistically or ethnically, sympathetic contexts in distant places. Many rap artists (see Chapters 4 and 8) have identified with minority groups elsewhere rather than with those of different class origins in the same place.

In the 'northern soul' scene, and others (see Chapter 7), particular credibility is attached to music that has 'come a long way' to the final consumer. In dance sub-cultures (more specifically house and techno) 12 inch singles are the crucial format for club DJs and DAT tapes are essential in trance scenes. The more rare a particular recording is, the more cultural capital accrues to it; Jamaican sound system DJs and producers scratched the labels off records to ensure mystery and exclusivity, part of the 'turf wars' that characterised inner Kingston (Bradley 2000). The friction of distance, and the role of DJs and others as cultural gate-keepers, slows the processes by which some particular sounds and scenes become commodified. Yet, simultaneously, the increasing use of Internet resources (see Chapter 11) has enabled many parallel sub-cultures to become connected, increasing the means of information flow across dispersed scenes. One effect of this has been to de-link the notion of scene from locality (although scenes continue to rely on fixed infrastructures within localities for their survival). Hence websites such as www.i94bar.com bring together niche garage punk scenes in Australia, the United States, Norway, Italy and Sweden, which share a common interest in Australia's Radio Birdman and The Scientists, and also such 1960s icons as MC5 and the Stooges, while providing a wider exposure to new (and otherwise obscure) labels such as Sweden's Low Impact label, home to groups such as Sewergrooves and the Maggots (see also Kibby 2000). For particular niche genres barely large enough to sustain a local scene, information flows generate a sense of 'imagined community' central to the idea of a scene, but disconnected from a single place.

In the contemporary popular music industry, little uniformity exists across geographical space, hence there is no cultural homogeneity that might standardise musical sounds and erase local uniqueness. The emergence of differences in popular music scenes has occurred in ways that establish social and cultural spaces for various styles, sounds and sub-cultures, and, importantly, these persist in relatively stable terms, alongside others in various geographical places. In North American 'alternative' scenes, musical sites have emerged for various styles and sounds that, rather than being replaced by newer, 'cutting edge' scenes in a rapidly cyclical process, are likely to persist not only in their 'place of origin' (as it is perceived from beyond) but also in a multitude of other urban settings:

> The development of alternative-rock culture may be said to follow a logic in which a particular pluralism of musical languages repeats itself from one community to another. Each local space has evolved, to varying degrees, the range of musical vernaculars emergent within others, and the global culture of alternative rock music is one in which localism has been reproduced, in relatively uniform ways, on a continental and international level.
>
> (Straw 1991: 378)

Straw thus challenged assumptions of 'local uniqueness' in rock mythology, arguing that a range of musical niches is more likely to persist across global markets – a logic more powerful than that in which local differences might emerge into distinct 'sounds' – and evident in the parallel university town scenes.

While music histories tend to identify 'authentic' origins and examples of regional uniqueness, music scenes have almost always been replicated across vast distances. For example, while the growth of punk was often historically attributed to the UK (made famous by the Sex Pistols and the Clash in the mid- to late 1970s), it is now increasingly acknowledged that the roots of the style and attitude were more geographically dispersed, incorporating and emanating from artists such as the Ramones, Television and Patti Smith in New York, MC5 and the Stooges in Detroit (McNeil and McCain 1997) and bands such as the Saints and Radio Birdman in Australia (Walker 1996). A sense of historical continuity in terms of style or attitude, such as the influence of late 1960s bands such as the Velvet Underground, and before that many influential garage bands, is underplayed, being replaced by a fascination with the anarchic, epochal sense of innovation with which UK punk was said to have swept the music world.

Similarly, dance music and 'techno' musical forms in the 1990s have their own histories that tend to emphasise 'authentic' origins, as in Chicago ('house' music), Detroit (electro) and the 'second summer of love' played out in the UK during 1988. Yet, experimental electronic music has been part of musical scenes across many continents, where the relatively anonymous repetitive beats and instrumental grooves of UK or American tracks are heard and enjoyed not as part of a 'passive' act of listening to overseas artists, but as the sounds and signifiers of a sub-culture that exists in distinct ways in each locality. One Israeli DJ has stated:

> Israeli trance in general tends to be the morning sound, soft and melodic with a lot of ethereal twists. Aussie trance, starting from the early days of Psy Harmonics, was into the deep twisted psy sound and as the years [have] gone by Aussie trance has gone through a lot of changes turning into breakbeats and the more progressive upbeat sound. In Europe the minimal progressive tone is taking over with a lot of the artists that used to make the European psy tone.
>
> (quoted in *3D World*, 24 September 2001: 32)

While a UK garage DJ suggested:

> An important reason for garage's popularity in the UK is because it's UK street music. It has a distinct attitude and style that is very UK. That's not to say non-UK people don't get it. . . [but] I think the nature of garage at the moment is too rough for Australians.
>
> (quoted in *3D World*, 24 September 2001: 32)

Different combinations of dance sub-cultures accompany those characterised as 'alternative', 'heavy metal', 'punk' and 'hip hop' in every country. While some 'sounds' are viewed as coming from particular points of origin – historical hubs of creativity – the actual sounds of the music emanating from these places (the notes, timbres, themes of songs) become less distinct over time as other influences intrude.

Cultural space can be metaphorically carved out of wider social space through musical praxis, through the 'affective alliances' (Grossberg 1984) constructed to support sub-cultures, scenes, performance spaces and events. Despite operating across a number of heterogeneous geographic contexts, these nodes of cultural production and consumption are linked within a 'network of empowerment' where 'pleasure is possible and important for its audiences; it provides the strategies through which the audience is empowered by and empowers the musical apparatus' (Grossberg 1984: 228). However, popular music remains eternally transient, emerging and dissipating as fashions change and generations pass – in the end neither providing a fundamental challenge to, nor even breaking away from, dominant mainstream social space (Grossberg 1984, 1992; Lawe Davies 1993), hence anxieties about longevity in particular scenes (Cohen 1999):

> [Rock 'n' roll] celebrates the life of the refugee, the immigrant with no roots except those they can construct for themselves at the moment, constructions which will inevitably collapse around them. Rock and roll celebrates play – even despairing play – as the only possibility for survival.
>
> (Grossberg 1984: 236)

Even in more 'central' locations, transience is inevitable, as techno producer Stacey Pullen highlights: 'People move on and there isn't as much to unify the Detroit sound anymore....People were hungry and motivated to make music because in Detroit there isn't much else to do. Now it's lost that edge a little' (quoted in Trimboli 1999: 21). Such fragility has been exacerbated by the incursion of new forms of leisure into household entertainment (including videos, computer games and Internet surfing and gambling), and competition from other forms of live music, such as cover and tribute bands, undercutting the viability of 'original' music. While superstars performed in global arenas, rarely performed at all (in the case of Kylie Minogue for seven years) or retired, manufactured cover bands played their music, mainly in pubs, to fans deprived of the 'real thing'. The phenomenon was particularly powerful in Australia, where the tyranny of distance reduced access to international stars. In Sydney alone, the 1990s saw such 'international' tribute bands as Bjorn Again (ABBA), Elton Jack, Beatels, Abbalanche, Meatball, Creedwater Revival, Nice Girls, Load Fire Reload (Metallica), Jagged Little Pill (Alanis Morrissette), Rattle and Hum (U2) and the Australian Doors Show. The logic of a multiplicity of scenes (central to the increasingly transnational distribution of dance, punk and indie pop sub-cultures) was accompanied by new social formations, in the ultimate example of mimicry and repetition. More generally, various scenes and places were associated with the 'good old days' in the sense that McLeay (1994: 44) has suggested that the widespread notion of a Dunedin sound was primarily nostalgia for a time (if it ever existed) of 'pure' and local music uninfluenced by the global reach of large corporations. In much larger cities, such as Sydney, appeals to a local (even national) tradition have 'been employed in order to hide the fact that local industries are dependent upon transnational capital' (Homan 2000: 33) and particularly so at a time when the 'pub circuit' – the mythologised sites of formative rock venues – was fast declining.

Mythologising 'the local'

Scenes are found in a multitude of locations, with kaleidoscopic mutations of global sounds and ways of consuming sub-cultural products. Within this diversity, the collective output of some locations has 'broken through' into international music distribution and secured a reputation for a particular 'sound'. In part this occurs because a few local music cultures are genuinely distinct and innovative, but specific 'sounds' are also bound up in wider processes through which places are mythologised: a fetishisation of localities (Appadurai 1990: 16; Mitchell 1996: 87). In the conventional narrative of music dispersal, styles are generally deemed to have originated from an individual or collective 'scene', generating new interest in the sites of production. The existence of music infrastructures explains part of this process – as the 'grassroots' of local styles from which music then disseminates through national distribution networks and (perhaps) through to global audiences. Cultural origins for a scene or style can often be traced to particular groups of musicians, producers and audiences – specific contexts from which a 'sound' develops and disseminates. The Motown sound relied on entrepreneurs like Berry Gordy and a specific set of songwriters and performers, as did Seattle grunge (with SubPop Records) and San Francisco psychedelia. Such 'authenticity' in music begins with individual musicians and performers, who are seen as credible if they can trace their roots back to organic, local scenes (see Chapter 2). Notions of 'paying your dues' or having emerged from a vibrant scene shroud the ways in which music products in contemporary mediascapes are always constructed (even the releases of the most 'authentic' artists are accompanied by marketing campaigns, press releases, cover artwork and production values that create and sustain credibility). The identification of musical difference through regional sounds is an integral component of the fetishisation of place – securing the 'authenticity' of local cultural products in particular physical spaces as they move through national and global economies. Connections to place emphasise roots and points of origin. Remaining 'true' to one's roots emphasises credibility. Thus Ian McCulloch, lead singer of Echo and the Bunnymen, after twenty years of performance 'is not just loyal to his roots. He has lived in Liverpool all his life and has no intention of leaving' (Shedden 2001: 17). Wider geographical variations (from city to city, region to region) serve to authenticate musical differences.

The authenticity of the 'local' is partly created from within: the activities of a few bands might lead to the growth of new venues and attract interest from record companies and the music press, and the city, town or region is entrenched in music folklore as a place of origin. This popular mythology has sometimes been appropriated by the state or local tourist promotions boards, as at Memphis or Nashville, as part of regional economic strategies. Local authorities have attempted to generate grassroots musical scenes, building on place identities, by converting manufacturing production facilities into musical venues, studios, practice and performance spaces. In Durham, England, the closure of the Consett steel works in the 1980s triggered new local development strategies, including the 'Make Music Work' programme that necessitated rethinking models of economic growth, and required difficult attempts to create and realise the value of the 'locality's cultural history…through which political issues relevant to its future could be raised' (Hudson 1995: 465). In Nashville, Tennessee, 'local' identities similarly played an economic function, as a distinct 'sound' formed the basis of the city's reputation as 'country music capital'. Its 'sound', emergent through the production of a blend of country and accessible rock ballads, was described as 'somewhat mellow…it is not nasal or twangy, and does not include a steel guitar, fiddle or banjo. What country music did retain was its emphasis on simplicity, heartfeltness and sincerity' (Blair and Hyatt 1992: 71). Yet it became typical, not so much because a swarm of musicians gravitated towards a particular style, or because features of Nashville's physical or cultural landscapes suggested certain musical accompaniments, but because a smaller group of session musicians were recorded on a string of famous songs during the city's boom years in the late 1940s and early 1950s:

> Actually the Nashville 'sound' was the jingling of money in the pocket – as Chet Atkins has wryly pointed out – because these musicians were economical. . . .The 'Nashville sound' became a buzzword to describe country music in general and though the term could never really be explained by those inside the industry, it served as a handy, convenient term for journalists and others in the media to hang on to when discussing country music. The result was a built-in promotion for Nashville every time country music was discussed.
>
> (Cusic 1994: 48, 53)

Although Nashville always maintained a major country music 'scene', it only really began to capitalise on its reputation as 'the home of country music' (and more recently 'Music City USA') after business interests and the major labels centralised their activities in the city during the 1950s (Cusic 1994). Popular music is:

> always producing (its own) mobility, always trying to differentiate itself from the logic of everyday life, to claim or reclaim authenticity, where authenticity marks difference from the commodified existence that perpetually encroaches. But because each new site of authenticity is always sliding towards inauthenticity, popular music must continually produce new sites of authenticity for particular bands to occupy and for fans to invest in, redefining what constitutes the authentic in the process.
>
> (Olson 1998: 280)

Hence there are always towns like Branson, seeking to challenge Nashville as being more credible (see Chapter 10). Assertions of local 'sounds' are never wholly created within places. While local authorities, media, corporations and musicians often entrench territorial claims or emphasise difference through identification with regional 'sounds', they also form part of a mythology of 'the local' created from beyond: a commercially constructed strategic essentialism of place.

Local musics and niche marketing

Mythologising 'the local' became particularly apparent as the internationalised recording industry sought niche marketing opportunities and different sounds. Since the reproduction of a performance in commodity form at the beginning of the twentieth century, music has been promoted in association with a particular name, style or place of origin. More recently, a range of ways of marketing music has accommodated the general shift towards targeting products to particular demographic groups and social scenes. The segmented market represents a number of processes: consumer demand for more diverse products, the rise in commercial status of sub-cultural affiliations and styles, and the decline in homogeneous mass marketing of products. Alongside these elements of corporate strategy, segmentation can imply greater connections (or the perception of greater connections) between products and particular cultural positions and locations. As Mitchell has argued, 'the production of "fierce concentrations of meaning" would more and more be inserted into the logic of capitalist development, becoming important "sites" for the generation of not just values, but *surplus value*' (2000:73; emphasis in original). Tying music to cultural histories and a sense of 'roots' in place becomes an important part of strategies to locate sounds in a cultural plane to buttress authenticity and maximise sales.

Many local recording houses such as Flying Nun (Dunedin, New Zealand), Island (Jamaica), Mushroom (Australia) or Factory Records (Manchester) were linked in to the global distribution networks of major infotainment corporations that avoided erasing the signifiers of localism and uniqueness associated with their labels, logos and personnel (see also Negus 1999; Mathieson 2000). Smaller labels were attracted to agreements with the majors because production costs dramatically decreased (Chapter 3) but distribution systems remained corporate controlled. Major corporations benefited from the credibility of these sources of musical creativity, dispersing the 'sounds of the street' through their global distribution and retail networks (Negus 1999). Hence:

> Those who promote such music emphasize the 'authenticity' of its condition of production, while seeking to make the 'product' commercially available for consumption by audiences who may be located in very different conditions, but who at the same time are drawn by the music's (actively promoted) claims to 'authenticity'.
>
> (Jackson 1999: 104)

Thus, for rap music:

> It became apparent that the independent labels had a much greater understanding of the cultural logic of hip hop and rap music, a logic that

> permeated decisions ranging from signing acts to promotional methods. Instead of competing with smaller, more street-savvy labels for new rap acts, the major labels developed a new strategy: buy the independent labels, allow them to function relatively autonomously, and provide them with production resources and access to major retail distribution.
>
> (Rose 1994: 7)

While scenes and 'sounds' necessarily emerge in certain places, maintaining the 'authenticity' attached to the 'local' is crucial within the global cultural economy. Thus 'grunge' emerged as the defining mode of a 'Seattle sound' – a version of post-punk guitar rock championed by Pearl Jam and Nirvana, and associated with particular modes of fashion (flannel shirts, long hair). These modes were quickly replicated in different rock scenes, advertising campaigns and even Paris catwalks. Yet as these signifiers of the 'local' became media hype, their 'authenticity' suffered:

> No-one predicted Seattle's ascent onto the world stage, and when it did happen some saw it as a triumph, others as a betrayal of its indie origins. Despite its anti-commercial beginnings, grunge has become a marketing success story. It's certainly no longer alternative – it's ubiquitous. You can pluck out countless imitation bands who have distilled the alternative brew into mainstream concoctions.
>
> (Triantafillou 1996: 20)

Increasingly, bands within such particular scenes distanced themselves from their mythologised 'sounds', as with Dave Dederer of the Presidents of the United States of America:

> In a lot of ways, that scene was a mythical creation anyway. Grunge is an anomaly – what we're doing is more in tune with what the Seattle music scene has been about for the past 30 years...most of the bands here have traditionally written catchy, over-the-top rock 'n' roll. As practised by the original creators, like Mudhoney, that's what grunge originally was – fast and loud, devil-may-care rock. Music writers then turned it into something else.
>
> (quoted in Usinger 1997: 63)

Rex Ritter, guitarist with Seattle band Jessamine, summed up this sentiment: 'There are ways to go about being in a band and making records that shouldn't have anything to do with being in Seattle' (quoted in Rose 1997: 14). In suburban Liverpool various performers deliberately sought to distance their style from any perception of a generic sound, much preferring a distinct style (Cohen 1991b:

182). Reality is much more complex that the simple environmental determinism that implied local sounds.

However, the local infrastructure of music production – concert halls, studios and clubs – and the response to that from bands and fans contributes to a local scene. People make music within communities without creating (or wanting to create) a distinct 'local' sound, but learn from and work with each other, sharing bills and practice venues, and even musicians and songs, not so much as a subculture, but simply as people who enjoy the same music. In North Wales, the growth of Welsh language bands was aided by complementarity: 'we all realised this was a new scene, and we were all mates, at each other's gigs, all gave each other support slots...and had a [common] fan base' (Rhys Alwyn, quoted by Owens 2000: 20; also see McLeay 1994). Certain similarities necessarily emerged. This was compounded by language and lyrics, where performers refer to, and relate to, local places, from the Beatles' Penny Lane to the Whitlams' identification with Newtown, Sydney (Carroll and Connell 2000), and where lyrics, as in rap, have a range of functions. Rap music exhibits musical variations, both in rhythm and lyrics, which usually emphasise local themes. In the United States this became a slower, more sung form in 'laid back' California, as opposed to the harsher sounds of South Central Los Angeles, the fast and varied rhythms of the Cleveland sound, or the softer beats and choruses of Atlanta rap, or, at the urban scale of New York, with performers 'representing' particular neighbourhoods and boroughs, without any conception of the city as a whole (Krims 2000:76–8, 124–51; Forman 2000). Ultimately rap produced a series of related yet distinctively localised sounds. Live performance, and lyrical connections to place, stimulate ties to fans, who identify with common issues, sentiments and words that may invest places with meanings. Audiences, 'far from being passive consumers of pre-packaged information...are active agents in the construction of meaning in a live performance' (Goh 1996: 63), through verbal interaction and audience validation. Through infrastructure and performance, 'local musics' – usually distanced from substantial commercialism – are a result of co-operation as much as competitiveness. Such local links engender credibility and pose problems for authenticity in the event of success, the migration of performers elsewhere or the choice of a more global language for lyrics. In the conflict of commerce and creativity, the most local music is likely to remain almost unknown.

Out of place

Yoko Ono complained in 1972 that popular music:

> is becoming intellectualised and is starting to lose its original meaning and function....It should not alienate the audience with its professionalism

> but communicate to the audience the fact that they, the audience, can be just as creative as those on stage, and encourage them to make their own music with the performers rather than just sit back and applaud.
>
> (quoted by Fowler 1972: 18)

Much contemporary music produced by major media corporations would still attract the same criticism, yet the vibrancy and multiplicity of music scenes across the world temper this pessimism substantially. In the initial stages, the production process (whether a work ends up as a recorded CD or a performance) still involves small-scale creativity – bands and songwriters making music in garages, recording studios, local pubs and cafés. However, the rich detail of local music scenes and sounds cannot be divorced from the wider economic contexts within which music production, marketing and dissemination take place.

Throughout popular music folklore, 'sounds' have emerged that have been attributed to particular places – either as a response to landscape, as expressions of local identity and difference or, as we have argued, as a representation of 'localness' in increasingly global music distribution networks. Artists have often talked in spatial terms – discussing places as sources of inspiration, as well-springs of styles and sounds, as home to encouraging 'scenes' and receptive audiences. In particular moments, these have coalesced into distinct 'sounds' representing a sense of place, a sense of historically and spatially unique cultural growth, but never being the entirety of that place. Particular locations have engendered music scenes, defined as 'that cultural space in which a range of musical practices coexist, interacting with each other within a variety of processes of differentiation, and according to widely varying trajectories of cross-fertilization' (Straw 1991: 373). Very different musics exist in very similar scenes (Grossberg 1994: 246), especially in the largest cities where diversity is obvious, with musicians even overlapping across genres (Cohen 1991a, b). However, even on quite small islands (Roberson 2001: 217) nascent scenes are seized on by music media and regional authorities alike, as potent marketing devices, and by audiences as providing credibility and nostalgia: 'to fix the meaning of particular spaces, to enclose them, endow them with fixed identities and to claim them for one's own' (Massey 1994: 4; Olson 1998). Credible places invest music with commodity value.

6

MUSIC COMMUNITIES

National identity, ethnicity and place

Popular music is an integral component of processes through which cultural identities are formed, both at personal and collective levels; moreover, 'the way people think about identity and music is tied to the way they think about places' (Wade 2000: 2). This chapter examines the role of music in shaping national and ethnic identities. Artists or even whole communities can represent themselves and their experiences of places through music, in much the same way as in literature or art. It was once assumed that communities and individuals were recognisable through relatively stable identities – traditions, cultural traits, opinions and practices that could be collected, mapped and objectified; in some instances, such perspectives were created from within communities as strategic moves towards particular outcomes. More commonly, these 'fixed' identities were imposed as clichéd images of place and culture in tourist promotions, political rhetoric and sensationalist media, essentialisms that had little resemblance to everyday, lived experience (Said 1978, 1995). Consequently attempts have been made to depict identities as process rather than state, as flow rather than fixed characteristic, as constantly 'becoming' rather than 'being' (see also Chapter 8). The ways individuals, communities, regions or nations see themselves involve engagements with a potential myriad options: ethnicity, religion, gender, occupational status, political beliefs and so on. Music remains an important cultural sphere in which identities are affirmed, challenged, taken apart and reconstructed. So a genre of music such as hip hop, which emerged from black and Hispanic neighbourhoods of American cities during the 1980s, simultaneously involved questions of ethnicity, political motivations, gender identity and discrimination, class status and awareness, cultural nationalism and place.

National identity and the musical construction of community

Nation-states have always been socially constructed, that is contingent to particular epochs, and the result of deliberate practices on the part of state leaders and strategic analysts to 'create' a sense of community within the nation's (often arbitrary) borders. The nation-state was an 'imagined community', which operated over geographical space, with a sense of unity created through national institutions (legal structures, police, bureaucracy) and through a variety of cultural means, which attempted to entrench the sense of unity within its borders and maintain the logic of its separate existence from neighbouring territorial units (Anderson 1983). Central to this were such features as national languages, though these were rarely uniform throughout the territory, but many other cultural forms have been employed to 'create' nations with a sense of community and unity.

Music, alongside national artistic traditions, common religions, ethnic identity and a range of visual symbols (flags, emblems, crests, currency, figureheads), is embedded in the creation of (and constant maintenance of) nationhood. Music has been used in a variety of political contexts related to the construction and maintenance of national identity, notably through some classical music, national anthems, state music policies and the more recent construction of national rock 'n' roll traditions and 'music ambassadors' for many countries. This was especially the case with a range of classical music traditions from the eighteenth century with the burgeoning Industrial Revolution, in which the modern nation-state was shaped. Sibelius wrote the evocative *Finlandia* as an 'expression' of nationhood, Bartók incorporated the folk music of Hungary into his style, Vaughan Williams and Gustav Holst's music was said to capture the pastoral essence of England (Revill 2000), while, by contrast, Wagner's music was used by Hitler in the 1930s to rally the German public behind fascist nationalism (and is consequently not performed in Israel). Composers such as Aaron Copland in the United States and Peter Sculthorpe in Australia continued to reflect muted nationalist sentiment in the twentieth century.

State policies and national music industries

In many countries state music policies were developed to encourage greater local (national) musical activity, usually by imposing quotas on radio stations for local content, requiring them to dedicate a percentage of time to recordings by local artists. Such policies were threefold: first, to censor sounds considered contrary to state interests, as in the Soviet bloc (see Box 6.1) or in Burma, where U2's album *All that You Can't Leave Behind* (2000) was banned because the song 'Walk On' was dedicated to the country's pro-democracy leader Aung San Suu Kyi. (Anyone importing copies faced penalties of up to twenty years imprisonment.) In Africa, Tanzania sought to ban 'soul music' at the start of the 1970s, while in Asia, North Vietnam feared the development of 'imperialistic records' that 'clandestinely popularise musical pieces fraught with profane, romantic feelings, stimulating the bestiality of man' (quoted by Ford 1978: 227). Such moral panics, widespread in both developing and socialist states, over threats to traditional or state-inspired values, continued for decades. Second, state policies sought to protect local music industries and artists, and provide some avenue for those artists to be recorded, broadcast and heard throughout their home country. Third, they sought the maintenance of 'local cultures' in the face of increasingly Anglo-American popular music distributed by major record corporations. This was particularly so where local artists wrote and sang in languages other than English. In Canada, during the recession of the 1980s, the Quebec recording industry lost the interest of major record labels, who dramatically reduced the number of local, French-language releases. The Canadian Radio, Television and Telecommunications Commission (CRTC) eventually imposed a 65 per cent French-language programming requirement for Quebec radio, a highly controversial decision fiercely opposed by broadcasters, which gained support from artists and local record labels. These policies were instrumental in bringing about structural change in the industry, resulting in the rebirth of Quebec *chanson* and ironically a more pluralistic, rather than nationalistic, popular music scene within Canada (Grenier 1993). Paradoxically, though local performers gained greater access to the airwaves, in some genres such as country and western they typically sang of distant places and minimised local lyrical content (Lehr 1983). By contrast, in New Zealand, attempts to pass a bill that planned to phase in a 20 per cent local content requirement for broadcasting were rejected, as a seemingly outdated state-interventionist mode of government, in an epoch dominated by the free market. Without such policies, mainstream New Zealand radio continued to play more commercially 'safe' imports and the 'dead certs of the international phonogram companies' (Pickering and Shuker 1994: 78), and the opportunity to develop a national cultural industry was lost. However, to suggest that state music policies can, or indeed should try to, create wholly insular national identities is unrealistic, denying the diversity of experiences of national populations.

While such cultural and economic policies remain highly problematic in terms of the promotion of local artists and labels (as opposed to entrenching the powerful position of local affiliates of overseas companies) and in terms of the essentialised notions of national identity they may presuppose, state intervention may provide mechanisms that enhance the formation of alliances, networks and distribution systems within local 'scenes' and, more obviously, provide elements of musical infrastructure such as national radio stations, cultural festivals and so on. The situation in Afghanistan, removed from Western contexts, demonstrates how national radio stations and nationalist music policies can provide for new and quite unintended alliances, influences and hybrid musical expressions with diverse outcomes (Box 6.2).

Box 6.1 Back in the USSR

Popular music penetrated the Iron Curtain, into Eastern Europe and the Soviet Union, more slowly than much of the northern hemisphere, as avowedly socialist states broadcast none of it, concerts were disrupted and musicians imprisoned, though little could be done to block Radio Luxembourg or, later, Radio Free Europe. Moral panic took on particular ideological overtones. However, precisely because of the 'subversive', anti-authoritarian image of rock music, it found a large audience – as jazz had previously done – though, ironically, it first took hold among a social elite, especially in Poland and Hungary, and was escapist rather than oppositional. Beatlemania provided a 'backdrop for everyday living under socialism' for millions of people (Ryback 1990: 51), yet, despite a Rolling Stones concert in Warsaw in 1967, rock 'n' roll was either 'battered into submission or seduced into complicity by the socialist state[s]', producing 'banal echoes of the western originals' (Ryback 1990: 101, 128), and music with ideologically correct lyrics, mainly in English. In the early phases of Soviet rock, musicians ignored 'ethnic' and 'traditional' instruments, and those who used accordions or balalaikas were scorned for 'inauthenticity' until the Beatles and other Western groups introduced Oriental instruments (Ramet *et al.* 1994: 197). By contrast, in Belarus, Ukraine and other republics within the USSR, using local languages, instruments and repertoires was more common as an expression of opposition to Soviet centralism and authoritarianism. There were enormous variations in the visibility and status of rock music within Eastern Europe; the Soviet Baltic states, especially Estonia, 'the California coast of the Soviet Union' (Ryback 1990: 111), were most liberal, while the more authoritarian Stalinist regimes of Romania, Bulgaria and Albania had little or no early rock music. The isolation of Romania and Bulgaria was broken down by the rise of tourism, with a rapid increase in the extent of visits and influence from Western Europe, but it was not until the very end of the 1980s that rock music filtered into Albania (Ramet 1994: 9). The slow diffusion was emphasised by state control of record industries, which released little rock music; the rise of cassettes in the 1980s, as elsewhere (Box 3.3) was a major boost. The Soviet invasion of Czechoslovakia in 1969 sparked renewed hostility and clampdowns on rock music, one of many institutional constraints to its diffusion. Music from the eastern bloc was rarely heard outside it, nor did it cross national boundaries within the bloc. Attempts to market successful Eastern bands, like Poland's Lady Pank, failed and the music was criticised as derivative and dated (Ryback 1990: 188–9, 233; Mitchell 1996). By contrast folk music, particularly *The Mystery of Bulgarian Voices*, was relatively successful following the vogue for world music. Rock stars – in their opposition to communism – were both the prophets and the muses of revolution; decaying communist regimes were said to fear electric guitars more than bombs or rifles (Ramet 1994: 2) and rock music provided youth with a vehicle for political and social discontent, perhaps 'the triumph of rock and roll…has been the realisation of the democratic process' (Ryback 1990: 233). Ironically the demise of communism in the late 1980s also contributed to the decline of many bands, who had drawn their vitality from challenging the system.

See: Troitsky 1987; Ryback 1990; Ramet 1994.

Box 6.2 Afghan music and national identity

Afghan popular music traditions are bound up in processes of constructing a sense of nationhood in situations where the population otherwise consists of divergent ethnic groups with competing claims and identities. Afghanistan was created as a nation through treaties signed between British and Russian superpowers that had fought over territory in the late nineteenth and early twentieth centuries; arbitrary borders were drawn up that divided several local ethnic groups. Thus, the nation of Afghanistan is made up of various local communities (notably the Pashtuns and Tajiks – ethnic groups with different languages) who have historically divergent cultures and whose music traditions reflected different backgrounds. These traditions were merged during the early twentieth century to form a national music repertoire – known as *musiqi-ye melli* ('national music'), which is made up of both 'art' music (*ghazal*) and 'popular' music (*kiliwãli*). Afghan *ghazal* is characterised by vocal performance of Persian poetry texts in couplets, while *kiliwãli* music combines Afghan, Indian and European instruments. Central to the construction of this national music repertoire had been Afghan radio services, notably Radio Kabul, whose explicit charter during this period was 'to spread the message of the Holy Koran, to reflect the national spirit, to perpetuate the treasures of Afghan folklore, and to contribute to public education' (Baily 1994: 57). The formation of a national identity was thus explicitly part of music broadcasting, resulting in new syncretic forms of music and a sense of shared heritage and tradition. However, this nationalistic use of music traditions served to marginalise local traditions, which remained site-specific and beyond the goals of broadcasters (Slobin 1993). For many people, the only contact with their own dialects and traditions was through radio broadcasts from neighbouring Central Asian countries (such as Tajikistan). These traditions only re-emerged as governments were faced with the possibilities of a fragmenting nation-state and began to broadcast local musics, again in the hope of gaining consensus under the rubric of the Afghan nation. Local popular styles then became more widespread: 'a fragmented mediascape allows for a lot of mischief' (Slobin 1993:22). This was exacerbated through the dispersal of inexpensive cassette recorders and the growth of an informal distribution network for local products, reinvigorating music production beyond the influence of the state. The revival of local music scenes came to a dramatic end in the late 1990s with the Taliban government's ban on all forms of music. State intervention took one much less subtle form.

See: Baily 1994; Slobin 1976, 1993.

Where popular music has been controlled in some way by the state, lyrics have been designed to support government positions, from Singapore's social engineering (Box 6.3) to Congo (Zaire) under Mobutu, where musicians 'felt compelled to denounce all deviant or delinquent proclivities of city dwellers' including single women who patronised bars, youthful materialism and the problem of AIDS (Gondola 1997: 78). In the 1960s the Ghanaian government sponsored 'high-life' bands, so called because they represented the kinds of high living to which Ghanaians aspired, and different government departments competed to produce the best bands (Coplan 1978, cited in Kaemmer 1993). In China nationalist and socialist songs, created and broadcast by an effectively state-run popular music industry, inundated the media (Jones 1992). State control of radio and television in many countries (irrespective of bans of performers and songs on moral and ideological grounds) contributed to popular music providing fewer challenges to society than it might have done. As popular music moved towards respectability and commercial success performers were co-opted by society, given decorations and awards (the most ironic probably being President Nixon's award of a Bureau of Narcotics and Dangerous Drugs badge to Elvis Presley in 1970) and moved closer to the 'middle of the road'.

Box 6.3 The creation of Singapore

The city state of Singapore, only created in 1965 through what amounted to 'ethnic secession' of the more Chinese city from the Malay hinterland (which became Malaysia), soon became one of the first 'Newly Industrialising

Countries' in Asia. Rapid social and economic change took place within an increasingly authoritarian political system. The government sought to stimulate national identity partly through popular song, beginning a 'Sing Singapore' programme in the 1980s to encourage both 'traditional' and 'national' songs, the latter group including anthemic songs such as 'We Are Singapore', 'Stand up for Singapore' and 'We Love Singapore' (Phua and Kong 1996), effectively ideological tools to legitimise government policies. Catchy jingles on television further emphasised government campaigns such as avoiding chewing gum, drugs and spitting, saving water and flushing toilets. While the ruling elite used commissioned popular music 'to achieve hegemonic ends and encourage socially acceptable forms of behaviour', this sparked resistance to the hegemonic rule of the elite class (Phua and Kong 1996: 215–16). Although most mainstream Singaporean bands had lyrics that focused on love and romance, alternative bands, whose names alone indicated difference – Opposition Party, Band of Slaves, Rotten Germs, Swirling Madness – berated government control and injustices within society and resisted cultural hegemony. Much of Singapore's alternative music portrayed the pursuit of material progress as having created unhappy, maladjusted people. One of Dick Lee's better-known songs, 'Internationaland' (1984), criticised the prevalent middle-class materialism. Lee, a largely mainstream performer, was once banned from the radio for being too critical of the coldness and high-handedness of the government (Wee 1996). Alternative music portrayed the image of a controlled and introspective place, dynamic yet problematic, a city that could be ugly in social and physical terms. Global Chaos' song 'Money Isn't Everything' attacked 'Uncontrolled greed/ Impetuous desire to be rich/ Immune to worldly issues/ Addicted to worldly pleasures' (quoted in Lockard 1998: 255). Both the officially sanctioned music of the state, and challenges to state control, created the image of a complex city, where social engineering stifled diversity and physical expansion removed both history and open space.

See: Phua and Kong 1996; Kong 1996a; Mitchell 2001.

A national musical aesthetic?

The music of nations is also mediated through much less formal means than state music policy or national anthems (see below) in the appropriation of local 'sounds' (similar to those discussed in Chapter 5) as national traditions. The 'nation' is nonetheless a field of meaning, a contested term, a concept that provides the basis for debates in society about cultural difference, uniqueness and attachments to territory (Regev 1997). In much the same way as musical differences were taken to be markers of uniqueness between places at the local level (and thus the basis of the social construction of unique 'sounds'), differences in the musical traditions of various countries helped to justify their distinctiveness as 'nations'. National 'trends' are sought after as evidence of a continuing sense of 'community', and national 'sounds' are often constructed by accentuating, celebrating and marketing local differences. Thus 'Irish' popular music is considered stylistically different from that of England and Scotland; U2, the Corrs, the Cranberries and Van Morrison have seemingly epitomised an 'Irish' sound in music (McLaughlin and McLoone 2000; Rolston 2001: 52–4); Abba and Roxette have been Swedish pop ambassadors; traditional enka is said to distil Japanese tradition (Yano 1997); and Caribbean carnival has been exported as a representation of Trinidad (Sampath 1997). In Mauritius sega is perceived as 'expressing the quest of a young Mauritian nation, in particular its Creoles, for an authentic expression of their identity', and is 'considered as a national cultural icon by most Mauritians' (Police 2000: 57). In Jamaica ska became nationally popular in part because it mixed Western musical forms with the particular demands of local sound systems, but, crucially, because this happened at the time of Jamaican independence (1962) when indigenous musical forms became more acceptable (Bradley 2000). Junkanoo festivals and artists such as Baha Men 'promote a sense of quintessential Bahamianness' (Rommen 1999: 90), qawwali acts as a 'sound idiom' that signifies Pakistan's public culture (Qureshi 1999) and 'Britpop' emerged in the 1990s as a deliberate media construction of national musical style in the 'new' United Kingdom of 'cool Britannia'. The list is almost endless but, just as the unique music of particular localities – such as Bristol or Iceland – has proved impossible to distinguish, so distinctive national styles are similarly marked by strategic essentialism, marketing and local boosterism. Differences also emerged from the varied ways in which cultural networks were built up in different national contexts (radio stations, venues, licensing laws, etc.). Indeed, not all nations had, or sought to have, their national identity reflected in a particular music tradition (see Box 6.4).

Box 6.4 Swedish pop – 'national' or 'international'?

Sweden is generally thought of as a semi-peripheral nation in terms of the global music industry. While Sweden has always had local music traditions, and cities such as Stockholm have played host to vibrant pop music scenes, Anglo-American domination of the domestic market has become profound. Meanwhile, Swedish labels such as Elektra and Sonet were bought by majors and incorporated into global distribution networks, with a growing emphasis on English-language recordings rather than Swedish lyrics; for many young people in Sweden, it is accepted that English lyrics are the norm, and part of a career path for artists that leads towards global acceptance. Yet despite this apparent marginality, Sweden's cultural industries continue to produce many international recording stars: Abba, Roxette, rappers Leila K and Papa Dee, Dr Alban, guitar-glam outfit Europe, Army of Lovers, Ace of Base and more recently the Cardigans, the Hellacopters and the Wannadies.

Moreover, Sweden became renowned for a particular brand of music – a 'national' heritage of pop melodies and playfully commercial 'hit singles' built on an almost placeless internationalism: '[Swedish musicians] don't mind leaving their national identity at the recording studio door in the quest for overseas success.... As a result, Scandopop is culturally anonymous. Perfect fodder for today's global, homogenised market' (Jinman 1997: 5). This tradition, established by Abba, was reflected in the music to follow – from Roxette's 'The Look' to the Cardigans' 'Love Thing'. For many artists, the use of English and 'global' sounds over Swedish-language traditions was a deliberate affront to nationalism, as explained by Ben Marlene, from Roxette's publishing company: 'The Swedish flag is mainly carried by the bloody skinheads and racists. We don't have healthy nationalism here, which is an advantage for pop. The bands don't come running and shouting "We're from Sweden!"' (quoted in Jinman 1997: 5). Thus, a national music emerged not as a product of overt nationalism (in the form of state music policies or through national radio stations, labels and anthems), but through the ways in which music was constructed for export and received by global audiences – a 'style' and 'sound' as a marker of distinctiveness (and ultimately, as the basis for a more effective marketing strategy). These sounds emphasised an absence of nationalism.

See: Burnett 1992.

Local musical differences have been appropriated and transformed into representations of the national, both within nations (by cultural elites and nationalists) and beyond the nation (as part of the trend toward exoticising culture). Thus 'Englishness' became associated with the Mersey sounds of the Beatles and, much later, 'Britpop'. Nostalgic images of small-town and suburban working-class life grounded the music in specific locales. The Kinks personified this with albums like *Muswell Hillbillies* (1971) (Muswell Hill being their home suburb in North London) and *The Kinks are the Village Green Preservation Society* (1969) (which traced, with influences as diverse as Dylan Thomas and Lewis Carroll, the broken dreams, fading hopes and inevitable losses of the inhabitants of a small English town), and singles like 'Waterloo Sunset' and 'Autumn Almanac' (1967): 'I like my football on a Saturday/ Roast beef on Sunday's alright/ I go to Blackpool for my holidays / Sit in the autumn sunlight' (Laing 1972: 47–8). In the 1990s several groups achieved commercial success by linking sounds and images to these notions of 'Englishness': Oasis modelled both their recordings and 'look' on the Beatles, while 'mod' styles underwrote the releases and marketing of a range of 'Britpop' bands including Blur, the Verve and the Stone Roses. For one record company, EMI, this style underpinned their repertoire, with labels such as Parlophone (home of the Beatles, and much later Radiohead and Coldplay) specialising in this sound. Yet, this terminology (and notions of 'Englishness') is highly selective. A certain style – an emphasis on Romantic ideals, major chord progressions and keys (distinct from the mostly blues-based structures of American rock, funk and soul) – has been claimed to define British pop traditions over others, including folk, punk, house, trip hop and heavy metal, all of which were heavily transformed (or created) in the United Kingdom. Englishness and Britpop were media-inspired creations that had little if any relationship to the musical diversity of the nation.

In Latin America, working-class musics (that were themselves hybrid musical forms) from the barrios of major cities were transformed into national emblems: tango in Argentina, samba in Brazil, danza in Puerto Rico, rumba in Cuba and ranchera in Mexico (Wade 2000), while in international mediascapes Latin dance sounds have been regularly homogenised and sexualised, as tropical soundtracks (such as 'The Lambada') featuring skimpily dressed dancers with high-flying colourful skirts and g-strings in promotional film clips. These appropriations and clichés, while never absolute, present simplified versions of nationhood and ethnicity that 'stick' in global mediascapes, 'postcard' images that are as much related to national tourism campaigns as they are to sustained local cultures. Achieving notional homogeneity, and a 'national music', demands crude essentialism. In Colombia, the 'imagined community' of the national government has emphasised hybridity, with the notion of *mestizaje* ('mixture') formalised in constitutional arrangements to 'enshrine certain rights for indigenous peoples and black communities' (Wade 2000: 1), in that country's complex post-colonial cultural landscape. Against this backdrop, costeño music, associated with the northern coastal Caribbean region, came to replace bambuco – an urban dance style from Bogotá in the interior – as the defining 'national' sound. Bambuco combined European influences with local traditions and was considered 'cultured'; as one journalist described it: '[there is] nothing more national, nothing more patriotic than this melody which counts all Colombians among its authors. It is the soul of our pueblo [people, nation] made into melody' (quoted in Wade 2000: 48). Yet, by the 1940s, it was rapidly replaced by costeño styles, and relegated to 'folk' status as new, commercialised musics swept the interior. Reactions to this transformation were varied – costeño musics were associated with blackness and promiscuity, 'lack of measure, lack of control, excessiveness of bodily movement, noise, and emotion' (Wade 2000: 127). Radio Nacional (Colombia's state broadcaster) refused to air new and 'invading' sounds that were seen to threaten national identity. By the 1960s, costeño's 'happy' sounds had been more fully commodified, standardised and dispersed through national and international distribution: it became the sound of Colombia, as 'música tropical'. Yet three decades later, these regional, black music forms were perceived as 'authentic' traditions, nostalgic roots in multicultural national identities. Meanwhile, in Nicaragua, similar processes elevated the Creole palo de mayo ('maypole') music to national consciousness during the late 1970s, associated with modernity, sexual liberation and multiple ethnicities. The revolutionary Sandinista movement attempted to 'fashion a constellation of elements that could constitute a new, more expansive conceptualization of the Nicaraguan nation through the deliberate accretion of cultural resources within the national borders that exist outside the dominant demographical center' (Scruggs 1999: 298). Yet, unlike in Colombia, by the 1990s successive economic and military attacks from the United States had destabilised cultural institutions; returning Nicaraguan migrants from the north (nicknamed 'los Miami boys') brought global sounds, and palo de mayo groups disbanded or went broke. The creation (or demise) of 'national sounds' was never immune to a variety of geopolitical and commercial needs and influences. In Colombia, Nicaragua, Afghanistan and elsewhere, national musical identities waxed and waned with the shifting balance between national aspirations and authority, regional cultural expressions and political shifts.

Anthems

The culmination of the role of music in constructions of national identity is the national anthem – the embodiment of nation in song. Many occupy an undisputed position as official musical representations of place, while others remain contested as nations and societies change. The Australian national anthem, 'Advance Australia Fair', adopted over traditional bush ballads such as 'Waltzing Matilda' (a popular favourite as an alternative anthem), is for some an outdated remnant of colonial sentiment in an increasingly multicultural society aware of its indigenous past. Indigenous Australians challenged the validity of the Australian national anthem, with its proclamations of 'boundless plains to share', youth, freedom and abundant natural resources – betraying the persistence of unresolved indigenous claims to sovereignty over land. The Aboriginal band Yothu Yindi's song 'Treaty' (1989) became for some an alternative anthem, whilst Tiddas, a Melbourne group (whose members were mainly Aboriginal), provided a challenge to 'Advance Australia Fair' in their song 'Anthem' (1996):

Don't sing me your anthem
When your anthem's absurd
We might have been born here
But we're not young and free...
This land may be beautiful
But it cannot be called fair
So don't sing me your anthem
till we've learnt how to share.

Similarly, the French national anthem 'La Marseillaise' has been the subject of debate concerning its militaristic lyrics (Eyck 1995: 52). Here and elsewhere, anthems are musical texts, historical documents and a means through which a sense of the contemporary nation is created and contested. Profound political shifts in the nature of existing nation-states (and new alliances beyond the nation-state) are reflected through the establishment of anthems. National tunes have emerged for countries such as Slovenia and Macedonia, the formation of the European Union was accompanied by an official European anthem (music from the last movement of Beethoven's Ninth Symphony), while the fall of apartheid in South Africa led to the adoption of 'Nkosi Sikelel' iAfrika' ('God Bless Africa') as the official national anthem of the new republic in 1994. The song, composed in 1897 by Enoch Sontonga and sung in Xhosa, was for decades performed by those oppressed through apartheid regimes, as an act of defiance and an assertion of African identity. Anthems are strategic devices to consolidate nationalist sentiment.

Solemn anthems have also been the subject of derision, subversive reinterpretations and spoofs: Jimi Hendrix played 'The Star Spangled Banner' at the 1969 Woodstock festival through a wall of swerving, mangled and psychedelic distorted sound; Serge Gainsbourg's (1979) reggae cover-version of 'La Marseillaise' (entitled 'Aux armes et caetera'), recorded in Jamaica with members of Peter Tosh's band, provided a twist on the otherwise confrontational lyrics; and the Sex Pistols' rendition of 'God Save the Queen' (released to coincide with Queen Elizabeth II's Silver Jubilee in 1977) all provoked critical appraisals of the representations of their nations. Hendrix's version of the American national anthem was widely interpreted as a savage critique of the United States' involvement in the Vietnam War:

> The ironies were murderous: a black man with a white guitar; a massive, almost exclusively white audience wallowing in a paddy field of its own making; the clear, pure, trumpet-like notes of the familiar melody struggling to pierce through clouds of tear-gas, the explosions of cluster-bombs, the screams of the dying, the crackle of the flames, the

> heavy palls of smoke stinking with human grease, the hovering chatter of helicopters....One man with one guitar said more in three and a half minutes about that peculiarly disgusting war and its reverberations than all the novels, memoirs and movies put together.
>
> (Shaar Murray 1989: 24)

Anthems thus evoke patriotism and nationalist sentiment, yet remain open to (re)interpretation in ways that subvert the dominant meanings of nation they usually convey.

Contesting the state and nation: ethnicity, identity and regionalism

Popular music has often challenged the authority and legitimacy of state systems, including repressive colonial and neo-colonial regimes, communist regimes (in the last years of the Eastern Bloc) and various forms of nationalism and capitalism; likewise, music has also contested systems of hetero-patriarchy (Denselow 1989; Harker 1980; Street 1986; see Chapter 9). This largely occurred through the various ways that individuals and communities negotiated representations of ethnicity, gender and place. National identities ordinarily implied articulations of ethnicity (as with the basis of nationalism in language), yet constructed identities of peoples of various ethnic backgrounds also served to legitimate subordination and reify European domination and colonial enterprises. Notions of the 'primitive', the 'exotic', the 'noble savage' – ideological constructions of ethnic identity imposed on peoples from Africa, the Americas, Asia, Australia and the Pacific – underwrote histories of slavery, conquest and domination, and were used as justification for genocide or for the activities of missionaries and 'pioneering' settlers. The remnants of these colonial identities continue to resurface in various ways in contemporary nations, despite official policies of decolonisation and multiculturalism supported by many governments, through the persistence of racism and economic disadvantage in countries such as the United States and the United Kingdom, and commonly held generalisations about migrant communities, indigenous groups, gays, gypsies – a range of peoples rendered as 'opposites' to the Euro-American centre, defined through a sense of 'otherness' (Goldberg 1993). In contrast, ethnicity has become an avenue for liberatory expressions, particularly apparent in music. Black nationalism acknowledged a 'pan-African' character that emerged in various cultural expressions (art, dance, music, literature) and provided a potent post-colonial demand for radical emancipation from oppressive social circumstances, in contrast to more repressive images and concepts of the 'exotic', where ethnicity was simplified and vulgarised for popular or commercial effect. Ethnic identities have been continually produced, contested and reproduced, at times resembling both 'essentialist' and 'pluralist' perspectives (see also Chapter 8). Popular music has been one cultural stage upon which these tensions are played out.

All music involves, to varying degrees, issues of race and ethnicity. The highly charged black nationalist rap of Public Enemy, the Rastafarian symbolism of reggae and the affirmative tones of Australian Aboriginal music are no more or less bound up in questions of ethnic identity than the Celtic twang of (white) American country and western, the heavy metal of Iron Maiden or Whitesnake, or representations of white ethnicity in literature and film (Dyer 1997). Such examples cannot be divorced from general debates about gender, music and identity, and particular expressions of ethnicity and gender (for example the exclusivity of male power in Rastafarianism). Hence, the American band Black Oak Arkansas used particular imagery in marketing campaigns and record covers to reinforce public perceptions of the rural south of the United States. Through depictions of the band members as rugged, simple rural folk, beside emblems of the United States' colonial, plantation history (in particular the confederate flag, a potent right-wing reminder of slavery mobilised by those seeking a 'mystic view of a glorious past', which commonly adorned stage sets and record sleeves), a white, male, southern ethnic identity was asserted. Music formed a conduit through which notions of the south, 'as a region of moonshine-guzzling rednecks, ignorant hillbillies, Dixie-glorifying racists, unambitious "wool hat" boys and/or submissive females' (Hutson 1993: 53), could be constructed.

Lines of division between ethnicities and other aspects of identity have always been blurred; simple associations between music, place and race are always problematic. As Longhurst asked, 'would a black musician performing a song by Lennon and McCartney count as black music?' (1995: 128). Some have found the African American Charley Pride's country music anomalous on racial grounds, while blackness has also been claimed as a means of asserting cultural authenticity, as in the white rapper Vanilla Ice's claim to credibility because 'he had grown up amongst African American poverty...and could identify with blackness because he himself had experienced its poverty-stricken lifestyle' (Boyd 1994: 292; see Chapter 8). Defining the boundaries of ethnic groups, or what constituted authentically 'black', 'Asian', 'Aboriginal' (or 'gay') music, became a matter of debate, but a pointless exercise, prone to redundant, often oppressive generalisations about race and biology (most commonly articulated in the idea that black musicians are somehow more 'naturally inclined' towards particular styles or expressions), and avoiding deeper issues of class and economic discrimination (Longhurst 1995; see also Garofalo 1993a; Gilroy 1991; Stephens 1999). These debates also shrouded issues concerning the geographies of identity and ethnicity that emerged through music.

Where separatist groups have attempted to break away from larger nation-states, or where nations have been created for the first time, music has played a significant role in the construction of an 'imagined community'. Music traditions and ethnic heritage have been invoked as part of separatist strategies, as evidence of cultural difference and justification for the logic of secession. In Scotland this occurred through a deliberate revival of 'traditional' Scottish instruments (pipes) and tunes as part of a folk scene centred on Edinburgh. A range of strategic essentialisms regarding 'Scottishness' was invoked, particularly evident when, under the Conservative Thatcher government, uneven development intensified between the nation's financial power-base in the south and Scotland. Rather than representing a 'mass movement' of the Scottish population, this particular folk revival, evident in the music of Jock Tamson's Bairns, was a somewhat elite 'urban and cosmopolitan movement, centred particularly on Edinburgh which received streams of musicians and singers interested in Scottish music and song from all parts of Britain, Ireland and abroad': part of a wider search for a 'sturdy vernacular culture' (Symon 1997: 205, 204), which rejected the myths of Scottish tartanry and 'shortbread tin' representations of Scotland prominent in tourism marketing campaigns.

Language is of obvious cultural significance, especially as English dominates popular culture and popular music. As music styles such as rock 'n' roll, rap and reggae diffused, those who first adopted them were usually also imitative in the parallel adoption of English but, over time, switched to local languages (and themes) with growing confidence, maturity and innovation, as in Japan (Mathews 2000: 64–5). Subsequently, where they sought access to a global audience and market, they reverted to English (e.g. Ramet 1994: 3). In the former Eastern Bloc countries, English-language popular music was initially highly popular among the young, not because of the lyrics, which were little understood, but because of the counter-cultural and political stance involved in listening to Western music. For some years Aboriginal rap music in Alice Springs, central Australia, simply copied African American music; efforts to encourage the use of local referents (let alone language) initially foundered on the delight that dispossessed Aboriginal youth had in using such words as 'motherfuckers'. Similarly, in Tanzania, rap groups initially took wholly derivative American names, such as Niggers with Power and Rough Niggaz, and wholly American dress, hairstyles and language use, before subsequently rapping in Swahili, and covering local themes, such as the rise of AIDS and promiscuity (Remes 1999). The diffusion of English has been resisted, particularly at the state level, especially in France where there are legal restrictions on the extent of foreign music on radio and on the use of English terms, but this is everywhere tempered with the desire to participate in global markets and trends.

Deliberate choice of language emphasises its cultural and political significance. This was perhaps best exemplified in the language and linguistic structures of rap

(Box 6.5) that both traced the origins of the genre and situated it within black America (despite its subsequent diffusion). In Germany, a growing number of neo-Fascist bands refused to sing in anything but German, taking up an overt political position: 'a self-conscious articulation of the Third Reich's attempt to ban all non-German music during the 1930s' (Bennett 1999b: 85; see Box 6.6). Language is a vehicle for nationalism at various scales. During the 1970s a Welsh 'Celtic rock' movement was initiated by musicians 'who felt they were losing part of their cultural identity with the demise of the language their parents spoke' (Wallis and Malm 1984: 140). Rather later, punk bands also adopted Welsh, in defiance of English, but in a very different manner to earlier bands, who were seen as being too close to the spirit of 'the Eisteddfod and shit folk music' (Rhys Mwyn, quoted by Owens 2000: 20; Llewellyn 2000). In Scotland the band Runrig abandoned a standard rock repertoire to pioneer Gaelic rock songs, from both their own loosely socialist and nationalist perspective, and some feeling that 'as a language, as a musical language...it's absolutely wonderful. You can interpret a Gaelic song because there's much more music behind it that an English song', yet their acceptance and success were limited outside the Scottish highlands and islands and their later albums had relatively few Gaelic tracks (Morton 1991: 161). Similar contexts are widespread: in Brittany a Breton language revival was linked to music by Alain Stivel, in Norwegian Lapland Mari Boine Persen achieved considerable success with Lappish songs and in Australia bands like Yothu Yindi, Warumpi Band and the North Tanami Band used Aboriginal languages for at least some of their songs (Dunbar-Hall 1997). In 'Recipe for a Master Race' (1997) Persen railed against the colonial 'use of bible and booze and bayonet' and one outcome: 'language and culture take their place in the museum, as research object and tourist attraction'. Motivations for using local languages have varied, from political and cultural to economic, but have often foundered on desires for wider success and the need for comprehensible lyrics.

Box 6.5 The poetics of rap

Rap is the product of several cultures but primarily those of migrant African Americans and Puerto Ricans in New York (see Chapters 5 and 8), and, to many, carries a 'message of disaffection and rage' associated with alienated youth in the inner cities of the United States. It has close links to other diasporic musics, going back to early call and response styles, including Jamaican dance hall and dub poetry (following a tradition of talking over recorded music), soca, British funk and ska. Not only does rap involve the 'continual citation of the sonic and verbal archives of rhythm and blues, jazz and funk forms', but this process occurs through sampling: 'its founding gesture: an incursion against the author function', incorporating not just music and lyrics but newscasts, sound effects, answering machine messages and political speeches. Consumption is transformed into production. Rap evokes the anger and violence of inner-city life, while the words – involving African American vernacular such as 'diss' and 'ho', and Jamaican patois – and the musical structures trace a particular socio-political history. Moreover 'as a vernacular practice, hip-hop depended on its audiences, its sites and its technologies to construct a zone of sonic and cultural bricolage', where musical competence was about performance: the ability to rap skilfully (Potter 1995: 9, 26, 36, 53). Distinctions between literate and oral modes of communication are fused. Rap is descriptive of everyday inner-city life – distinctive styles of dress, shared plights, violence and sexism. Resistance to capitalist and racist structures is marked by language, distilled from what Niggaz With Attitude (NWA) have called 'the strength of street knowledge', though there is no single black American vernacular (and speaking in the vernacular has not always been regarded as a valued or empowering mode of resistance). The power of the language resides in its violence and aggression, and its close association with the urban disadvantaged (both elements that have encouraged its widespread diffusion). One commentator has even argued that:

> The hums, grunts and glottal attacks of Central Africa's pygmies, the tongue clicks, throat gurgles and suction stops of the Bushmen of the Kalahari Desert, and the yodeling, whistling vocal effects of Zimbabawe's mbira players all survive in the mouth percussion of such 'human beat box' rappers as Doug E. Fresh and Darren Robinson of the Fat Boys.
>
> (Marc Dery, quoted by T. Rose 1994: 197)

In rap too voices become musical instruments with phraseology and pauses that 'are not merely stylistic effects [but] aural manifestations of philosophical approaches to social environments' (Dery, in T. Rose 1994: 67). Words and structures, especially the breaks, emphasise repetition and rupture, building on long-standing black cultural forces (Dery, in T. Rose 1994: 70), creating a distinct musical genre that through its form and historical roots resists and challenges institutional structures.

See: Potter 1995; T. Rose 1994; Toop 1992.

Box 6.6 Neo-Nazi music in Germany

During the 1990s an underground, white, right-wing political movement grew in Germany, tracing its lineage to the pre-war Nazi era, and supported by as many as seventy established rock groups, touring the country, giving concerts and releasing CDs and tapes, with an audience of around 45,000 far right-wing radicals and neo-Nazis. Concerts operated under a shroud of secrecy, as word of mouth gave details of meeting points far removed from the final venue. Stewards using mobile phones directed audiences to the venues only when they were sure they could proceed undetected. Many supporters were male skinheads adorned in Nazi regalia. CDs and cassettes were manufactured abroad and directed back into Germany from Denmark and Britain to make use of postal privacy. Much of the music contained either incitements to racial hatred or the glorification of National Socialism, both crimes in German law. One group on a track entitled 'Blood Must Flow' openly exhorted listeners to go out into the streets and murder Jews. Frank Rennicke, one of the leading performers, recorded a ballad dedicated to the Nazi war criminal, Rudolf Hess. The five members of another group, Landser (an old-fashioned German word for low-ranking soldier or private), whose songs called for attacks on foreigners, Jews, gypsies and political opponents, were arrested in 2001. No political movement, however extreme, has been without mechanisms of cultural support such as music.

Just as the use of Celtic languages drew attention to regional distinctiveness so variations of standard English (or other languages) emphasised local spaces, as in rap and country and western music, stressing difference and often marginality. Particular forms of slang, such as *verlan* (the reversed syllables used by French youth) have been incorporated into popular music, from French suburbs to Congolese towns, to create and represent localised sub-cultures (Gross *et al.* 1994; Gondola 1997), while Creole has increasingly been used as a symbol of identity and thus the language of songs in the French Caribbean (Berrian 2000: 39–40). Zouk music from Martinique and Guadeloupe challenged colonialism through its championing of Creole, in opposition to colonial French, emphasising the particularities of the African legacy in the islands. Music and language remain potent and ubiquitous markers of ethnic identity. Until recently in the world's 'largest non-state nation in the world' (Blum and Hassanpour 1996: 325), that of the Kurdish peoples who inhabit Iraq, Turkey, Syria and Iran, the distribution of recordings featuring either of the two main Kurdish dialects remained an illegal act, as the rights of Kurds to their language were systematically denied in all these countries. Numerous separatist political movements have been explicitly aligned with sub-cultural music styles that emerged originally in Western contexts, as with punk music in Northern Ireland (Rolston 2001) and in the Basque country of Spain (see Box 6.7). In certain contexts particular claims to ethnicity, combined with claims to regional identity, were reflected in deliberate musical constructions.

Box 6.7 The Basque nationalist movement and punk rock

Punk rock emerged as a highly potent political music within the Basque region of Spain, closely aligned to radical separatist activities in search of a new *Euskadi* (Basque) state. ETA mobilised support not only through appeals to a sense of distinct ethnic identity, but also through more conventional class-based struggle, attempting to unite workers, unemployed young people and students against processes of uneven development and modernisation that were seen to discriminate against this region of Spain. ETA fought, literally and philosophically, for the legalisation of all political parties, the release of 'political' prisoners, ethnic self-determination and the withdrawal of Spanish military forces from the Basque country (Lahusen 1993). Punk rock (or 'Basque radical rock') emerged as an equally violent musical embodiment of this, and bands such as Eskorbuto ('Scurvy'), Barricada ('Barricade'), Kortatu (named after an ETA member killed by Spanish police) and Negu Gorriak ('Hard Winter') appropriated the imagery of revolutionary movements in album covers, stage sets and the lyrics of their songs, providing a twist on the nihilistic, pessimistic flavour of punk. Music could be appropriated and incorporated within radical nationalist movements in an active way, an emotional tool to elicit support and patriotism. This included not only musical texts, but the sub-cultural practices of punk – anarchic live performances, aggressive and confrontational dance styles and a 'symbolic universe and argumentative repertoire' of anti-authoritarianism and militancy.

See: Lahusen 1993.

In contrast to the link between music and separatist movements, the role of music in opposition to the actual nature of the state has been widespread. Particularly marked in the communist world, this has been matched by complex state attempts to control opposition. In China, as in Eastern Europe (see Box 6.1) rock music functioned as one arena where different voices attempted to speak for 'the people'. In the early 1990s Chinese popular music was divided into two broad camps: officially sanctioned popular music (*tongsu yinyue*) and underground rock music (*yaogun yinyue*, literally 'rock and roll'), each produced, distributed and performed in different ways. *Tongsu* music was broadly propaganda, circumscribed by the goals of the Chinese Communist Party to retain power, bolster its ideological supremacy and make money. *Yaogun* music, relegated to sub-cultural margins, united musicians and fans in a more coherent ideology of cultural opposition, in support of individualism and in opposition to feudalism, and sought to claim a wider space in society; the underground cassette industry took it beyond government control. Cui Jian, its most famous exponent, used subtle lyrics that found an audience among intellectuals and young workers who recognised themes of discontent. Chinese popular music was not merely leisure but 'a battlefield on which ideological struggle is waged' (Jones 1992: 3; Stokes 1997). *Yaogun* musicians sought to cut through what they perceived as the hypocrisy of *tongsu* to reclaim the 'authentic voice of the Chinese people' (Jones 1992: 3), but, simultaneously, urged the embrace of Western conceptions of modernity (democracy and science) in a new 'poetics of authenticity' (Jones 1992: 146). Its concerts and performers were frequently banned because of their significance in articulating voices and strategies of political discontent. Music was not merely symbolic; it was an active component of a lengthy political struggle.

In a wide range of contexts popular music has opposed both political repression and cultural imperialism, usually welding these themes together as in Chile (Box 6.8). Similarly, in Haiti, in the early 1990s, the music of Boukman Eksperyans, blending indigenous voodoo and rara rhythms with imported funk rock and dance music, played a part in the popular revitalisation of voodoo, united town dwellers and peasants in opposition to political corruption, and contributed to the movement that brought a democratically elected government: 'The group's historical references, Creole lyrics, and voudou metaphysics speak to distinctly Haitian realities, but its finished products also circulate as nodes in a network of global cultural commerce' (Lipsitz 1994: 11). The international success of reggae was related to symbolic and real forms of resistance and politicisation; global themes of poverty, oppression and hunger gave the lyrics (and music) widespread relevance and appeal (King and Jensen 1995). At a national scale this was also true of Nigerian juju music (Waterman 1990b: 224–7). Similarly Thomas Mapfumo's Zimbabwean music became most popular in the 1970s, the years prior to independence, through 'his recognition of the value of Zimbabwe's traditional musical instruments, and the importance of a people's heritage especially at a time of conflict' (Herman 1990; Turino 2000), accentuated by lyrics that called for the overthrow of colonialism and the construction of a united nation. Thus music, both circulated as a commodity in the world market and through the appeal to history and local or regional cultural forms and traditions, contributed to and influenced powerful local political movements. Cultural battles were constantly interwoven with struggles for political and economic liberation, as popular music proved a vehicle for developing populist nationalism.

Box 6.8 The Chilean struggle for democracy

In the 1960s many Chilean musicians, collectively known as the New Chilean Song Movement (NCSM), opposed the North American incursion into national popular culture by writing songs reminiscent of traditional folk music, using Andean instruments to accompany them. During the 1950s and 1960s folk singers, notably Violeta Parra and Victor Jara, collected folk music throughout rural Chile while also writing songs dealing with social, political and economic problems, notably the struggle of the 'common man' against the wealthy and powerful, and, by extension, the struggle of Chileans and other poor people against the United States. The NCSM wrote campaign songs and performed at political rallies in support of Salvador Allende's socialist government, elected in 1970. Allende appeared with a group of musicians under a banner that said: 'There can be no revolution without songs'. Jara himself had argued that 'the authentic revolutionary should be behind the guitar, so that the guitar becomes an instrument of struggle, so that it can also shoot like a gun' (Taffet 1997: 97). After the election new songs glorified labour, but a sense of doubt entered the lyrics. Allende was overthrown by Pinochet's military coup in 1973 and Victor Jara was tortured and killed by the military. Political repression effectively ended the NCSM movement within Chile, but groups like Inti Illimanni, who had gone to Europe in 1973 as 'cultural ambassadors of the Allende government', remained in exile, introducing Andean music to the world and maintaining the political struggle through songs like 'Cancion del poder popular' ('The Power of the United People') and 'Venceremos' ('We Will Triumph'). In Chile the songs of Jara, Parra and other supporters of the socialist government constantly circulated underground, but only Parra's love songs could be played on the radio. The military government instituted a policy that favoured Western classical music, and foreign instruments were cheap and easily available, whereas Chilean folk instruments were virtually impossible to obtain because they could not be bought on instalments and shops thought it unwise to sell or display them.

See: Taffet 1997.

Music and black nationalism

Unlike separatist nationalism, black nationalism in the West, especially the United States, conveys both expressions of cultural linkages and unity throughout the diaspora, alongside demands for a more prominent place within the nation. Until at least the war years the 'negro market' did not exist in a national sense; some records were produced, mainly by small independent companies, and distributed within particular regions. Though regions still constituted distinct markets into the 1970s, with particular singers depending primarily on local audiences, a national network of distribution effectively began in 1945, much influenced by DJs across the country (Gillett 1970: 11). Black popular music evolved largely separately. Despite the presence of African Americans in vaudeville theatres throughout the country, there were also separate theatres, dance halls and tours, and distinctive musical themes – incorporating work songs and dance tunes, and fragments of African musical forms, which eventually emerged as blues and jazz. In the inter-war years much of this was known as 'race music' or 'race records', until in 1948 the principal trade journal, *Billboard*, reclassified the genre as 'rhythm and blues', a designation that remained in place until 1969 when *Billboard* officially renamed it 'soul' (Ennis 1992: 25), a mark of its movement into the mainstream.

African American history exemplifies the manner in which music and ethnicity may combine to convey experiences of slavery, forced displacement and continued survival in the face of marginality: '[music] demonstrates the aesthetic fruits of pain and suffering and has a special significance because musicians have played a disproportionate part in the long struggle to represent black creativity, innovation and excellence' (Gilroy 1991:132). Musically related to African traditions, blues provided an early sphere of solidarity and expression for musicians and audiences alike, and, for the former, an occupation that allowed respite from dreadful labour conditions. Many of the earliest blues songs, by artists such as Robert Johnson, Son House and Skip James, articulated intimate attachments and reactions to physical places; migratory and transient experiences were littered throughout blues songs dedicated to themes of escape, songs of wandering and leaving home. Meanwhile, cartographies of places emerged from the texts, as destinations became mythologised, lines of exodus became clearer and homelands and transitory stops were imbued with emotional resonance:

> I'm tired of being Jim Crowed, gonna leave this Jim Crow town,
> Doggone my black soul, I'm sweet Chicago bound,
> Yes sir, I'm leavin' here, from this ole Jim Crow town.
> I'm going up North, where they think money grows on trees...
>
> (Charles 'Cow Cow' Davenport, 'Jim Crow Blues' (1929), quoted in Oliver 1990: 46)

Spaces of the colonial United States, of frontiers, white wealth and plantation control, were symbolically transformed into spheres of desire, promise and escape. Geographical mobility provided strategies of resisting domination; yet these patterns of transience, constant moving and resettling strained family relations, maintained economic deprivation and intensified feelings of exile, as the immensity of migration became apparent (Oliver 1990: 44). Between 1910 and 1940, nearly 2 million black Americans migrated from the south to northern cities (Thomas 1990), a population shift reflected in new sounds, as rural, acoustic blues made way for an overwhelmingly urbane, electric blues developed by such figures as Howlin' Wolf, Sonny Boy Williamson and Muddy Waters. Blues music cultures represented 'rhythms of resistance' through subtle evasions of control, developing a dialect that articulated stories of itinerant life in words and phrases alien to colonial classes and white elites, maintaining old tunes and evocative oral traditions (Pratt 1990). Locations such as Chicago and Detroit, the principal northern cities at the end of these mythical and actual migration routes, became new centres of the black diaspora and of black musical creativity.

Early black American musical forms, at once examples of local music traditions (creating new sounds in localised contexts, expressing attachments to and reactions to places), also suggested links to 'homelands', in the case of the blues through aural reminders of African beats and scales:

> The contemporary musical forms of the African Diaspora work within an aesthetic and political framework which demands that they ceaselessly reconstruct themselves time and again to celebrate and validate the simple, unassailable fact of their survival.
>
> (Gilroy 1993b: 37)

Jazz exemplified the process of continually reproducing tradition: the notes, points of emphasis and musical structures of African roots were present throughout early styles and on into post-war be-bop. Quincy Jones has said that 'the essence of African music as we know it is the sound of the drum' (Jones 1977). Be-bop jazz drummer Max Roach described it as 'politics in the drums' (Lipsitz 1994: 38) – aural reminders of a common African heritage, absent in political rhetoric but captured in the music. Celebrations of survival and solidarity through musical traditions have remained powerful. The concern of jazz musicians, reggae and dub DJs, hip hop crews and others with 'the beat', 'breakbeats', 'grooves' and creative use of 'flow' and 'rupture' in the structure of songs can be seen as political in this way, as they 'create the mood, set the beat, and prompt the engagement' (Walser 1995: 194; T. Rose 1994). Trends and themes established in earlier blues and jazz scenes were reworked in more recent contexts as funk, reggae, rap, rhythm and blues, while gospel recordings captured new pan-African expressions and mythologies of an African homeland. This was apparent in the work of soul/funk stars such as Marvin Gaye and Curtis Mayfield, who sang about race politics in America to both black and white audiences (as on

'Inner City Blues (Make Me Wanna Holler)' (1971) and 'We the People Who Are Darker than Blue' (1970). This African connection was later expressed in various hip hop musical collectives and activist groups, such as the Native Tongues project initiated by A Tribe Called Quest, Queen Latifah and the Jungle Brothers (whose explicit aims to maintain and respect black music traditions included hip hop samples of James Brown and Marvin Gaye). Many black artists engaged with the Nation of Islam project (where African Americans sought to reclaim heritage and ancestry in a distinctive version of Islam that emphasised ties to Africa), and Afrika Bambaataa's Zulu Nation gathered street gangs, DJs and graffiti artists together to create the first examples of an explicitly political hip hop culture.

Outside the United States, similar musical evolutions and emphases on pan-African consciousness were evident in the music of Manu Dibango, Angelique Kidjo and others. The liner notes to Manu Dibango's *Wakafrika* (1994), recorded in France, which brought together such performers as Ladysmith Black Mambazo, King Sunny Ade and Salif Keita, stressed 'Africa, the land of all origins, must now heal from the wounds of colonialism or it will collapse in a near future… for pan-Africanism should be a guarantee for democracy through its ambition for African people' (Bigot 1994; see Mitchell 1996: 86–7). In Jamaica particularly, but also in the Caribbean diaspora, Rastafarianism emphasised the validity of an African

Figure 6.1 Manu Dibango, *Wakafrika* (1994)
Note: Manu Dibango's was one of the earliest records to evoke Pan-African identities, symbolised in this cassette cover design.

homeland, specifically in Ethiopia (see Chapter 8). A more committed musical stance reflected the continuity of marginality and rejection, interpreted over time in a variety of musical genres. Pan-Africanism in music involved mobilising a range of metaphors and unique rhetorical systems, including articulations of unity and liberation that were traditionalist and experimental, confrontational and even phantasmagorical (Box 6.9). While much black music had relatively little concern with ideological issues (as with any other musical tradition), particular stylistic and lyrical concerns mapped out displacements and trajectories based on strategic affirmations of ethnicity. Identifications with some sense of 'Africanism' in music and lyrics created a diasporic identity.

Box 6.9 Pan-African identity and science fiction

A vivid example of surrealist and playful engagements with pan-Africanism involved the link between music and images of 'otherworldly' escape and transformation. A musical trajectory can be traced back to John Coltrane, and later, Sun Ra, the black American jazzman who recorded hundreds of albums from the 1950s until his death in 1993 with his big band (known among other names as the Intergalactic Research Arkestra and the Astro-Galactic Infinity Arkestra). Sun Ra created the image of himself as a space-travelling jazz missionary – a child from another planet sent to generate unity among humans, and transcend the racial politics of his time. This concern with the 'cosmic' was deeply rooted in an acknowledgement of Africa as homeland:

> With sound, light, words, colour and costume, Ra concocted a moving, glittering hallucination of ancient Egypt, deep space, the kingdoms of Africa – Nubia, Ashanti, Hausa, Yoruba, Mandinka, Songhai, Sudan, Mali, Malinke, Xhosa – a history of the future with which he battled for the souls of his people against the legacy of slavery, segregation, drugs, alcohol, apathy and the corrupting powers of capitalism....Every record was an apocrypha, a vibrant cosmic map of unknown regions, lush solarised rainforests, cold domains of infinite darkness, astral storms, paradisical pleasure zones, scenes of ritual procession, solemn ceremony and wild celebration. Space was the place.
>
> (Toop 1995:28–9)

The 'mothership' metaphor of Africa, made explicit in Sun Ra's 'ark of space', provided a course to liberation through the construction of a symbolic vessel of black unity, progression and escape (Szwed 1997), a rhetoric that was to resurface later in different settings. In the 1970s and 1980s, George Clinton's space-age funk costumes, elaborate set designs and long, improvisational jams of speedy blues and jazz riffs created similar metaphors of liberation. In his group Parliament, Clinton created stage dramas and special effects – the P-Funk Mothership Connection, a glittery flying saucer, figuratively landing at the start of the set, from which Clinton would emerge to mobilise and unite the crowd (Keyes 1996).

Related themes ran through rapper Afrika Bambaataa's *Planet Rock*, and the electro sounds emerging from Detroit's techno artists such as Juan Atkins, Mad Mike and Underground Resistance, who took the connections between technological futurism and black expression to their most extreme. Atkins released records under a series of names, representing 'distant points of the universe': Infiniti, Output, Channel One and Model 500; Underground Resistance released *X-102 Discovers the Rings of Saturn* (1992), which featured tracks cut into the record corresponding to the size and relationship between Saturn's rings. Black 'Afro-futurist' musicians 'take space iconography seriously and turn it into a platform for playful subversion, imagining a productive zone largely exterior to dominant ideology....[T]hese are the new, unreasonable vessels for travel in discursive space' (Corbert 1994: 8, 18). With heavy emphasis on collage, improvisation and 'groove', a sense of radical fluidity, of appropriation and reinscription, pervades this music which uses both European electronic music and black traditions, creating a sense of spatial flow, of tapping into old worlds and creating new expressions of ethnicity.

See: Corbert 1994; Keyes 1996; Szwed 1997; Toop 1995.

Towards the transnational?

Music is ubiquitously 'in place' yet relationships with place are ambivalent and constantly changing. Deliberate attempts to create or nurture national musics demand essentialism, often crudely so, alongside repression of dissident and alien musical traditions and lyrical forms. Countries like Afghanistan have gone through various different attempts to create (or reject) a national musical identity. Politics intrudes into commercialism in creating and challenging national identities. Nonetheless, frictions of distance sustain regional variations, especially in the context of ethnicity, language and other forms of difference. However, the range of music accessible in most places provides a 'contested terrain' where 'dialectics of class, gender, ethnicity, age, religion and other aspects of social identity are symbolically negotiated and dramatized' (Manuel 1993: 259). Hence, the commercial also intrudes into the political; just as a certain degree of artifice produced the 'Dunedin sound' and also 'world music' (see Chapter 7), similar strategic essentialism, and commercial selectivity, has linked nations and regions with identifiable market concepts. Nonetheless, nations exist; their languages and media stimulate national identity, even if contested, while musical traditions (whether lyrical, harmonic or associated with particular instruments) cannot easily be brushed aside. Yet just as nations are difficult to construct, in theory or in practice, linking music to physical places or ethnic groups can only be a contested enterprise. Boundaries are porous, constantly being broken, necessitating new national anthems and new attempts to sustain imagined communities in the face of transnational flows. National sounds are retrospective and nostalgic, markers of a less fluid, more local era: a wider scale of the fetishisation of locality.

7

NEW WORLDS

Music from the margins?

Global flows of music have become more rapid and numerous as movements of people, whether voluntary or not, have become more widespread. Diasporic networks now connect metropolitan communities across continents; migration maps out lines of cultural flow between cities and homelands. Meanwhile, marketing of cultural products (including distribution and tours) makes similar global connections. Globalisation is both multifaceted and selective. Some music has achieved a global reach, while other performers and genres remain distinctively local; some places continue to be isolated, and tangential to global trends, while others are centres of creativity, reception and transformation. This chapter examines the rise of 'world music', and the manner in which it was linked to particular places; perhaps better than any other style it exemplifies how music is simultaneously an agent of mobility and a cultural expression permanently connected to place. Equally the chapter traces the 'pathways' of musical flow from places perceived as marginal to the centres of Anglophone musical production, and the movement of music away from developing countries to meet the needs of the West for new sounds, sources of creativity and expressions of authenticity.

'World music': from appropriation to hybridity

Many countries, even at the end of the twentieth century (especially those most distant from wealthier Western countries) had little exposure to Western popular music or no real taste for it. Conversely, while jazz and the blues were seen to have roots among African Americans of the southern United States, there was little interest in the music of African areas from which the ancestors of black Americans had come as slaves, and even less in the music of most other developing areas (despite occasional dance crazes, such as the tango and other forms from Latin America). While Anglo-American companies came to dominate global flows of recorded music, they nonetheless incorporated exotic sounds: a deliberate absorption of non-Western forms and traditions into Northern popular music. Globalisation did not merely entail one-way traffic.

Every phase of the evolution of popular music displayed 'alien' elements, though often substantially modified to suit Western tastes and prejudices, notably in the rise of minstrelsy in the nineteenth century, the fad for 'Ethiopian' music in early British music halls, Palm Court orchestras and the success of a small number of individual performers from other regions. Throughout the colonial era, shows, exhibitions and festivals were established in major European cities to display the unusual and the idiosyncratic discoveries of the New World, while museums placed the skulls of distant peoples in glass cases; artists such as Gauguin and composers like Debussy absorbed influences from Asia, the Pacific and Latin America into their European canon. Discussing the late nineteenth-century expositions in Paris, Toop described the logic of collecting and fetishising objects from 'other' cultures:

> These expositions were models of the shopping mall, as well as being the precursors of themed entertainments, trade shows, post-Woodstock rock festivals and the kind of fast-foods-of-the-world 'grazing' restaurant clusters that can be found in Miami malls. In Paris, world cultures were laid out in a kind of forged map as choice fruits of empire, living proof of conquest and self-aggrandisement in the face of imminent decline.
>
> (1995: 20–1)

Music was part of this. Instruments such as banjos, once African, were incorporated early into American music. Musical forms spread rapidly, often far beyond their apparent origins. By the 1930s, Hawaiian music and the rumba were well established in Japan (Hosokawa 1994, 1999), Indonesian gamelan had become an object of fascination (largely through ethnomusicologists) and similar eclecticism occurred elsewhere (Toop 1999). The more successful introductions became occasional exotic additions to the lower levels of hit parades, culturally distinct, but ultimately little different from many other popular music performances; the Weavers sung 'in the jungle, the mighty jungle, the lion sleeps tonight, Wimoweh' (1962), which took its place alongside Israeli songs and the Cuban 'Guantanamera'. In 1963 the cute Japanese song 'Sukiyaki' (performed by Kyu Sakamato) was a global hit, as was Antonio Carlos Jobin's bossa nova smash 'The Girl from Ipanema' (1963), but these novelties were exceptional.

Harry Belafonte's two major calypsos of the 1950s, 'Kingston Town' and 'Banana Boat Song', with obvious African influences, sung of a place 'where the nights are gay' and banana loading was enjoyable – the height of romanticisation of a town viewed quite differently a decade or so later when reggae became Jamaica's largest export. Similarly the South African Manhattan Brothers brought themes from a continent that was 'a romanticised, mythological continent, a timeless Africa of animals and jungles and hunting rituals and mystery: the West's exotic other' (Ballantine 1999: 16), often cast in simplistic, gendered terms. Such appropriations underpinned early easy-listening music of the 1960s, from Bert Kaempfert's *A Swinging Safari* and Martin Denny's *Exotic Love*, to Morton Gould and His Orchestra's *Jungle Drums* and the Ei'e Band's *Polynesian Playmate* (see Figure 7.1) or Manuel and His Music of the Mountains' *Shangri-La*. Latin dance styles were particularly popular (as with Edmundo Ros, Sergio Mendes, Mantovani, Herb Alpert's Tijuana Brass). Such exotica formed a genre where 'the elaborate fiction of the tropical paradise functioned as an exoticised complement to American suburbia: a colourful, dangerous, mysterious, heterogeneous Other which contrasted with the safe, predictable, homogeneous and sexually repressive environment at home' (Leydon 1999: 48; see also Chapter 9). Gender representations aided constructions of exotica. 'Semi-naked natives' on album covers were reminders of the distances music had travelled to reach lounge rooms of the West, particularly in 'ethnomusicological' recordings that sought authenticity. In these appropriations, beyond the obvious sexualisation of 'the other', little credit was given to indigenous origins and performers.

Flirtations with Eastern mysticism in the 1960s brought new influences; the success of the Beatles, and George Harrison's fascination with the Indian sitar, increased exposure to Indian music and to Ravi Shankar, probably the first distinct 'world musician' (Box 7.1), unquestionably promoting musical sounds and structures quite different from those in the West. Prior to the successes of Miriam Makeba, Ravi Shankar and Manu Dibango, the first African musician to have an international hit, and whose music helped usher in the disco era (Garofalo 1993b; Mitchell 1996), musicians with exceptional local and regional popularity were otherwise largely unknown in the West, because their music was unfamiliar and inaccessible, and the words incomprehensible (hence Western recording companies took little interest). Writing of one of the more legendary regional performers, Umm Kulthúm (Box 7.2), the musicologist Virginia Danielson observed 'I had to be taught why Umm Kulthúm's singing was good singing' (1997: 3) since her previous musical and cultural experiences and conventions provided little guide. All too often music from developing countries seemed no more than strange sounds and obscure structures that, without modification, were too challenging for Western audiences.

The Beatles' quest for mysticism, enlightenment and innovative sounds (which could be incorporated in Western musical structures, rather than being given a life of their own) was the forerunner of other Western performers' similar searches for authenticity and difference. Paul Simon's *Graceland* (1986) recorded English lyrics over tracks performed by black South African bands and the vocal group Ladysmith Black Mambazo. A host of other Western musicians

Figure 7.1 Morton Gould and His Orchestra, *Jungle Drums* (*c*. 1962) and Ei'e Band, *Polynesian Playmate* (1962).

Source: © RCA Records; © Viking Sevenseas NZ Ltd.

Note: Early easy-listening albums combined exotica, ethnicity and lurid gender representations to create images of sexualised 'others'.

Box 7.1 Ravi Shankar

Ravi Shankar was one of the best known popularisers of Indian music, and the first 'world musician' to achieve international success, though it took elements of patronage from Yehudi Menuhin and the Beatles to enable that success. Shankar, who partly grew up in Europe, initially studied Indian classical music, and later observed of the late 1940s: 'At that time there was no-one else [in India] who could communicate, could speak English fluently and articulate the nature of our music to a foreign audience.' He toured the United States in 1957, enabling himself and Indian music to become better known. Shankar noted that great European musicians, such as Heifetz and Segovia, did not appreciate Indian music. To them 'it was just a monotonous voice'; consequently 'when we came to the West I cut back the opening improvisational passages to get to the virtuosity and rhythmic excitement that people could more easily appreciate' (*Daily Telegraph*, 17 April 1999). Though he recorded *West Meets East* with the classical violinist, Menuhin, he became famous in the West after the Beatles used the sitar in their music. Shankar gained status in being credited with the introduction of the sitar to a Western audience, and became associated with the flower-power era, an association with which he was not entirely happy. The echoing sound of the multi-stringed sitar enjoyed vogue status in popular music for some years. Shankar taught Peter Sellers how to hold the sitar for the film, *The Party*, appeared at the 1967 Monterey Pop Festival and the 1969 Woodstock Festival – in partnership with tabla players – and was signed to the Beatles' Apple recording company. He was heavily involved in the 1971 Concert for Bangladesh, but within a few years the 'raga-rock genre' faded from prominence. Critics accused Shankar of commercialising and Americanising Indian music, and even of becoming a hippie.

See: Shankar 1969.

Box 7.2 Umm Kulthúm: the voice of Egypt

Umm Kulthúm, whose performing career stretched over fifty years, 'was unquestionably the most famous singer in the twentieth century Arab world' (Danielson 1997: 1). She recorded about 300 songs, had a monthly radio concert broadcast throughout the Middle East, was on numerous Egyptian national cultural committees but, although 'the cultural symbol of a nation' (Danielson 1997: 1), was virtually unknown in the West. Umm

Kulthúm was born in 1904 in the Nile delta of Egypt and, despite being a female peasant villager, had become an Egyptian star in Cairo theatres by the 1930s, and was already touring other parts of the Middle East, at a time when both radio stations and talking films (in which she performed) were becoming popular. Her music was acclaimed as *asil* (authentic) in her ability to master the complex repetitive forms and cadences of Arab music, alongside her song compositions that brought together traditional Egyptian instruments, notably the mizmar, with such themes as Sufism and recitations of the Koran into a wide repertoire. Her widespread popularity 'testified to the vast appeal of constructions that could be viewed as indigenous, the deeply felt pride in Arab history and culture and the support for Arab or Islamic alternatives to Westernization' (Danielson 1997: 198). Though she performed widely in the Arab world, from Tunisia to Pakistan, she performed just once outside it, in Paris in 1970, where her three songs lasted over three hours. After her death in 1975 her recordings acquired a transcontinental audience much larger than any in her lifetime; she had her own web page, and remained popular in Egypt, as the music enabled recollection of the value, strength and potential of local culture and society, whether as a small group of listeners in a coffee shop, the urban neighbourhood or village, the community of Egyptians, Arabic speakers or even the wider Muslim world (*umma*). Through her musical performance, she reinforced cultural identities and social life throughout a massive area, but that culture and its musical expression was inaccessible to a global audience.

See: Danielson 1997.

and 'cultural gatekeepers', such as Peter Gabriel and Ry Cooder, similarly introduced new musicians, sounds and structures from a diversity of regions and styles. Collaborations became increasingly unexpected, including that of Pearl Jam with Nusrat Fateh Ali Khan and of the metal band Sepultura with the Xavante Indians of Brazil (Harris 2000). *Graceland* was highly successful commercially, though the whole project was criticised for cultural and economic imperialism (Box 7.3). Its success however opened the way for many, like Ladysmith Black Mambazo, to achieve success in their own right in the West, alongside other non-Western individuals and bands, such as some from Papua New Guinea (Hayward 1998), who would otherwise have remained of merely local significance. The music of those performers who became successful in the West, with distinct new styles, eventually became known as 'world music' (or 'world beat').

Box 7.3 Paul Simon, *Graceland* and cultural imperialism

The production of the album *Graceland* in 1986, despite its introduction of many largely unknown African performers to Northern audiences, was criticised as manifestation and embodiment of the dominant power relations in the music industry, which mirrored global power relations. *Graceland* won international awards, was celebrated as a blending of global pop and African 'folk' music and as an important event in consciousness raising over the evils of apartheid. Yet Simon's supervision of production, copyright for the finished product and superimposition of 'lyrics about cosmopolitan postmodern angst' (Lipsitz 1994: 57) over songs about the lives and struggles of marginalised African communities indicated his superior power in the scheme, as a wealthy American with access to capital, technology and marketing resources. He had defied UNESCO boycotts to make the record and implied that black–white co-operation in South Africa was normal rather than exceptional. Simon was even more strongly criticised for the appropriation of the music of a Chicano band from Los Angeles and a Louisiana Cajun band for the final two tracks of the album (Feld 1994). Simon vigorously rebutted the various charges, claiming that art (music) and politics were separate entities, but his apparent good intentions were offset by negative consequences in South Africa, such as South African whites portraying his role as one of civilising 'rough' black music and the apartheid government celebrating the success of *Graceland* as a measure of South Africa's acceptance in the wider world. Paul Simon also noted that when he first heard South African township music, 'It sounded like very early rock and roll to me, black, urban, mid-fifties rock and roll' (quoted in Feld 1994: 241). Black musicians involved in *Graceland* rejected the notion of cultural imperialism as a denial of their own agency, and supported its validation of urban African musical styles. Similar criticisms were levelled at other examples of appropriation, such as of African pygmy music by Deep Forest and others (Feld 1996a, 2000a), without the complex political situation of South Africa.

See: Feld 1994; Garofalo 1993b; Goodwin and Gore 1993; Lazarus 1993; Lipsitz 1994; Meintjes 1990; Mitchell 1996; van der Lee 1998.

Numerous other Western performers turned to other musicians to play (usually) supporting roles, and imbue their own music with new sounds and some supposed elements of authenticity, in a growing complexity of music making and collaboration (e.g. Connell 1999). One of the most successful of such collaborative ventures was that of Neneh Cherry (of Swedish-Ghanaian origins, singing in

English) who combined with the Senegalese musician, Youssou N'Dour, on 'Seven Seconds' (1994), a song which sold several million copies. Similar cultural fusions and creations had long been celebrated, especially on record sleeves. Thus on Rochereau's *Tabu Ley* (1984) liner notes, the emergence of Congolese popular music was attributed to a diversity of influences.

> Congolese popular music began taking shape during the 1950s when imported Caribbean (especially Cuban) records ignited a passion for rhumba rhythms and lyrical melody lines. At the same time, acoustic guitars became popular in much of Africa. The resulting interaction of Afro-Cuban music, acoustic guitars and traditional African rhythms gave birth to a new style of music, dubbed 'Congo' – and that region dominated other African styles (electric guitars, choruses and horns emerged in the 1960s and in the 1970s). Elements of salsa, American rhythm and blues, reggae and disco drumming were injected into the Congo style.

Performers from the non-Western world were repeatedly heralded as being both exotic and accessible. In Zimbabwe Thomas Mapfumo was argued to be the most important musician 'in transforming and updating Zimbabwe's traditional music...a crystallization of styles' that transposed the sound of the mbira (thumb piano) to guitar lines, and that of the hosho (shaker) to the cymbals. Early in his career Mapfumo's repertoire included songs from the Beatles, Elvis Presley and Sam Cooke, South African township music and rumba from Zaire, but he was later the first major national artist to sing in Shona (rather than English), in live shows that were proclaimed 'a triumphant blend of the traditional with the modern. The group's irresistible sounds distil the deepest soul of Zimbabwe's music' (Herman 1990). While simultaneously representing national and regional traditions, non-Western performers had invariably absorbed a range of stylistic influences from the West. There was no necessary close relationship between identity, ethnicity, place and music, even though particular genres were usually strongly tied to regions (and time periods), but a particular spatial origin was seen to imbue a degree of authenticity.

Some degree of hybridity was central to world music, not least zouk, which emerged in the French West Indies (Guadeloupe and Martinique) and was exported, in the 1980s, where performers absorbed and combined diverse influences:

> These range from local drumming song-dance music, quadrille, biguine and mazurka, to Haiti's compas direct, Dominica's cadence-lypso, the Dominican Republic's merengue, French-Africa's soukous, Latin

> America's salsa, Jamaica's reggae and dub, the United States' soul, funk music and rap – and other genres.
>
> (Guilbault 1997: 33)

This hybridity of zouk (see Figure 7.2; cf. Warne 1997: 138) can be repeated indefinitely for other genres of world music (though rarely with such diversity). Eclecticism and cosmopolitanism have also been emphasised in advertisements, as in Real World's description of Myriam Marsal's *The Journey* (1998): 'Myriam Marsal mixes the intoxicating Arabic rhythms and instruments of her homeland [Egypt] with the dynamic, hip-swivelling pulse of Afro-Pop. Shades of electronica give this album a unique glow, courtesy of producer Simon Emerson of the Afro Celt Sound System', while Sheila Chandra integrated Indian and Celtic traditions, which had commonalities with Islamic and Andalucian vocals, and the music of Bulgaria (Taylor 1997: 149). Diverse ethnic and regional origins, as long as they were distant from the West, seemingly conveyed extra authenticity.

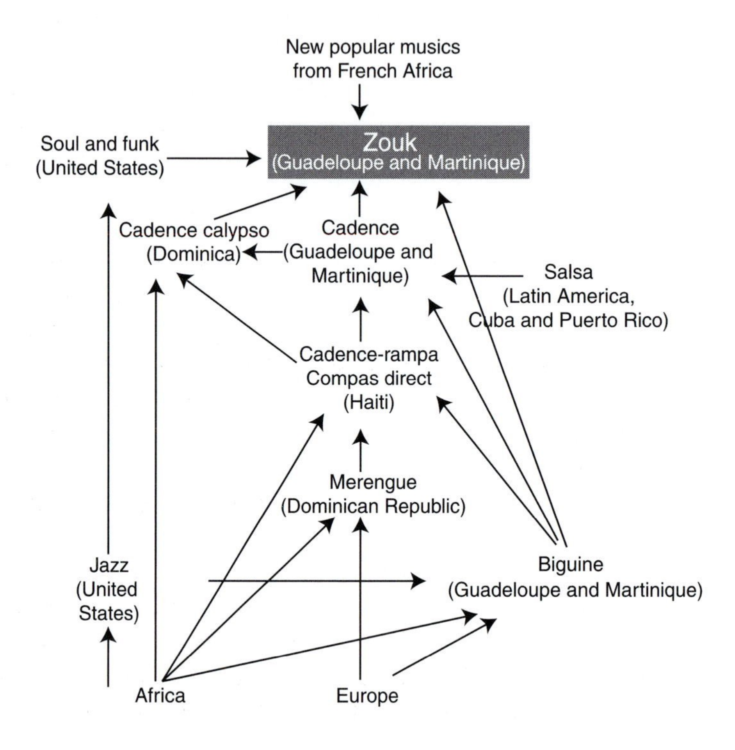

Figure 7.2 Musical influences on zouk
Source: Guilbault 1994: 167.
Note: Zouk's influences incorporated the music of at least four continents alongside various Caribbean islands.

Commercialising 'world music': exotica and essentialism

Early successes, alongside the rise of reggae (see Chapter 8), resulted in greater interest by metropolitan music companies in performers from developing countries. Hitherto such performers usually had particular audiences, with markets restricted by languages, limited production of records and cassettes, and, above all, little exposure beyond their home areas. World music is not a musical genre but constitutes, at best, a marketing category for a collection of diverse genres from much of the developing world. Its definition depends on the social, political and demographic position of certain minority groups in a particular country. Reggae is rarely characterised as world music (and reggae never italicised as alien). In the United States, where there is a substantial Hispanic population, salsa is not usually considered world music, whereas in Britain it is (Guilbault 1993b: 45). Australian Aboriginal music may be both viewed as world music, if seen as 'traditional', and excluded if 'modern', categories that are inevitably comprehended in quite different ways in different contexts. In music stores in Italy, country and western music is usually displayed under 'world music', hence even the simplistic notion that world music is 'simply the music sold in the world music section of record stores throughout the Western (and partly non-Western) capitalist world' (Roberts 1992: 231) provides nothing like a stable category (Barrett 1996: 238–9; van der Lee 1998). More accurately, but in a limited sense, it has been:

> a marketing term describing the products of musical cross-fertilisation between the north – the US and Western Europe – and south – primarily Africa and the Caribbean basin, which began appearing on the popular music landscape in the early 1980s [through] the emergence of new, interlocking commercial infrastructures established specifically to cultivate and nurture the appetites of First World listeners for exotic new sounds from the Third World.
>
> (Pacini-Hernandez 1993: 48–50; Feld 2000a)

Erlmann thus noted that 'the term displays a peculiar, self-congratulatory pathos: a mesmerising formula for a new business venture, a kind of shorthand figure for a new – albeit fragmented – global economic reality with alluring commercial prospects' (1996: 474). World music flourished from the 1980s, because of distinctive features of that decade's social, political and economic situation, including international migration,

> the breakup of the communist block; the resurgence of many ethnic groups; the realignment of various communities and the formation of new alliances; increasing problems of multiculturality and polyethnicity;

> the consolidation of the global media system; and the reconfiguration of the world economic order with a more fluid international system – all marking the end of bipolarity.
>
> (Guilbault 1993b: 36)

In short, an increasingly post-modern world of much greater mobility, transience, urbanisation and rapid technological change was marked by musical diversity and eclecticism. The rise of world music marked capitalism's relentless need for new sources of inspiration and innovation, and new areas for production and consumption.

Despite eclecticism and hybridity, the success of such performers as Youssou N'Dour or the Benin singer, Angélique Kidjo, required that they remain, to some extent, 'musically and otherwise premodern...culturally "natural" because of racism and western demands for authenticity', what Taylor therefore described as 'strategic inauthenticity' (1997: 126). Yet the music of many famous 'world' musicians had 'passed into the commodity stage and secured for itself a firm position on full-grown national markets long before the new global musical culture was even dimly perceived...it was already thoroughly modernised before it came to the global bazaar' (Erlmann 1996: 475). Moreover N'Dour performed songs reflecting the pressures of modernity – the impact of tourism and environmental degradation, migration and nostalgia for the ancestors and their wisdom – using a distinctive new drum-based musical style (mbalax). Kidjo, too, has been 'a kind of postcolonial critic', drawing attention to social change and development problems (Taylor 1997: 136). Both absorbed Western influences, from Jimi Hendrix to John Coltrane, and other syncretic African musics, sung in several languages (Kidjo in at least four), lived in Europe for extended periods and hence had to travel 'home' intermittently to (re)capture local influences and sounds; both faced criticism for being too commercial and popular. Much like Christine Anu or Salif Keita (Lipsitz 1994: 44) both claimed to be modern, and viewed Western demands for authenticity and purity as concomitant with demands that they, their countries and cultures, remain pre-modern – the other – while the rest of the world moved towards a 'postindustrial, late capitalist, postmodern culture' (Taylor 1997: 143). Rather, performers like N'Dour saw themselves as 'global citizens', with Kidjo embarking on a project in the late 1990s to produce a trilogy of albums that 'build a bridge between the Black diaspora', from Benin to Brazil and beyond, combining a diversity of musical influences (Thomas 1999: 15). At the very least musicians were making conscious choices over content and style yet, in the musical world, 'reifying culture and simplifying the identities of people has been standard practice...for a long time' (Keil 1994: 177). Strategic inauthenticity represented the other commercial face of cultural imperialism. For these musicians, popular music had become an avenue for diverse musical diffusions and post-colonial expressions. Yet record companies' perceptions of Western market demands and contemporary themes often remained subordinate to the marketing of 'exotic others'.

World music became increasingly diverse, sometimes pitched as 'yuppie-directed exotica', combining elements of 'quality' art rock, dance craze, mystical mind expansion, scholarly folklore studies (Goodwin and Gore 1990: 67) and a form of 'aural tourism' (Cosgrove 1988), taking fragments of other cultures for the benefit of jaded Western tastes. Creative Vibes were advertising in 1999: 'Travel the World with Creative Vibes', while a year later Putumayo, a major world music label, urged of a new CD:

> *Gardens of Eden* is an exquisite selection of acoustic music from some of the Earth's most beautiful places. Since time immemorial, people have been on a quest for remote idyllic hideaways where life approaches perfection. The music of *Gardens of Eden* has an organic, ambient quality that conjures up images of a magical, tropical paradise where humanity's day to day stress disappears...a musical journey to the world's Shangri-La's. Be transported!

Consumers of rai music in France were identified as 'young, upwardly mobile, urban consumers of travel and leisure products on their way to becoming decision-makers in French society' (Schade-Poulsen 1997: 67; cf. Ellison 1998, Turino 2000). Tuvan (Mongolian), Tibetan or Cape Verdean music was often more a mark of the cosmopolitan pretensions of the purchaser, than a reflection of musical tastes. Even more pointedly, in some respects:

> a recourse to the category of 'world music' often occurs in the context of a search by well-meaning cultural tourists for pre-capitalist authenticity, for signifiers of rootedness...a projection screen for the anxieties and neuroses of ethnic majority Westerners dealing with the guilt of a colonial past.
>
> (Warne 1997: 135)

Underlying the diversity of world music was the 'paradox inherent in the transnational recording industry' where Third World performers could gain more effective access to global markets when they conformed to the 'use of preponderant Euro-American scales and tunings, harmony, electronic instruments now seen as standards, accessible dance rhythms and a Euro-American based intonation' (Guilbault 1993a: 150), inserting music such as the Pakistani qawwali 'into a trendy, cosmopolitan world music culture' while dragging the music away from its textual base (Qureshi 1999: 94). Thus Real World (whose name implies authenticity and stability) recordings of qawwali music were argued to have 'virtually ignored the crucial religious and socio-critical elements of the music, and...attempted to reduce the music to an aesthetic form [simultaneously producing the] elevation and erasure of Nusrat Fateh Ali Khan' (Sharma 1996: 24–5). The performers themselves played their part; Xavier Cugat commented 'To succeed in America I gave the Americans a Latin music that had nothing authentic about it....Then I began to change the music and play more legitimately' (quoted in van der Lee 1998: 50). Strategic inauthenticity, alongside the convergence – often accidental – of different musical genres, brought an emerging sense of 'commodified otherness, blurred boundaries between exotic and familiar, the local and the global in transnational popular culture' (Feld 1991: 134). Indeed ethnomusicological recordings of local performers in developing countries, such as Feld's recordings in the Southern Highlands of Papua New Guinea, have not usually been commercially successful. Where world music has been particularly successful, whether rai in France or Hispanic music in the United States, it is often partly linked to migration and migrant markets (Chapter 8). That too confers some degree of authenticity.

A premium is attached to the exotic. In the United States, 'the sounds of salsa and merengue may simply not be as appealing to world beat consumers as the music of exotic – and less numerous, or more safely distant – others' (Pacini-Hernandez 1993: 57). In Brazil musica sertaneja, a Brazilian parallel to American country music, was ignored through preference for Brazil's more 'exotic' and 'exportable' musical genres, associated with the Afro-Brazilian heritage (Reily 1992: 337). Conceptualisations of the exotic were everywhere similar. Notions of discovery, authenticity, heritage, uniqueness and a sense of place are rarely far from the covers of world music albums, as on *Paranda*:

> Three generations of legendary Garifuna composers and musicians from Belize, Honduras and Guatemala together for the first time....Imagine African drumming, American blues, Cuban son and West African guitar all wrapped into one....[Previously] we knew very little of the Paranderos and their music...[but] Nabor's incredible passion and raw emotion made one appreciate Garifuna music like nothing I had heard before....Unfortunately Paranda is a dying art, just a handful of composers are still alive and very few young musicians still practise the style.
>
> (Duran 1998)

In a more commercial vein, Ry Cooder's 'unearthing' of Cuban music (and subsequent revivals) capitalised on the apparent 'purity' of Latin sounds hitherto hidden within a socialist state, and 'kept alive' by a distinguished older generation of forgotten musicians.

Strategic inauthenticity was not only applicable to musical genres, styles and lyrics, with indigenous languages seen as 'authenticating' their overall performances (Evans 1997: 39), but it was firmly attached to place. In world music, more than most forms of popular music (other than folk and country music) ethnicity and locality became 'a fetish which disguises the globally dispersed forces that actually drive the production process' (Appadurai 1990: 16) but, much more than that, they became the cradle of authenticity – despite the lack of coherence between culture and space. More generally world music exemplifies a 'fetishisation of marginality' and an essentialist identification of cultural practices in developing countries with otherness itself (Erlmann 1996; Mitchell 1996). This fetishisation is part of a broader trend to seek out cultures that are relatively untouched by processes of commodification, most evident in some forms of tourism, which exaggerate, reify and romanticise the extent to which any culture, and place, is isolated from others. Exceptionally this is part of a tendency to eulogise Africa, and African 'roots', above and beyond that of other non-white people, such as Asians, let alone minority white groups; racism has often resulted in blacks being seen as more 'authentic' in terms of musical (and sexual and sporting) expressions of the body, whereas Europeans have often been associated more with the mind and less spontaneous types of musical performance (Gilroy 1993b). In more extreme form, it resulted in the discovery of the 'healing' sounds of particular kinds of (usually drum or ambient) music, the 'primitivist fantasy' that resulted in Western musicians adopting the music of central African pygmy peoples (Feld 1996a; Lysloff 1997; Gibson and Connell 2002). Thus, Anthony Copping's pan-Pacific compilation album *Siva Pacifica* 'is an emotive voyage signposted by the passion and majesty of the unique musical ambience of the South Pacific region. A place where nature itself abounds with rhythmic glory' (1997). Such eulogies were commonplace.

Yet, paradoxically, for all the desire for authenticity, world music was decontextualised most obviously on compilation albums and at world music festivals, such as Womad (World of Music and Dance), where an array of exotic sounds from a diversity of locations was brought together. Here world music became 'a kind of commercial aural travel consumption, where the festival...assembled from "remote" corners of the world, could be a reconstructed version of the Great Exhibitions of the nineteenth century' (Hutnyk 2000: 21). While origins conferred local authenticity, tensions remained in the placelessness of certain forms of consumption: world music constituted 'the ubiquitous nowhere of the international financial markets and the Internet' (Erlmann 1996: 475), while, rather differently, ' we no longer have roots, we have aerials' (Wark 1994; Clough 1997). In most extreme form it has been claimed both that world music constitutes 'the soundscape of a universe which underneath the rhetoric of roots has forgotten its own genesis' (Erlmann 1996: 477) or, in the context of a review of a recording of sacred flute music from Papua New Guinea, it has 'a real capacity for getting inside another culture, then extracting the creative essence of that culture to whet the insatiable curiosity of people who listen to their own inner muse' (*Diaspora*, 3, 2000: 33). Despite inevitable changes and compromise, alongside commercialism, world music's popularity reflected widespread interest in new, engaging sounds and rhythms. Simultaneously the operation of more political agendas, in opposition to perceived economic and cultural imperialism, gave world music particular credence for anti-racist movements and in support of multiracial societies. Success empowered local musicians and peoples. In the French West Indies zouk's international success helped Antilleans 'lose their inferiority complex and to feel comfortable competing with others on the market' (Guilbault 1993a: 41). In Jamaica a museum was dedicated to Bob Marley (see Chapter 10), while Cesaria Evora became a national hero in Cape Verde. At one level, the rise of world music emphasised the importance of 'the local' in global commodity flows; at another it drew little-known places and performers into global markets and culture.

From local to global?

The rise of world music, or perhaps more accurately 'third world music' (Feld 2000a) through its connotations of exotic difference, remoteness, poverty and simplicity, emphasised ethnicity, initially as curiosity and eventually as commodity. Marketing championed difference: local and regional sounds, obscure performers and nations, strange instruments, creativity and energy, and unusual rhythmic and vocal structures. Ironically, performers such as Thomas Mapfumo were 'actually marketed to worldbeat audiences on the strength of [their] localist and nationalist imagery' (Turino 2000: 12). As world music entered global market-places, and was constructed and defined by those market-places, it became evident that it was not so much 'a new aesthetic form of the global imagination' (Erlmann 1996: 468), as a product of fusion that therefore necessitated strategic inauthenticity, selectively appropriated, even romanticised, from more 'traditional' world regions. Authenticity was thus 'implicitly a spatial model, positing a [usually] neo-African, sacred, rural peasant tradition at the center…and secular, foreign, urban elite space at the margins' (Averill 1997a: 44). Such authenticity required the West. Yet, ultimately, world music labelled its places of origin as exotic and 'third world' – with its aura of inferiority – while attributing virtues to their musical traditions (though many, in those states, were seeking 'modernity', including karaoke and country and western). As migration, especially from poor countries to rich, increasingly became a critical element of globalisation, so music moved away from 'bounded, fixed or essentialised identities' (Feld 2000a: 152), a product of more rapid fusion and change in different socio-economic, political and geographical contexts. Artists resented being categorised as 'world music', separate from a wider body of popular music; thus Nitin Sawhney argued that:

> world music is a form of apartheid...you go into a record store and they have 4 CDs from Egypt, 4 from Africa or wherever...it's like they read the Indian name on the cover and didn't understand it, and therefore shoved it on a shelf somewhere where it's marginalised.
>
> (quoted on Triple J, Sydney, 13 August 2001; see Chapter 8)

The fetishisation of place, persistent in the marketing of world music, could no longer be intellectually sustained.

8

A WORLD OF FLOWS

Music, mobility and transnational soundscapes

While technological changes, marketing strategies and changes in taste and style have been crucial in the geographical distribution of music, other forms of music diffusion are based largely on the movements of people rather than products or capital. Everywhere music is played and consumed it contains multiple networks. Cities are nodes in international mediascapes – centres of production and retailing – and hosts to multicultural communities and their diverse musical texts and spaces. This chapter examines the complex relationship between migration and music, and the manner in which music has been transformed in a variety of destinations. Cities such as Paris maintain connections to such Francophone centres as Montreal, Algiers and Abidjan, while providing the stage for African musical transformations; New York is a central node in Latino traditions (with connections to Miami, Havana and Puerto Rico) and a major part of pan-Caribbean links (with Kingston and London), while Indian diasporic networks link cities as diverse as Mumbai, Birmingham, Chicago and Kuala Lumpur. Cities also host diverse domestic recording industries, while several styles of music (such as hip hop, R&B and techno-pop) have become transnational urban soundtracks, as likely to be heard in Seoul as Sao Paulo.

The geographies of music evident through migration are more limited and selective: rai and mbalax are little known outside major French cities and Haitian konpa was concentrated primarily in New York (Averill 1994); hybrid musical forms such as bhangra, produced by Asian migrants and their descendants, were concentrated in the larger English and North American cities. Music followed both colonial and post-colonial migration routes. While the most familiar and successful 'non-Western' music in developed world cities has been that from Latin America (in the United States), the Caribbean, Africa and to a much lesser extent South Asia, that from East and South-East Asia has been largely absent. Diffusion required both migrants and infrastructures for production and performance including venues, specialist retail outlets, ethnic newspapers and radio programmes (though music might also be diffused commercially, independent of social networks).

In the post-war era, and particularly since the 1970s, contemporary migrants from developing countries have transferred their musical traditions to new urban destinations (much as eighteenth-century migrants created new rural 'folk regions' in the United States), a process involving both conservation and dissolution:

> Music is central to the diasporic experience, linking homeland and hereland with an intricate network of sound. Whether through the burnished memory of childhood songs, the packaged passion of recordings, or the steady traffic of live bands, people identify themselves strongly, even principally, through their music.
>
> (Slobin 1994: 243)

Yet other ties and memories (such as kinship, sport, landscape and food) may be at least as important. Migrants, refugees and their children all experience, to varying degrees, senses of displacement and dislocation, mediating memories of the people and places of home with the realities of their new surroundings. Music is one element of this experience. It provides a mechanism by which the 'cultural baggage' of 'home' can be transported through time and space, and transplanted into a new environment, assisting in the maintenance of culture and identity.

Leaving home

Outside the West, with its tradition of mobility, migration was more likely to be permanent and a response to uneven development. The music of migrants typically involved such themes as solitude, homesickness, nostalgia, unemployment, racism (and other hardships) alongside the longing for their converse – kin, community, familiar foods and sights, and a place in society. They were the harsh, more pointed versions of country music displacement. Portuguese *fado* (fate), long associated with sailors and the sea, offered a sorrow that was almost hope; it had begun with Portuguese fishermen leaving for distant fishing grounds uncertain over how their homes and loved ones would change in their absence. Canto-pop songs from Hong Kong, in the years before reintegration with China, and Okinawan migrant songs traced remarkably similar uncertainties, fears and frustrations (Lee 1992; Roberson 2001: 226–9). Migration was not necessarily traumatic; Barbadians expected wealth from what was a temporary expedient rather than an absolute loss, hence those who came back empty-handed were derided (Chamberlain 1995: 269). Different destinations provided different outcomes and expectations.

Irish popular music, whether within or outside Ireland, is replete with references to loss and longing, exile and emigration. Every album of the Pogues, and of their singer-songwriter, Shane MacGowan, refers to exile; London is prominent in songs such as 'Transmetropolitan' and 'The Dark Streets of London'. 'Thousands are Sailing' reflects on Irish migration to the United States: 'Where e'er we go we celebrate/ The land that makes us refugees/ From fear of priests with empty plates/ From guilt and weeping effigies/ And we dance', and provides a telling expression of the ambiguity of the Irish culture of migration (Boyle *et al.* 1998: 226). That ambiguity is present in every migration experience, in the contrasting images and lifestyles of two places: homes and destinations. For more than a century labour migrants, moving sometimes unwillingly from villages to organised and regulated workplaces, reflected on this duality (Box 8.1). In recent times Vietnamese migrant music in the United States has reflected the cultural dreams of the community, including a desire for elegance (especially dress), aspirations to upward social mobility, emulation of what is perceived as the romantic style and individualism of American culture but alongside idealised memories of Vietnam (Lockard 1998: 27). Rai music in France expresses a similar yet fluctuating duality. Migrant music epitomises dream and nostalgia combined.

Box 8.1 Sound tracks

Migration from Lesotho to South African mines and from Melanesian islands to plantations in Fiji and Australia was well established by the end of the nineteenth century. The songs that migrants composed sought to resolve the contradictions between village and mine or plantation as the domains of experience, and the challenges and problems of divided individual and family lives. Songs usually involved commentaries on religion, politics, work and the strangeness of new landscapes. In South Africa especially the narrative structure was relatively straightforward – a reason for migration, the journey from home to recruiting office, an impending sense of doom and the loss of personal autonomy and control: specifically the contrast between a humane rural existence and the impersonal recruiting process. The dangerous and harsh conditions of the mines and the threat of the crowded colonial city are juxtaposed with an idyllic and highly fictionalised and romanticised rural existence. Images of home were both fluid and contradictory: 'romantic longing for the freedom of home contrasts with fear of the disintegration of family and the unfaithfulness of spouses. Soaring images of the stark beauty of the physical landscape contrast with harrowing depictions of rural poverty and privation' (Crush 1995: 241). The cover notes to the Venanda Lovely Boys' *Bo-Tata* (1989) observed that male chorus songs (*isicathamiya*) typically 'lament broken up family and love relationships, the decline of parental and chiefly authority and the collapse of rural-traditional value systems'. By contrast women's songs reflected autonomy and independence, and their experiences of migration as a means of escape from poverty, abusive relationships and demanding relatives, alongside critiques of male attitudes and actions – they spoke loudly and eloquently of the 'catharsis of affliction' and displacement (Coplan 1995). Migrants in South African mines and townships, Melanesians on coastal plantations on distant islands, created and maintained alternative visions of existence and 'imaginary geographies lodged within the everyday experience and subterranean spaces of the migrant world' (Crush 1995: 243). Performance of the songs allowed migrants and home communities to experience and retain the intense emotion, and sacrifice, of critical historical and contemporary events.

See: Crush 1995; Coplan 1995; Erlmann 1998; Thomas 1992; Ballantine 2000.

Lyrics do not always dwell nostalgically on the past or reflect successful transitions. Many are fraught with the uncertainty and challenges of new, distant lives: the songs of Brazilians in New York often stressed the fear of being caught by immigration officials and deported (Margolis 1994: 176) while Mexican-Americans deplored oppressive economic and cultural structures (Lewis 1992). The Chicano singer Robert Lopez (El Vez) transformed the lyrics, but not the tune, of Elvis Presley's 'Suspicious Minds': 'I'm caught in a trap, I can't walk out/ Because my foot's caught in this border fence/ Why can't you see, Statue of Liberty/ I am your homeless tired and weary', and the song was particularly well received amongst Turkish migrants in Germany (Habell-Pallan 1999). At 'home' villagers feared for their loved ones' well-being and lamented those who failed to return. Yet return was rarely simple: Block 47's 'American Wake' (1994) – 'Always remember at the end of the day/ You can always go home, you just can't stay' – captures the ambivalence, sense of displacement and loss. In urban South Africa and Papua New Guinea the songs of male migrants frequently disparaged female migrants as prostitutes, where women are alternately objects of derision and conquest; their supposed 'easy virtue' was a measure of migration being allowed to men, despite their loss of power in urban society, and denied to women whose 'true place' was in the village (Webb 1993; Crush 1995). Public performances in bars, nightclubs and churches intensified both displacement and the difficulty of assimilation.

Migrant songs anchored and validated the present, providing a means for migrant communities to sustain critical elements of identity while simultaneously changing them (Shelemay 1998). Identity was fixed in distant times and places, which necessitated new generations transforming lyrics and musical styles to provide more idyllic notions of the past that was left behind. Romantic nationalism thrived in migrant songs, such as those of Haiti (Averill 1994) or Ireland, invoking populist visions of idyllic lifestyles and harmonious communities, with only tenuous relationships to the contexts that created diasporas. In such visionary worlds homecomings were impossible.

The way in which music is used by migrant groups and received by the dominant culture varies considerably. In rare cases, the musical form remained largely unchanged, fixed in the style of that community at the time of the migrants' decision to leave (hence Mikis Theodorakis sought 'authentic' Greek music in Melbourne, just as the Chieftains brought back 'traditional' Irish music from New York). Yet, simultaneously, other migrants (especially the young) absorbed local forms and established new hybrid identities. Moreover, migration cannot be conceived as simply the definitive movement of individuals from one place to another, but is a constant and never complete process of mobility, linking individuals and their communities, however defined, into transnational circuits and diasporic networks at various scales. Like migration itself, the processes acting on migrant musics – pushing, pulling, shaping and protecting it – are intense, varied and complex. As a result 'no one has formulated a worldwide viewpoint on the music in diaspora' (Slobin 1994: 243) and no straightforward view is possible.

Some simplistic models of the relationship between music and migration have been proposed; Hirshberg identified three outcomes for music that result from the movement of migrants: compartmentalisation, synthesis (transculturation) or unidirectionality, but failed to specify how and why such variations might occur (Hirshberg 1992; Hirshberg and Seares 1993). Criticisms of such models emphasised the difficulty of generalising complex social processes that were often unique, subjective and reversible. The specifics of location – economic, social, political and cultural – were almost wholly absent, yet were critical to both links with home and broader processes of cultural stability and change. In California, the diversity of urban neighbourhoods, and the differential impact of media infrastructure (recording industries, radio, film and TV) within them, resulted in a parallel diversity of black musical genres (DjeDje and Meadows 1998). Where music has changed little in the destination, stability was linked to limited duration, negative response to migrants and their own perceived lack of success. The context of Mexican banda music in Los Angeles, where it became a major symbol of Mexican identity, encouraged stability as 'an expression of immigrant consciousness, and a strategic and symbolic source of unity in the face of outside attacks' (Lipsitz 1999a: 230). The mainly instrumental music, dominated by horns, associated with an acrobatic and difficult *quebradita* dance, came from the western coastal states of Mexico. Performers flaunted their rural Mexican roots and carried leather straps engraved with the name of their home state, so affirming 'an intense affiliation with regional and national identities' (Lipsitz 1999b: 202) and symbolising rejection of assimilation. Banda musicians consciously addressed Mexican (Spanish-speaking) audiences and did not seek to penetrate wider markets, enabling them to 'imagine and create their world', enhanced by a mainstream 'cultural backlash' against Hispanic migration (Simonett 2001: 319–20). As migration changed the structure of power relations between men and women in the new workplaces, banda did change, as 'a resurgent hyper-masculinity' came to the fore, further emphasising the 'self-pitying romanticism and sexism' of banda lyrics (Lipsitz 1999b: 206; Haro and Loza 1994). Concerts often included techno, rock and other Latin American styles while banda music came to incorporate both North American melodies and English lyrics.

Some degree of change, most often in lyrics, instruments and the social context of performance is inevitable. Equally, music in the homeland of migrants is not static but, as in the diaspora, is a product of synthesis and hybridity. Genres such as rai and bhangra have substantially changed; others, such as salsa and hip hop, have been largely created in the cities of diaspora, but have links with more or less distant 'sources'. Some genres have either not been transplanted, or have died out in the diaspora; yet others, such as Japanese taiko drumming (Fromartz 1998) and Yiddish klezmer music (Sapoznik 1997), have been revived, through cultural resurgence (or 'ethnic revivalism') often on the part of long-established migrants or second generations. Japanese-American youth revived group drumming (*kumi-daiko*) when they were challenging 'the stiff assimilationist outlook of their parents' generation and the prevalent stereotype of the "quiet Japanese"' (Fromartz 1998: 46) though the particular form of drumming that developed was quite different from that in Japan. In the Malaysian city of Malacca Eurasian descendants of Portuguese settlers created an 'invented tradition' of music and dance derived from Portuguese folk music, some 400 years after settlement, which asserted their simultaneous Portuguese roots and Malaysian ties (Sarkissian 1995). Indeed dissolution was often matched by ethnic revivalism, not least in the music of migrants such as Christine Anu and Sheila Chandra, encapsulated in the title of the latter's album *Weaving My Ancestors' Voices* (1992). Migration nonetheless clearly emphasised the new global fusion of styles, especially in the diaspora, reflecting both new, often post-colonial geographies (as in the rise of Tex-Mex/Tejano in the increasingly Hispanic Texan borderlands of the United States) but also the creativity and pleasure of music composition and performance. Where migration was limited, musical change was usually greatest. In London, with little Hispanic migration, salsa music evolved as some British musicians were absorbed into bands, economic constraints reduced the number of band members and the use of technology increased while performance styles adapted to new venues and expectations (Roman-Velazquez 1999; cf. Hirshberg and Seares 1993). Beyond the extent of and response to migration, the critical variables influencing conservation and dissolution include the extent of assimilation (in terms of access to employment, social and political institutions), geographical concentration and cultural distinctiveness, all of which are a function of the duration of migration. Yet, as 'ethnic revivalism' demonstrates, there are no simple correlations. Finally the particular form of 'migrant music' is itself important, and responses vary enormously within a dominant culture (as they do to those musical genres that are diffused primarily electronically) in every destination.

Some of the most influential and successful genres of music (such as rai, zouk, reggae and hip hop) have roots in Africa yet were consolidated in the diaspora, both in the capital cities of developing countries such as Dakar, Kingston or Abidjan, and around migrant communities in large Northern cities such as Paris, Los Angeles or Marseilles. Similar parallels exist for salsa, and other Hispanic forms, linking New York and Miami, for example, with Caribbean and Latin American cities. Migration has brought 40 per cent of all persons of Puerto Rican ancestry to New York, and made the United States the fifth-largest Spanish-speaking nation in the world (Lipsitz 1999a), taken vast numbers of Caribbean islanders to London and Paris, and made Auckland the largest Polynesian city in the world. Many are more affluent and educated than those at 'home'; not surprisingly, migrants influence home cultures. Music simultaneously evolves in both places. Even these generalisations are simplistic. Bhangra, for example, has changed in many diasporic contexts, yet in Newcastle, England, where there are few migrants from former colonial territories, it is largely unchanged – adopting none of the 'Caribbean' influences that have been transformative in other urban contexts. It is perceived by many Asian youth there 'primarily as a form of folk music', and thus analogous to the Irish folk music performed around the world that 'serves to stimulate collective memories of traditional Irish culture, thus becoming a crucial link with notions of heritage' (Bennett 1997: 111). There is therefore no one version of bhangra in the diaspora. In North America bhangra took quite different forms in New York, Chicago and Canada (Maira 1999); in Britain, at the same time, the band Cornershop (whose name deliberately played on a stereotype) took their much altered bhangra music to the top of the British charts, with the single 'Brimful of Asha' (1997), a paean to an Indian actress. There was no necessary direction that all performers chose, or were constrained, to follow, in any place or at any time. Performers were active producers of musical culture, not passive respondents to globalisation, creating something 'qualitatively new, with its own dynamics, rather than just a dilution or corruption of something formerly authentic' (Barber and Waterman 1995: 240), through intricate processes of fission, fusion and creativity.

Transformations in diaspora: rai, bhangra and salsa

In a variety of contexts music was a vehicle for the maintenance and expression of cultural (and national) identity (see Chapter 5); for Caribbean migrants in Britain, ska and then reggae informed 'the struggle for a sense of identity, to keep their own voice alive in a tough environment' (Stapleton 1990: 88). In French cities rai became one primary form of cultural practice for Maghrebi (North African) migrants and their children (*beurs*), of whom there are more than a million in France. Like reggae, the Argentinean tango, Greek rebetiko – and, rather later, hip hop – rai emerged from the more depressed urban environments, specifically in Oran (Box 8.2).

Box 8.2 The rise of rai

The basic form of rai music comes from music performed on social occasions – from weddings to parties, even in brothels – that evolved in cabarets in the Algerian city of Oran in the 1970s. Oran's geographical position and role as a port led to rai musicians absorbing an array of musical styles: flamenco from Spain, gnawa (a musical form of Sufi musicians) from Morocco, French cabaret, the music of the Berbers from the Kabyle region, the rhythms of Arab nomads and others (Gross *et al.* 1994: 6). Its best-known contemporary exponent Cheb Khaled has said:

> Music is a mixture, no matter where it comes from…where I come from we were influenced strongly by Moroccan music; you heard the flamenco, and the music of the Jewish people. And then there was the western stuff. Nobody sung about sex so we all went wild when we heard James Brown on the radio singing Sex Machine…we all tried to sing like him. This is how I grew up with everybody mingling. Oran is a port and everybody brings their own traditions and gets together.
>
> (quoted in Myers 1997: 16)

The essence of rai is the development and refinement of catchy key phrases. Songs from the 1970s, in the local Arabic dialect, reflected men's problems with 'free women' and crime and alcohol abuse. Recordings maintained the character of live performance through improvisation. At the same time, in the cabarets, acoustic rai was being fused with the new technology (drum machines, synthesisers, etc.) of Western music, while cassettes (opening up new possibilities for widespread production and diffusion of recorded music through a parallel 'black' market) replaced records. Rai increasingly tapped into the concerns of Algerian youth, including issues of sexuality and lifestyle, but because of its reputation was neither broadcast nor performed in concerts until the 1980s. After public concerts in Paris in the mid-1980s, rai spread throughout Algeria and into Morocco and Tunisia, and by the 1990s had become the main pop genre across the Maghreb. By then it had also incorporated songs and rhythms from American disco, Julio

Iglesias, Egyptian instrumental interludes and Moroccan wedding tunes. In 1985 rai singers were performing in Oriental cabarets and at weddings in France. By the following year leading singers had managers experienced in both the Arab and Western music business, and backing bands (rarely found in Algerian rai), while Island Records had begun to sign recording contracts. By 1990 the main performers had given concerts in Europe, North America and Japan. The musical components of rai were altered for Western audiences: Arabic was far from the 'ideal' language for audiences to comprehend or sing along to, and rai's bass rhythms were difficult to dance to, hence on tracks like Khaled's hit 'Didi' (1992), the first record by an Arab to reach the French Top Ten, 'the only thing left of Algerian rai consists of its stock phrases, the sense of phrasing melodic lines and primarily the voice of Khaled'. A funky bass and choirs had been added, and dedications left out (Schade-Poulsen 1997: 78). Similar forms were not created in more impoverished Algeria. Khaled was one of the first to introduce the accordion (now replaced by the synthesiser) into a music originally composed for flute and percussion. Cheb Malik, born in France, pioneered a strand of techno, labelled *l'after-rai*. After such transformations some critics argued that music like that of Cheb Khaled could no longer be considered rai (Gross *et al.* 1994: 21; Virolle 1995: 163). Indeed, attempts to market his first American-produced album used posters proclaiming 'Ceci n'est pas un disque arabe'/'This is not an Arab record' – perhaps implying that no Arab nation or individual could produce 'anything as modernist and progressive as popular dance music' (Warne 1997: 144), yet such attempts to achieve commercial, mainstream success largely failed.

See: Virolle-Souibès 1989; Schade-Poulsen 1995, 1997, 1999; Warne 1997.

In French cities music was also 'an important means of cultural expression for a minority struggling to carve out an ethnic identity and space in an inhospitable, racist environment' (Gross *et al.* 1994: 9). Rai was widely played on local radio stations (such as Radio Beur dedicated to Maghrebi migrants), the most popular programmes being those during the nights of Ramadan, when evenings were given over to celebration during the Islamic holy month; such programmes became 'a nostalgic return to an ambience resembling what they had heard of or remembered about Ramadan celebrations in their home country [which] reduced the burden of exile by establishing a mood of community closeness' (Gross *et al.* 1994: 10). The most common way to exert their kinship and ethnic identity was to dedicate a song, usually of Cheb Khaled, to a relative or friend. Ironically Khaled sang about women and alcohol, sentiments inappropriate to the observance of Ramadan, but loosely reflecting the more secular, modern character of rai lyrics that appealed to younger people. Lyrical preoccupations shifted from the more oppositional, inter-generational Algerian forms to concerns with race and religion, and rai was increasingly fused with funk beats, reggae and flamenco styles. It was danceable (to the young), and hybrid, yet recognisable as Algerian, but of a certain kind of Algeria: 'a contemporary, relaxed, sophisticated, tolerant, urban Algeria – the vision of the homeland selectively privileged by the rai audience in France' (Gross *et al.* 1994: 30). Moreover, despite the transformations, second and third-generation *beurs* gradually rejected rai in favour of rap, since rai was increasingly perceived as retrospective and nostalgic rather than contemporary.

In much the same way bhangra was closely tied to the experiences of Punjabi migrants in distant cities. The music sustained connections with Punjabi folk music, but was transformed in Britain (particularly Birmingham, the 'capital of Bhangra') and North America by young migrants, who retained the traditional instruments but added guitars, synthesisers and drums, and sometimes English lyrics, creating a new genre that 'was as genuinely Indian as it was recognizably disco' (Banerji and Baumann 1990: 142; Baumann 1990), enabling Asian youth to affirm their own ethnicity (Gilroy 1993b: 82). It has been stated that bhangra 'displaces the "home" country from its privileged position as originary site and redeploys it as but one of many diasporic locations' while creating a 'hybrid notion of "Indianness" produced within and deployed from the former colonial power' (Gopinath 1995: 304, 313). Such changes resulted in bhangra not merely being accepted as the music of migrant Punjabis, but being accepted by other South Asians as 'their' music – a pan-Asian identity – while also becoming of some relevance to migrant groups of different ethnicities and from different continents. The significance of rapping, sampling and electronica took bhangra far from notions of Punjabi folk music, though the complexity and creativity of the new continued to be asserted and constructed with reference to the old (Hutnyk 2000: 11), however tenuous such associations became. It nonetheless served as a vehicle through which young Asian migrants (or their children) 'of differing religious and cultural backgrounds are able to overcome their own particular differences, thus forging a common form of identity in order to symbolically counteract the marginalization, exclusion and hostility to which they are often subject' (Bennett 1997: 108). Bhangra thus went far beyond its ancestral origins to enable first- and second-generation migrants from throughout South Asia to develop a new common identity.

This wider acceptance and intensified hybridity went even further where quite distinct musical genres emerged within migrant communities. Salsa, for example, began in the Puerto Rican colonial diaspora during the first half of the twentieth century, and evolved among migrant communities from Puerto Rico, Cuba and other parts of the Caribbean, with significant black American input, in and around Manhattan, New York. In the wake of the Depression, and in the face of racism, salsa flourished. In the post-war years, especially the 1960s, it again emerged as a distinctively New York–Puerto Rican style, with the 'rise of a new class of Puerto Rican musicians' who deliberately sought to produce a new style more in touch 'with the Puerto Rico *barrio* [village] reality' (Padilla 1990: 89; Aparicio 1998). This was tempered by Cuban and wider Hispanic elements, while simultaneously being encouraged by an American recording industry, then in an oligopolistic phase, which sought to export Latin music to other American cities and throughout Latin America, to the extent that it was described as a 'publicity stunt and marketing ploy' (Brennan 1997: 287). The new salsa was thus linked to its 'Puerto Rican roots', but created in New York and increasingly designed to appeal to a broad public. It alternated between being a hybrid of all Latin American music – but especially Cuban rumba and son, mambo and chachacha – and a more obviously Puerto Rican form, using instruments such as the maracas, which reflected Taino Indian culture, and lyrics describing Puerto Rican experiences in New York or Puerto Rico. Despite commodification (especially in exoticised club venues), attempts to achieve crossover success (generally unsuccessful) and 'cultural voyeuristic looting' (with elements of salsa incorporated into other musical genres), it retained a distinct identity and cohesiveness to the extent that:

> Salsa music within the Puerto Rican diaspora helps us maintain an intimacy with our pasts and a living connection to our present, orchestrating a coherent embodied sensibility that can only be fully understood somewhere between the purely discursive dimensions of Puerto Rican and Afro-Caribbean cultural history.
>
> (Sanchez-Gonzalez 1999: 237–8)

For the jazz musician Willie Colon, salsa (whose primary meaning is a spicy sauce) was a 'validation [of] home, a flag and grandma to displaced young Latinos all over the world' (quoted in Steward 1999: 12). Salsa, a style of music invariably linked to Latin America, thus emerged in New York, but became associated with a transnational, Hispanic identity.

The production of migrant musics in 'world cities' has enhanced the sense of 'local' in other contexts. In several cases the music of migrants – especially folk music – returned to the 'homeland' where it had partly disappeared, as in parts of Ireland, Greece and Poland. Elsewhere the rise, or revival, of particular forms of 'local' music marked 'ethnic revivals' of various kinds, some deliberately engineered (Reck *et al.* 1992: 434). One of the first groups to popularise bhangra in young British South Asian communities started in order 'to bring them back into their own culture.... We noticed that young Punjabis in London knew little about their own culture and language' (quoted in Gopinath 1995: 309; cf. Diethrich 1999). In Chicago South Asian music became more, rather than less, popular over time though many South Asian youth 'do not understand the Hindi or Punjabi lyrics of the songs, and many may rarely if ever go to India, the song lyrics, like the musical sound itself, convey primarily non-representational...and affective meaning' (Diethrich 1999: 55) – crucial senses of nostalgia (even for 'homes' never directly experienced) and identity.

Music, like other cultural forms, is selectively integrated into social worlds that are 'driven by the frozen furrow of memory and the politics of nostalgia' (Maira 1999: 51), but the obsession with a more authentic, pure, original tradition occurs at exactly the time when ethnic (or other) identity is most in question (which may partly account for the vast number of CDs of Irish ambient music, and even the spectacular success of a volume of 'traditional' Irish hymns, *Faith of Our Fathers* (1997), excluded after Vatican reforms). Music therefore represents 'enumerations of loss, of a yearning and longing to recover and recuperate that which is also simultaneously and implicitly acknowledged to be irrecoverable and irrecuperable' (Gopinath 1995: 310) with both music and life involving local and transnational neighbourhoods: what Appadurai (1990) termed the 'ethnoscape'. Perhaps, most startlingly, 'one of the central ironies of global cultural flows, especially in the arena of entertainment and leisure is that elements of youth culture are partly shaped by a "nostalgia without memory"' (Appadurai 1996: 30). The past, and memories of distant places, are never entirely lost.

Transnational urban soundtracks

While the diffusion of music has obvious links to migration (and commercialisation), the most successful examples were those that extended beyond migrant communities into the mainstream, eventually even losing part of their identification with migrant communities. Dominant populations in migrant destinations both regarded migrant and minority groups as socially and culturally remote from their own lives, and often ridiculed aspects of migrant culture, including music, as in Newcastle, England (Bennett 1997: 114–15). However, selected migrant musics have become associated with ideas of transnationalism as they have been transformed in the diaspora. The emergence of styles and scenes that spawned successful international products also created new lines of cultural affiliation between youth cultures in very different contexts. In the hands of contemporary black producers of hip hop, world music and dance hall reggae:

> 'transnationalism' becomes a utopian symbolic space between the culture of the racial diaspora and the political economy of the American nation-state. For them transnationalism is not a simple transcendence of the national, but rather an expressive project that samples and mixes the identities of immigrant, citizen and refugee.
>
> (Stephens 1998: 163)

Certain individuals and groups such as Transglobal Underground and the Afro Celt Sound System deliberately transcended and fused a variety of genres, identities, places and even ethnicities, to challenge conventional stability and certainty. Hence, Gopinath has argued of bhangra that not only do '*multiple* diasporas intersect both with one another and with the national spaces that they are continuously negotiating and challenging' but, somewhat dramatically, 'diasporic popular cultural forms such as bhangra are always many texts at once; they are never purely enabling, resistant or celebratory, but are simultaneously available to recuperation within hegemonic constructions of identity, culture and community' (1995: 304), none of which are static. A sense of transnational 'sound' and cultural politics emerges as global cultural industries commodify and distribute the materials (CDs, music videos, etc.) necessary to create such urban soundtracks. Reggae and hip hop both demonstrate how youth sub-cultures develop means of identity construction in frequently adverse circumstances, which are then further commodified by media corporations, often also located in the same cities. Music recordings in global mediascapes consequently embody a central tension – as both the products of strategic branding campaigns (designed to capture youth audiences across nations) and the artefacts of transnational cultural identification for migrant audiences.

Despite the broad processes of globalisation (including migration, commercialisation and the rise of global media matrices), shifting trends in local and global music are embedded in 'a complex, tangled web of modern, mass-mediated transcultural communication' (Bilby 1999: 273). Particular trends vary spatially and temporally, with simultaneous fusion, diffusion, decline and revival, such that developing models of change is impossible. The rise of the Internet ensured both that the pace of change accelerated, and the potential for hybridity became overwhelming. Chude-Sokei provides an elegant example of how such shifts operate within one specific diaspora network:

> The legendary team of Steely and Cleevie in Jamaica, or maybe Bobby Digital in Kingston, may send a floppy disk with the basic rhythm track to Daddy Freddy, who is in London with the up-and-coming production team of Mafia or Fluxy (or maybe Fashion, today's dominant U.K. sound). This track may feature the latest craze in dancehall rhythms – sampled Indian tablas mixed with Jamaican mento patterns from the 1950s. After a brief vocal session, that same information could go to Massive B in the Bronx for hip hop beats or to Sting International in Brooklyn where R&B touches are added again. Again, all of this is by modem or by floppy disk. Within a few days this mix is booming down the fences at the weekly 'sound clash' between Metromedia Hi-Fi and the mighty Stone Love Sound System somewhere in a crowded field in

> West Kingston. Or in a community center in Brixton. An 'authentic' Jamaican product! And this trade goes both ways, circulating through diaspora. (Even in Lagos, Nigeria, I have sat listening to Igbo rude boys and Yoruba dreads rap in Jamaican patwah about the virtues of Eddy Murphy!)
>
> (1997: 221)

Such is the current and evolving complexity of integration and disintegration, and the tangled personal, physical and electronic web that links the local with the global. Yet, while this kind of hybridity may be successful on the dance floor and in the record store:

> though fashionable in theory and also literally in 'ethnic chic', it is not always easy to live, for families and communities demand loyalty to cultural ideals that may be difficult to balance for second-generation youth. Theoretical valorizations of hybridity do not always hold up on the ground, where the contradictions of lived experience challenge binaries of authentic/syncretic cultural identities.
>
> (Maira 1999: 44)

Such contradictions were made explicit by Nitin Sawhney on *Beyond Skin* (1999), with the British-Indian musician questioning both allegiances to prior ethnic nationalisms and new identities. Sawhney's music, electronica and soul with splashes of tabla, qawwali, drum and bass and rapping, criticised Indian government policy on nuclear proliferation, yet articulated the experiences of racial vilification in the migrant context:

> I am Indian. To be more accurate, I was raised in England, but my parents came from India – land, people, government or self – 'Indian' – what does that mean? At this time, the government of India is testing nuclear weapons. Am I less Indian if I don't defend their actions?...Less Indian for being born and raised in Britain? – For not speaking Hindi? Am I not English because of my cultural heritage? – Or the colour of my skin? Who decides? – 'History' tells me my heritage came from the 'Sub' continent – a 'third world' country, a 'developing' nation, a 'colonised' land – So what is history? – For me, just another arrogant Eurocentric term....I learned only about Russian, European and American history in my school syllabus – India, Pakistan, Africa – these places were full of people whose history did not matter – the enslaved, the inferior.
>
> (Sawhney 1999)

Sawhney's music, which incorporates such voices as those of Nelson Mandela (in a sample stating 'we are free to be free'), Mandawuy Yunupingu, the Aboriginal lead singer of Yothu Yindi, and a Soweto primary school choir, explicitly moved away from any sense of strategic inauthenticity, or the cliché of world music, by incorporating a diversity of beats, sounds and emotions into tracks that escaped ethnic or place categorisation, and that questioned cultural identity rather than embalmed it. The hybridity of musical forms was scarcely reflected in contexts where performers such as Sawhney were simplistically labelled as 'Indian' and their music as 'bhangra', and where families, and communities, were divided by the extent of assimilation and change. Indeed 'renegotiating your identity purely through the expressive cultural forms of diaspora is at best naive, at worst opportunistic' (Banerjea 2000: 76). Even within particular migrant or transnational communities there is both conservation and dissolution; hybridity and change are tempered by family loyalties, generational shifts, employment, class structures and personal preferences.

Reggae: innertainment and outernational

By far the most familiar diasporic music has long been reggae, primarily centred on Jamaica where it has often been the most valuable export. Like other genres of world music, there is no exact definition of reggae, which has many elements of other Caribbean music forms. The word 'reggae' is said to mean 'comin' from the people' or 'regular', notions coined, like the name itself, by Fred 'Toots' Hibbert, of Toots and the Maytals, in his first single 'Do The Reggay' (1968), to refer to a new dance and a new sound (Cooper 1998). Reggae ultimately combined Western technology with African and African American culture, and evolved from earlier local musical forms such as mento, ska and rocksteady, which in turn had developed from the music of African slaves. Mento, slow and rhythmic and closely linked to calypso, gave way in the 1960s to ska, which, like its successor, reggae, was most popular in depressed urban areas of Jamaica. Ska was strongly influenced by both modern jazz and American rhythm and blues, partly a legacy of the Second World War when many black American soldiers were stationed in Jamaica, and partly a response to exposure to black American music and rock 'n' roll (Winders 1983: 67). Reggae emerged from its ska and rocksteady origins in the slums of west Kingston, with some early songs glamorising crime and rebellion, and glorifying rude boys – unemployed, angry youths and gangleaders, with a distinctive dress style, motorcycles and knives – to incorporate a variety of forms, the most commercial of which – such as Johnny Nash's 'I Can See Clearly Now' (1971) – became internationally successful. It also came out of a longstanding tradition of sound systems and highly territorial dance party spaces held in clubs, or in open streets and 'lawns': turf wars through noise. In Jamaica a harder form of reggae, involving Rastafarian themes, became significant, and it too was popularised and spread, notably by Bob Marley and the Wailers. Reggae combined African rhythms with European melodies and harmonies, though there are 'pure African survivals' in the call and response structures and references to old African songs. Much reggae music, especially that with Rastafarian intent, concerned issues of black pride and identity, deliberately drew on African experience and hence involved social criticism, political protest and the particular problems of urban slums. Drumming, religion and resistance to authority were closely linked (Hebdige 1987: 43–5; see Box 8.3). For many years it was almost impossible to hear live reggae music in Jamaica: concerts were banned and the Jamaican Broadcasting Corporation broadcast reggae only at certain times, often after midnight, rather than calypso, derided by Rastas as 'rum culture' for tourists who never saw places such as Trenchtown (Winders 1983). As reggae evolved, expanded and became commercialised, each of the early distinctive elements became less evident.

The growth of reggae was largely synonymous with the influence of Bob Marley and the Wailers, who discarded their earlier rude boy image, adopted Ethiopian colours and dreadlocks, composed several militant songs (such as 'Get Up, Stand Up') and acquired a producer, Chris Blackwell, who had set up the highly successful Island Records. The first album, *Catch a Fire*, was an international success, but after modifications from the original version recorded in a Kingston studio:

> Blackwell considered it a little too heavy for the white audience, so he re-mixed it in London. He brought Marley's voice forward and toned down the distinctive bars. He also added some flowing rock guitar riffs recorded by British session men to the original tape

while the lyrics were reprinted on the album sleeve in case white listeners could not understand the Jamaican patois. In the United States the speed of some reggae records was increased for that market (Hebdige 1987: 80, 82). The success of the British star, Eric Clapton, with a cover version of Marley's 'I Shot the Sheriff' (1974) ultimately gave Marley international recognition. Local music was being transformed for a global market.

Box 8.3 Rastafarians, resistance and identity

The first Rastafarians appeared after Haile Selassie was crowned Emperor of Ethiopia – then the most recent manifestation of a black king in Africa – in 1930. In Jamaica a group of preachers claimed that Haile Selassie was more than just a secular king and black hero, but was nothing less than the Living God. Rastafarians quoted biblical prophecies to justify this claim, identifying Selassie as Prince Ras Tafari ('Jah'), the Conquering Lion of the Tribe of Judah, a direct descendant of King Solomon. At much the same time an exiled Jamaican, Marcus Garvey, promoted a Back to Africa movement, and this was linked to the story of Moses leading the 'suffering Israelites' out of slavery to a promised land (which eventually stimulated Desmond Dekker's enormous 1969 hit 'Israelites'). Rastafarians (Rastas) developed their own beliefs and rituals. Orthodox Rastas refused to cut their hair, grew dreadlocks and smoked ganja (marijuana). Many refused to pay taxes or work in the commercial world ('Babylon'), but looked towards black Africa ('Zion') for some form of salvation, inspiration or even 'repatriation'. The number of Rastafarians increased in the 1940s and 1950s – both in remote rural areas, where ganja was grown, and in depressed urban areas, such as Trenchtown. The most militant, distinctive and socially outcast Rastas – hounded or ignored by authorities – had the most impact on the music, as they sought out African themes and traditions, especially drumming, and emphasised a particular variant of Creole, but particularly in dress and style, not only in the Caribbean, and in the red, green and gold colours of the Ethiopian flag, or the red, green and black of Garvey's Universal Negro Improvement Association (UNIA). As reggae music spread, the colours and symbols have dominated the music scene, rather than the religious messages of peace and love, or the political ideal of a return to Africa. Although the notion of return to Africa was largely impractical, 'It was not the literal Africa that people wanted to return to, it was the language, the symbolic language for describing what suffering was like...a metaphor for where they were...a language with a double register, literal and symbolic' (Hall 1995: 114). That language dominated Rastafarian influenced music.

See: Hebdige 1987: 49–61; Clarke 1980: 36–56; Davis 1983; Hansing 2001.

Beyond the Caribbean, reggae music first became successful in England, the principal destination of West Indian migrants during the boom immigration years of the 1950s and 1960s. Throughout this period it was ignored by the main radio programmes, being perceived as 'repetitive' and 'moronic', but dominated Caribbean clubs, and was later adopted by mods, punks and skinheads (which created further disdain among music critics and radio stations). The combination of major groups incorporating versions of reggae rhythms into more mainstream popular music (such as Blondie's 'Heart of Glass' (1978) and The Clash's 'London Calling' (1979)) resulted in reggae moving from the music of an immigrant minority towards mainstream status. Initially, the distinctiveness – in rhythm, style, language and ethnicity – emphasised that it was a transnational music with a particular migrant market. Later however:

> A sizeable white reggae market developed, and some of these record buyers began tracking down the more 'ethnic' roots product. Despite the Rastafarian trappings, roots reggae appealed to many young white people. It was 'rebel music'. It hadn't become smug and indulgent like so much of the rock produced by the rich, successful white stars. Reggae stars in general remained in touch with their roots. They tried to stay close to their fans. Above all, they sang songs about poverty, desperation and revolt – songs which related to the everyday experience of many white kids.
>
> (Hebdige 1987: 95)

or at least what white teenagers would have liked it to be. Even though their own lives rarely bore even tangential reference to the urban Caribbean, 'the white concept of the authentic is linked with the raw and unembroidered musical expression of the stark poverty and social decay that characterise a certain aspect of Kingston life' (Clarke 1980: 169). In British cities such themes found a social context. By the end of the 1970s British groups, mainly white, such as the Specials, UB40 and Madness, began to play ska and reggae music almost exclusively, as reggae moved 'closer to the mainstream of British society' (Hebdige 1987: 98; Alleyne 2000). Black British reggae groups also appeared, taking up familiar Rasta themes in their songs, but usually linked to urban British issues, especially as unemployment and social stress increased. British-born or raised black performers felt removed from such Rastafarian themes as repatriation to Africa while the dance hall rhythms of Jamaica seemed too slow for the faster cosmopolitan life of urban England. With different sounds, by the end of the 1970s, bands were beginning to have hits in Jamaica: reggae was crossing the ocean in a different direction. A decade later it was possible to argue that 'British reggae is now more alive, more interesting to listen to and more in touch with the fans than the music which is coming from Jamaica' (Hebdige 1987: 119). At the same time reggae was becoming further commodified and, as it diffused, being transformed 'from a form of cultural criticism into a cultural commodity' (Cushman 1991: 38). The processes begun at Island Records were extended further, turning reggae performers into international pop stars 'to achieve broad-based multi-market crossovers' (Alleyne 1994: 78), and Marley may have colluded:

> This foray into pop stardom was a calculated development in which he was intimately involved, having realised that the solidification of communicative networks across the African diaspora was a worthwhile prize. The minor adjustments in presentation and form that rendered his reggae assimilable across the cultural borders of the overdeveloped countries were thus a small price to pay.
>
> (Gilroy 1987: 169–70)

Marley's image shifted from Rastafarian outlaw of the 1970s, to family man of the 1980s and natural mystic of the 1990s, representing ideologies of national liberation and black power, multiculturalism, universal pluralism and, eventually, transnationalism (Stephens 1998). By the time that the best-selling compilation *Legend* (1984) was released, both 'reggae' and 'revolutionary' had largely been removed from the conception of the album, in favour of an emphasis – especially visual – on Marley as an individual, and on 'one love' as a pluralist vision that might legitimate the national narrative of a multicultural United States and, by extension, of the world. This, and the later compilation, *Songs of Freedom* (1992) became identified with one of the enduring myths of black music, 'one nation under a groove', where music had the power to unite: the ultimate incorporation of what was once dangerous, subversive and counter-cultural (Stephens 1998: 149–56). Consequently, as one West Indian author subsequently wrote:

> We usually think of Bob Marley as our own and of the recorded representation of his work as authentic. While one can argue that in some sense this remains so, the persistent reformulation, amendment and commercialisation of his music suggests that we have been listening to an alien, inauthentic representation. Paradoxically since it is the only one to which we have access, it automatically assumes a level of 'authenticity'.
>
> (Alleyne 1994: 83)

The legacy remained in the presence of a 'reggae aesthetic' making bass and drums the most important part of the mix of many variants of house music, and 'the dub revival' of the 1990s alongside the commercialisation of a faster dub/techno form, known as jungle (Blake 1997: 108). This further emphasised the diverse ethnicity of those involved, however marginally, with reggae. Massive Attack and the Dub Pistols used reggae samples, alongside soul, jazz-funk and British folk influences, Washington DC's Thievery Corporation sampled Rastafarian elders as part of their dub electronica; while even more complex fusions were evident in the music of Apache Indian, Asian Dub Foundation (Box 8.4) and Transglobal Underground.

Box 8.4 Apache Indian and Asian Dub Foundation

Apache Indian (Steve Kapur) was born into an Indian migrant family in Birmingham, England, and adopted the performance name derived from that of Jamaican reggae star Supercat (the 'Wild Apache'). His first album *No Reservations* (1992) was a contemporary Anglo-Indian fusion of dance hall and Punjabi bhangra music with a reggae beat. The lyrics, relevant to Indian youth identity in Britain, challenged racism, bridging black–Asian hostilities, using English, Punjabi and Jamaican patois interchangeably, frequently referencing the sonic geography of his influences ('Chok There' (1992): 'Them a ball from Bombay City/ Chok there/ From the deepest parts of Delhi.../ So me hear from Karachi, New York, Kingston and London City'). Asian Dub Foundation, another Anglo-Asian group, explored similar thematic territory, with their combination of ragga vocal techniques, jungle beats, guitar riffs (based around South Asian scales) and sampled sitars and political speeches. Both their releases *Rafi's Revenge* (1998) and *Community Music* (2000) adopted explicitly radical political stances on the entrenched racism and class relations affecting Anglo-Indian migrants ('Colour Line' (2000): 'Today, the colour line/ Is the power line/ Is the poverty line'), celebrations of peasant socialist uprisings ('Naxalite'), critiques of racist legal systems ('Free Satpal Ram') and declarations of Indian diasporic self-determination, alongside a celebration of one earlier Pakistani musician's contribution to cultural survival:

'New Way New Life' (2000)
Nusrat Fateh Ali Khan
Kept our parents alive
Gave them will to survive
Working inna de factories...
Stayed an we fought an now de future's open wide
New Way New Life...
Inna de dance our riddims rule
But we knew it all along
Cos our parents made us strong
Never abandoned our culture...
Inna de battle we belong
When we reach the glass ceiling
We will blow it sky high

Asian Dub Foundation found critical acclaim not only in migrant communities but in global 'alternative' and dub scenes, while Apache Indian, through his visits to and success in India, contributed to some diffusion and a greater appreciation of reggae there.

See: Sharma *et al*. 1996; Hutnyk 2000.

Jamaicans themselves have referred to reggae both as 'innertainment' and 'outernational', a music of both pleasure and protest, unconstrained by national borders and able to adapt and respond to new conditions (Clough 1997; Hansing 2001). Reggae linked the spiritual and mystical elements of Rastafarianism with protest, in Jamaica and across the world, the resultant popular political identification giving Bob Marley and his music 'a profound galvanizing impact throughout the black diaspora and the postcolonial community' (Stephens 1998: 143). This offered what later became 'the classic Caribbean package of bitter social commentary wrapped up in a light, refreshing rhythm. In effect Marley was making the world dance to the prophecies of its own destruction' (Hebdige 1987: 81). The rhetoric of anti-authoritarianism (although originally a reaction to the specific legacies of the Caribbean's plantation slavery past) allowed other groups to invoke similar comparisons, while the pattern of using tricolour designs and symbols was easily interpreted as a sign of subaltern identity in other contexts. Furthermore, reggae's stylistically and semantically 'open' musical form revolved around specific rhythmic patterns (syncopated beats, heavy bass-driven sound), which were more easily adopted and transformed in local circumstances (such as adding vernacular instruments as layers on top of the beat, or using local languages). Reggae's explicit support of the use of drugs to achieve a state of spiritual awareness buttressed its radical edge and subcultural reputation, making it the international music of marijuana and a standard soundtrack in backpacker haunts from Amsterdam to Kathmandu.

Reggae's 'global template' (Garofalo 1993b: 28) gave rise to numerous local variants such as the Afro-reggae of the Ivory Coast's Alpha Blondy and South Africa's Lucky Dube, and much lesser known bands, including New Caledonia's Flamengo and Solomon Islands' Toxie, many of whom sung in languages other than English (or Creole) but copied rather than transformed the originals. In other contexts transformation was enormously complex, even allowing for the reggae tradition of 'versioning': making new tracks from deconstructed instrumental mixes of well-known rhythm patterns and melodies:

> *Mantra Mix* (1996) by the Sydney studio aggregation, Sacred Sound System...was created around a sample bass-line from a 1979 track *Fattie Boom Boom*, by the Jamaican M.C. ('mischanter' or reggae rapper) Ranking Dread, itself a rearrangement of a bass rhythm made earlier in the 1970s at Studio One in Kingston. The first version was by a Jamaican domiciled in England. The second was constructed, in part, by a Jewish-English woman, Yvonne Gold, a Maori reggae/hip hop singer and M.C., Kye (David Heta) and a former member of pop groups Split Enz and Crowded House, Pakeha (non-Maori) New Zealander, Tim Finn. The latter three live in the Sydney suburb of Bondi.
>
> (Clough 1997: 12)

The final track incorporated Tibetan Buddhist mantras and bells, the sampled laughter of the Dalai Lama and a 'Polynesian kind of riff' on the ukulele (Clough 1997: 12), and one of its creators emphasised 'We all watch international films, we all listen to international music, we all read international books. I mean it's an international society. So why wouldn't music produced out of here reflect this?' (quoted in Clough 1997: 13). In Surinam, on the north coast of Latin America, local versions of reggae incorporate not only Jamaican influences, but also melodies from the Ivory Coast's Alpha Blondy, such as 'Bori Samory' (originally a West African song in Dyula and French), with new lyrics constructed in the local Sranan language (Bilby 1999: 272). Reggae influenced the 'Jawaiian' tunes of Hawai'i (see Weintraub 1993), the musical traditions of places as different as Colombia, Nigeria and South Africa (see Manuel 1995), and was adopted in various indigenous communities in Polynesia, New Zealand and Australia (Gibson and Dunbar-Hall 2000) and in the cities of the Islamic world – from the Palestinian alleyways of Old Jerusalem to Jakarta or Algiers:

> In the smoky dives known as *les bars* the disaffected young men of Algiers gather to drink, talk and listen to live music. But close your eyes for a second and you could be forgiven for imagining you were in South London. Blended with the traditional *rai* music are reggae basslines. 'Reggae Islam' has taken off in the Algerian capital. The inspiration comes not from Jamaica but Brixton, where 'reggae Islam' is also said to be thriving. . . . In Muslim neighbourhoods, Bob Marley posters and marijuana icons jostle for position with graffiti in support of the Muslim guerrilla group GIA.
>
> (*Guardian*, 6 May 1999: 8)

In these diverse contexts, the meanings of the original pan-African metaphors of reggae have been both mobilised and transformed into new expressions of ethnicity and political motivation, and 'watered down' for tourist pleasure. As Baulch put it, 'the Bali reggae scene is much less about the struggle for Rastafarian liberation than about creating a Caribbean atmosphere and promoting Bali as a beach culture' (1996: 25). Yet elsewhere, in Nepal, reggae was commonly used as 'a persuasive musical setting for seductive love songs' (Greene 2001: 173). Its sound, meaning and symbolism were constantly in flux.

Meanwhile, at the core, in Kingston, Bunny Wailer, the last survivor of the original Wailers, was literally stoned in 1991, by a new generation whose notions of identity and cultural survival had changed. Ragamuffin (and dance hall), descendants of reggae, evolved in a privatised urban market-place, and reflected a world where the African diaspora had been realigned. No longer was it linked to the Rastafarian vision of an African past and future, and its 'pious moralisms'

(Chude-Sokei 1997: 221). Reggae's new adherents were more likely to express individual faith than apocalyptic messianic messages, and rail against the International Monetary Fund rather than slavery (Ross 1998). In the dance halls, reggae returned to early themes, and crude lyrics about gun culture and sexuality (incorporating homophobia and sexism) linked into new urban sound systems. These privileged rhythm at the expense of lyrics and melodies, and traced complex relationships with – and within – diasporic black communities. Reggae had diversified, developed and become open to a series of interpretations and agendas. Neither music nor message were static.

Hip hop and the global 'hood

Like so many other musical genres, hip hop (rap) emerged from a fusion of elements brought by migrants (in this case from the Caribbean to the United States) with local musical forms of residents of deprived inner-city neighbourhoods. Hip hop emerged in the South Bronx (New York) in the 1970s, and was quickly taken up by residents in similar American urban neighbourhoods, notably in south-central Los Angeles, but also 'other urban ghettos in major cities such as Houston's fifth ward, Miami's Overtown, Boston's Roxbury' (McLaren 1995: 14). The South Bronx was an inner-city, public housing development largely populated by black and Hispanic (mainly Puerto Rican) families and newer migrants, mainly from the Caribbean. Clive Campbell (later better known as Kool Herc) came to New York from Kingston, Jamaica, in the late 1960s, constructing the largest and loudest sound system in the Bronx and using the Jamaican style of 'toasting', calling out catch-phrases to the crowd through a microphone over the top of the record. Since most crowds were uninterested in reggae rhythms, Herc, and other emerging DJs, such as Grandmaster Flash and Afrika Bambaataa, played faster funk music and began to extend the 'breaks' (those parts of popular records where rhythm took precedence over vocals) by cutting backwards and forwards between the two copies of the same record on twin turntables. The success of the music turned on the skill and dexterity of the DJs, their use of original and obscure record sources, the ability to provide breakbeats and mix several different records using twin turntables and volume faders.

Since reggae, hip hop has been probably the most widespread transnational urban soundtrack, reflected in its intense territoriality and focus on 'the ghetto' as a real and mythical space. As Rose argued:

> rap is a contemporary stage for the theater of the powerless. On this stage, rappers act out inversions of status hierarchies, tell alternative stories of contact with police and the education process, and draw

> portraits of contact with dominant groups in which the hidden transcript inverts/subverts the public, dominant transcript.
>
> (T. Rose 1994: 101)

The visual and aural elements of hip hop cultures have articulated a mesmerising array of urban American experiences, from contradictory sexisms and anti-Semitic comments to stories of pleasure and violence, alongside militant expressions of black nationalism.

The Caribbean sound system format made famous with reggae and dub outfits in Jamaica (where portable, bass-orientated public address systems were set up in the open air, on street corners or in informal club environments) was to play a vital role in developing hip hop. Otherwise derelict or degraded urban spaces were transformed into spaces of pleasure and expression (Toop 1984). Street corners, basketball courts, abandoned warehouses and clubs became central spaces in the process of making meanings and rituals out of commodities. The commodities in this case – dub mixes of funk hits, old second-hand records of Led Zeppelin or Kraftwerk – formed the basis of creative mixing, cutting and scratching by DJs and toasting by MCs:

> The artefacts of a pop industry premised on the individual act of purchase and consumption are hijacked and taken over into the heart of collective rituals...which in turn define the boundaries of the interpretative community. Music is heard socially and its deepest meanings revealed only in the heart of this collective, affirmative consumption.
>
> (Gilroy 1993a: 38)

The cultural meanings of a single commodity (which in itself represented the global music industry and distribution reach of transnational corporations) were challenged and subverted. Urban sub-cultures constructed social arenas in which music products of big business were reused and reinterpreted as artefacts of locally based cultures that sustained African American oral traditions (Gilroy 1993a: 38–9). In such cultures, the ways in which music filled space became important. As Jackson put it, 'there is a world of difference between listening to music performed live in a communal setting and listening to recorded music in the privacy of one's home' (1999: 103). Rather than simply being contained within the enclosed spaces of private living rooms (or the even more secluded space between Walkman headphones), hip hop and dub beats reverberated through public space, either from sound systems or passing cars in acts of self-assertion and attempts to gain recognition from passers-by and other motorists; while:

> the 'rap' might seem the most recognisable element of rap music, the bass is what turns your head when a car drives by, it is often used...as an attention-getting device, drawing gazes, as well as more than one angry look, to the one who 'brings the noise'.
>
> (M. Quinn 1996: 80)

The bass, then, served an important function in urban space, tied into deliberate attempts to construct self-identity, while graffiti artists, another crucial element of such sub-cultures, spray-painted public property in an overt process of 'signifying' a presence and control over territory.

While hip hop has generally been seen as a vehicle for giving voice to the minority inner-city African American community, enabling black ethnic pride (Potter 1995), Puerto Ricans and whites were also involved in its development (Flores 1994; Lipsitz 1999b). Hip hop also appeared in urban areas in many other parts of the world, often those associated with recent international migration, such as Sweden, France and Germany. In France, hip hop developed in the outer suburbs of Paris and, to a lesser extent, in other migrant cities such as Marseilles. Most performers, and their audiences, were of West Indian, African or Maghrebi descent; the leading performer, MC Solaar, was born in Senegal of Chad parents, and had moved to Paris from Cairo (Cannon 1997). Marseilles group IAM reflected the diverse origins of the city's population with five of the group's members from Senegal, Madagascar, Algeria, Spain and Italy, while they rapped in Marseilles accents using the constantly changing slang of the housing estates (Box 4.1). Similarly in Germany, rappers were often young people in migrant guest-worker families from Turkey and Morocco. Everywhere rap increasingly became localised by using national languages (or local variants) and lyrics concerning specific cultural, economic and political issues or by incorporating the languages – and some ethnic themes – of particular migrant groups, as in Germany (Box 8.5), New Zealand (Mitchell 1996; Wall 2000; Zemke-White 2001), Australia (Maxwell, 1997a, 1997b) and elsewhere. Initially there were few if any musical links to the migrants' (and their children's) home countries, but over time lyrics, styles and delivery were adapted to particular migrant circumstances.

Box 8.5 Hip hop in Frankfurt

German rap groups first formed in Frankfurt, a large multiracial city, in the 1980s, much influenced by the American groups featured on local and national radio and on MTV Europe. The original groups briefly tried to copy their African American counterparts, being initially motivated by a similar 'identity of passions' in which the appropriation of hip hop 'became bound up with a sense of imagined cultural affinity with African-Americans' (Bennett 1999b: 81). The first prominent rap group was called White Nigger Posse, who often claimed 'we are the blacks of Germany' (Caglar 1998: 248) but, quite quickly, rap culture moved away from parental and personal heritage. Frankfurt groups realised that, as in North America, theirs was a particular form of 'lived' ethnicity, which necessitated its own local and particular mode of expression (Gilroy 1993a: 82). Initially German lyrics were incorporated, then specific themes – primarily the fear and anger caused by racism, and insecurity over issues of nationality and citizenship (not automatically given to children born in Germany) – as rap became a more locally politicised discourse. In Berlin the same processes provided:

> diasporic youth with a ground where they can use their ethnicity as a strategising tool to articulate their identities in response to the majority nationalism...[and] the unpleasant present pervaded by racism, unemployment, exclusion, poverty and exclusionary regimes of representation.
>
> (Kaya 2002: 43)

However the relationship between German rap and the desire to be seen as German prompted a 'backlash of hip-hop nationalism' among some ethnic groups in Frankfurt. A small independent rap label began to specialise in producing Turkish rap music, combining traditional Turkish melodies and rhythms with rap; its manager observed 'a lot of this German rap is all about coloured guys saying look at us, we're like you, we're German ...[but] I'm not German, I'm a Turk and I'm passionately proud of it' (quoted in Bennett 1999b: 84–5). Rap culture bridged the gap between the displaced Turkish diaspora community and the 'imaginary homeland', constituting and creating 'an imaginary journey back home' (Kaya 2002: 48). The process of localisation of hip hop thus went from slavish imitation of an overseas genre, to the development of a national form and finally its localisation in terms of the concerns and aspirations of particular ethnic groups.

See: Bennett 1999b, 2000; Elflein 1998.

Hip hop also appeared in places where there was no significant international migration but, usually, as in the United States, where there were significant ethnic minorities. Rap enabled migrant groups and others who perceive themselves as 'brown' and 'black' to live with and express their racialised identities. In Auckland, young Maoris valued American rap music above all:

> can you imagine being Maori and not listening to hip hop....It's part of being brown. Like I know the traditional stuff's good but....Like with the American stuff we can relate...it's a brown thing. Like they know what goes on for us aye.
>
> (quoted by Wall 2000: 79)

Maori rap groups performed lyrics in opposition to racism and police harassment (Mitchell 1996: 244–50), and in post-apartheid South Africa hip hoppers used ethnicity as a symbol of strength and resistance (Watkins 2001). Hip hop emerged in southern Italy, in Ireland, where lyrics focused on issues of high unemployment and the cost of living, and in Japan, where rap performers used their musical style to differentiate themselves from the 'mainstream' youth of Japan (Bennett 1999a: 5). In Micronesia there was even Yap rap. Variations occurred within countries. In Mexico, the group Control Machete, like IAM in Marseilles, were using local words: 'We are not trying to imitate Chicano...we don't use words Chicanos would use....We're doing Mexican rap, and even more specifically Monterey rap' (quoted in Kun 1998: 57). In southern France, the Fabulous Trobadors rapped their nostalgia for a medieval, less centralised world, using old Occitan French and even medieval texts, but in support of the anti-racism and Afrocentrism of hip hop (Gross and Mark 2001). In many of these places rap existed where there was virtually no established black population; in Newcastle, England, a hardcore of hip hop enthusiasts believed that their intimate understanding of hip hop's essential 'blackness' was the key to its relevance for their own marginalised, white working-class experience (Bennett 1999a: 10–15). Rather differently, in the Netherlands, white Dutch preferred 'Nederhop' – the rap produced by white performers – while black consumers, mainly from the Dutch Antilles and Surinam, ignored Nederhop, which was seen as 'skateboarder' music, and preferred imported African American rap, indicative of how musical style, sub-cultural allegiance, social class and ethnicity are combined in imagination and performance (Krims 2000: 163–4). In every destination, even Newcastle (where there were local lyrics and accents), it was restructured to account for local issues and sensitivities. In many respects, its emergence paralleled the formation of what were seen as hybrid 'glocal' subcultures (Mitchell 1999), but rap also represented the separation of migrants and their children from the music of distant places and homelands.

Webs of cultural connections

Music has sometimes been the medium for large-scale expressions of unity, which transcended national borders and acknowledged similar cultural positions held by quite different migrant groups. In Peru the music of migrant Aymara panpipe players in Lima came to be symbolic of a newly imagined and culturally united

Peru and even represent a wider highlands Andean identity (Turino 1993); in the West Indies Trinidadian steel bands and Jamaican reggae contributed to broad Caribbean identity (Averill 1997b); in most parts of the African diaspora, including Jamaica, Surinam and England, music suggested an emergent pan-African consciousness (Bilby 1999: 275–6). In France rai music stimulated a pan-Maghreb (Algerian–Moroccan–Tunisian) identity, just as bhangra had contributed to a wider South Asian identity. In the United States, the santeria music (which accompanied a syncretic North American religion) united Hispanic migrants from various countries (Velez 1994), though pan-Hispanic identity was most actively represented in the salsa music that had also emerged in the north. More popular performers such as Julio Iglesias, Ricky Martin and the Buena Vista Social Club contributed to this broader identity.

The rise of a pan-Hispanic identity and the wider commercialisation of Hispanic music was fundamentally shaped by a 'salsa matrix': a web of commercial ties connecting music company offices and studios in the triangle of New York, Miami and Puerto Rico, where most salsa composing, arranging, production, recording and manufacture takes place. Further ties linked this web to other regional offices in Venezuela, Cuba and elsewhere in Latin America, and, partly through the Canary Islands, into Europe and West Africa (Negus 1999: 146–7). The transmission of salsa thus mirrored Hispanic culture – shaped by ethnicity, region, language and religion – but also cultural ties beyond, mediated in Europe and elsewhere by the commercial preoccupations of the music industry. Such flows challenged the notion of cores and peripheries in music, where globalisation usually proceeded from the West to the more impoverished countries elsewhere. The complexity of musical (and human) migration points to the fragility of international borders, and further questions the congruence of place and culture. National boundaries are increasingly irrelevant to the location of cultures that have become transnational (Box 8.6). Many musical cultures, like cultures more generally, are

> rooted in specific histories of cultural displacement, whether they are the 'middle passage' of slavery and indenture, the 'voyage out' of the civilising mission, the fraught accommodation of Third World migration to the West after the Second World War, or the traffic of economic and political refugees within and outside the Third World.
>
> (Bhabha 1994: 172; cited by Guilbault 1997: 31)

As new geographies have been created through human movements, so they are paralleled in musical cultures. In some contexts, whether within Aboriginal Australia or in the French West Indies, musicians have acted as cultural brokers – even political geographers – in promoting new national and transnational identities.

At least in urban Nigeria: 'Highly mobile and positioned at important interstices in heterogeneous urban societies, they forge new styles and communities of taste, negotiating cultural difference through the musical manipulation of symbolic associations' (Waterman 1990b: 9). Music changed in the diaspora, as it reflected new contexts, technologies, opportunities and performing situations, empowering migrant groups by 'staking out a unique cultural space in the host nation...an entirely new space, one that asserts and affirms both aspects of their hyphenated identities' (Diethrich 1999: 36). Music also affected the national identity of host nations. Thus, in France, Maghrebi migration influenced French culture – from the more prosaic presence of couscous and tabouli in supermarkets and restaurants, to the novels of Tahar Ben Jalloun and the music of bands like Carte de Séjour (literally 'residence card') or of Amina, of Tunisian origin, France's entrant in the decidedly mainstream 1991 Eurovision song contest (Gross *et al.* 1994: 29). Music is one component of ever-evolving cultural, national and international identities.

Box 8.6 Aboriginal reggae and hip hop

Growing connections between Aboriginal Australian reggae and hip hop and black music politics from North America exemplify the transcultural possibilities opened up through music. Reggae and hip hop have become important music genres for many indigenous musicians and consumers across Australia, both in rural and urban settings. In more remote areas, reggae became a common soundtrack to community life (along with country and western, rock 'n' roll and traditional song) during the 1980s. The largely Aboriginal (and non-European) regions such as Arnhem Land (the 'Top End' of the Northern Territory) and the Kimberley region of Western Australia have witnessed the growth of vibrant reggae scenes, from which bands such as Yothu Yindi, Blekbala Mujik and Sunrize Band have gone on to achieve varying levels of national and international exposure. Musical scenes maintained intricate connections to the Aboriginal land rights movement, which accelerated after the granting of citizenship rights in 1967 and during the term of the Labour Party government during the 1970s. This movement involved the generation of a national indigenous consciousness, captured in the red, black and gold colours of the Aboriginal flag, which also adorned record covers, were worn by reggae musicians, fans and activists, inscribed on T-shirts, badges, political posters, electric guitars, and tattooed on bodies, as symbols of resistance and cultural pride. Lyrical concerns mirrored issues in wider national political arenas, articulating a politics of cultural survival, human rights and post-colonial liberation. These themes were vividly expressed in a classic reggae track,

'We Have Survived' (1981), from No Fixed Address:

> You can't change the rhythm of my soul,
> You can't tell me what to do,
> You can't break my bones by putting me down
> Or by taking the things that belong to me.
>
> WE have survived the white man's world,
> And the horror and the torment of it all
> WE have survived the white man's world,
> And you know you can't change that.

During this time, Bob Marley and the Wailers toured Australia; on stage Marley alluded to the plight of indigenous Australians and called for greater solidarity with his own pan-African cause. Many indigenous musicians, including members of No Fixed Address, attended this series of concerts, absorbing reggae styles and attitudes into their own repertoires. During the 1990s hip hop also became an important influence for indigenous music, particularly in urban areas. Aboriginal hip hop groups emerged (such as Raven, Native Rhyme Syndicate, Nokturnl), indigenous rock bands linked rapping with black nationalist politics, while touring American artists including Public Enemy, Ice-T and Michael Franti all established connections with local performers, Aboriginal radio stations and communities, solidifying lines of cultural exchange initiated by consumption of black American recordings. However, some Australian Aboriginal political leaders suggested that hip hop culture, despite its unique and distinctive Australian flavour, had a debilitating effect on the maintenance of local traditions, as it and other forms of popular entertainment drew indigenous teenagers away from customary law towards urban areas and pastimes. Hip hop music and fashion had become yet another form of American imperialism alongside hamburgers and soft drinks, though participants rarely shared this perspective. As in New Zealand, Australian Aboriginal engagements with hip hop commonly mobilised the rhetoric of black pride and unity apparent in Afro-American rap, translating earlier Aboriginal political movements centred on land rights and cultural survival for younger generations. Selected elements of reggae and hip hop (and, by inference, black music traditions and politics) were transformed into indigenous Australian circumstances – particularly their political focus and use of visual symbols to affirm local cultures and attachment to land.

See: Gibson 1998; Neuenfeldt 1993.

Figure 8.1 Club flyers in Leeds
Note: In the 1990s many clubs consciously evoked international references in flyers and other promotions.

Cities have become the sites where social and cultural transformations have taken place, whether through migration, the growth of youth sub-cultures or the mediascapes opened up by entertainment corporations. Cities are places of migration, and host a range of musics that have travelled with migrants who partly differentiate and simultaneously characterise particular social areas and networks within cities. Whatever has become of the music, migrants have retained distinctive ethnic identities through that music. At one level this is banal. Migrants sing the songs and play the music of the regions from which they have come, whether African American migrants coming as slaves to North America or, much later, as economic migrants to northern cities, or Irish emigrants in Australia. Lyrics often encapsulate resistance, nostalgia and diverse senses of identity. Music is one audible example, along with language, of group solidarity and common ancestry. Yet music, just like migration itself, has been 'restless, ever-evolving – through absorption and transformation...disrupting the supposed certainties of ethnic and national cultures from a position in the margins of these cultures' (Gilroy 1993b: 89). Cities are also sites of production and exchange for new sub-cultural styles that emerge from homeland or diaspora. Reggae music (mythologising African roots) emerged from Jamaica, and travelled to London (where it was mobilised by West Indian migrant communities, eventually having a strong influence on punk bands such as the Clash) and to New York (where sound systems developed in Jamaica would become the basis of hip hop performances on street corners, parks and basketball courts). Over time music has been transformed in a variety of ways (in lyrics, instruments, rhythms, format and so on) into something quite new, especially when it was taken up by other minority, and migrant, groups, without ever losing some link to 'home', whatever the degree and duration of displacement.

Despite globalisation, transnationalisation, international migration and the commercial underpinning of music, each musical genre, in every place, required at least some local identification, and had its own internal musical structure, its particular technology, performative contexts, and social and political environment. Not only have local musics occasionally resisted, or failed to be swept up by globalisation – as in parts of Asia – but transnational cultural products, in whatever direction they appear to be travelling, do not simply replace local ones, but are refashioned and given new meaning. Rather than eliminating specificity and creating homogeneity, capitalism absorbs and works through difference, resulting in multiple capitalisms and multiple modernities, though it remains centred in the West. Societies can no longer be seen as 'self-contained, authentic, meaning-making communities; rather most communities are derivative and mutually entangled, enmeshed in the complex, power-laden relations between local worlds and larger systems' (Lockard 1998: 266). In a world of decentralisation, fragmentation and compression, 'all cultures are involved in one another; none is single and pure, all are hybrid, heterogeneous, extraordinarily differentiated and unmonolithic' (Said 1995: 15). Popular music reflects this combination of hybridity and differentiation.

9

AURAL ARCHITECTURE

The spaces of music

The sound of music

Music, through its actual sounds, and through its ability to represent and inform the nature of space and place, is crucial to the ways in which humans occupy and engage with the material world.

Sound is invisible; music cannot be seen. Yet music plays an important role in defining our behaviour in certain locations, creating a mood or atmosphere, eliciting reactions and responses, and reinforcing roles in particular geographical situations. From music at funerals or in aeroplanes, the fragments of music used between innings of baseball games, the regal music of official state affairs, to the plethora of spaces dedicated to live music (bars, clubs, entertainment centres), there are cultural expectations of what music to encounter in particular places, as associations are created between sound and space. Music does not exist in a vacuum. Geographical space is not an 'empty stage' on which aesthetic, economic and cultural battles are contested. Rather, music and space are actively and dialectically related. Music shapes spaces, and spaces shape music. In various ways sounds have been used to create spaces and suggest and stimulate patterns of human behaviour in particular locations. Corporations have taken advantage of music's emotive qualities to stimulate productivity in their workforces, piping music into factory floors, office blocks and shopping centres, reinforcing a desired corporate image and marketing strategy, or building 'walls of sound' between shops, shoppers and unwanted minority groups. Individuals have transformed their bedrooms into places of recreation or recording, or their lounge rooms into concert venues.

Sounds themselves can also occupy sites of political tension. Political music, from reggae and hip hop to the penetrating bass of 'underground' techno, have all been associated with certain movements and ideological concerns, and employed in ways that define certain spaces and events. Music cannot be thought of as 'pure' sound, divorced from the social and economic contexts in which it is produced and distributed, or from the range of cultural settings in which people make meaning out of music: 'the treatment of music as purely a kind of sound (as opposed to a whole ensemble of practices such as dancing, playing and so on) is a specific cultural construct, and not universally valid' (Sterne 1997: 24). Understanding how the sound of music can alter spaces, and people's interactions with them requires differentiation of the ways in which music is consumed and experienced in different locations and contexts. In addition to recognising the wider economic and political prerogatives behind the production of certain 'sounds', it is necessary to understand that listening is 'an ambiguous term that shuttles between activity and passivity. In part, this...is an effect of the social organization of music in a capitalist mass media environment' (Sterne 1997: 25). The extent to which people switch between modes of listening varies with personal preference (we may choose to 'tune in' to music in the background at a restaurant or bar, or alternatively 'tune out' at a concert), but is also influenced by the ways in which music is deliberately played in certain spaces.

Music can be used in certain contexts to create a sense of space, to reaffirm various social identities or challenge or subvert power relationships; thus the music of kd lang can challenge the ways in which everyday, urban spaces are gendered in particular ways (Valentine 1995). While few may listen actively to a kd lang song being broadcast as background music in a bar, restaurant or fashion boutique, others may positively identify with the music. In city streets the booming basslines of hip hop spilling from car windows may act as an affirmation of male bravado (or, depending on the context and interpretation, be seen as an affront to pedestrians in otherwise busy, commercial areas). Buskers on street corners, cathedral steps, subway stations and on the trains themselves create public sounds and informal economies. A sometimes seasonal phenomenon, buskers rely on certain physical sites that satisfy both aural needs (reverberating fragile acoustic sounds) and commercial intent (maximising passers-by); buskers may annoy shopkeepers and urban planners, yet suggest a vibrant cityscape welcomed by tourism promoters that challenges the anonymity (and alternative noises) of public space. Underground railway platforms, passageways and sometimes the trains themselves are temporary concert venues, while particularly popular tourist venues, such as the steps below the Sacré Coeur in Paris, are contested arenas for buskers of different kinds. The best create 'transitory communities' of spontaneous human contact, across ethnic and linguistic boundaries (Tanenbaum 1995: 105–6), invoking moments of nostalgia, pleasure and diversity: unexpected underground harmonies. Concerts change the nature of any space – from pubs to the open fields of raves and festivals, while carnivals may invert the urban order at least temporarily. The idea that sounds can influence behaviour in spaces is nothing new – the classical neo-gothic architecture of cathedrals and churches in part accommodated certain styles of choral singing and religious music (creating sombre, heavenly effects with sustained reverberations in large, open spaces); Victorian concert halls were designed to maximise the acoustic reach of individual performers' voices. Music may not always shape the spaces we inhabit, but in the media-saturated environments of contemporary cities and towns it has become a ubiquitous, and often deliberate, presence.

Music, commerce and the control of space

In the first three decades of the twentieth century, various 'sound pioneers' experimented with ways of controlling human behaviour in particular spaces. In 1911 industrial economist Frederick Taylor developed a system of 'managing' workers and their activities to maximise efficiency and reduce labour costs. This highly mechanised, impersonal system of management involved intense time and motion studies to minimise the time taken for each basic function and eliminate activities unrelated to the production process. 'Taylorism', as it became known, was accompanied by similar sorts of research in less obvious areas including 'piping' music into particular spaces, including homes, as a 'soundtrack' to daily life: 'It was the dream of any corporation to plug electronic lines into homes and businesses, transmitting not only music but advertising and public service announcements' (Lanza 1994: 27), and so into workplaces, where studies had already suggested that factory efficiency could be improved through broadcasting certain types of music at particular times of the day.

The North American Company (based in Cleveland) established Wired Radio Inc. (later to be changed to Muzak, by combining the words Music and Kodak), which began to transmit specialist programmes that could be subscribed to by businesses. During the Second World War Muzak transmitted 'uplifting' sounds to soldiers, alongside military information and warning signals. By the late 1940s 'travel Muzak' was broadcast on trains and passenger ships, and Muzak had become commonplace in workplaces; major corporations, including the Federal Bank and Bell Telephone, were customers of Muzak channels: 'the idea was to combat monotony and offset boredom at precisely those times in a work day when people are most subject to these onslaughts' (Lanza 1994: 49). Muzak workplace channels broadcast fifteen-minute cycles, alternating subdued and more stimulating songs in order to manipulate workers' 'fatigue curve'. While this may have enhanced the quality of the working day, its primary objective was to maximise worker output, as explained by Richard Cardinell, a researcher examining the productivity gains of Muzak in the 1940s:

> Factors that distract attention – change of tempo, loud brasses, vocals – are eliminated. Orchestras of strings and woodwinds predominate, the tones blending with the surroundings as do proper colors in a room. The worker should be no more aware of the music than of good lighting. The

> rhythms, reaching him [*sic*] subconsciously, create a feeling of well-being and eliminate strain.
>
> (quoted in Lanza 1994: 48)

Many types of pre-programmed music have subsequently emerged, including pre-recorded tapes of music by retailers and factory managers; genre-specific channels on airlines; music video programmes, sold on pre-recorded laser discs; and stock-music libraries, which can be used to suit a particular function – for an advertising jingle, to provide an atmosphere of excitement and tension at art shows or product launches, or to re-create a sense of space and place in film. In libraries such as those developed by Chappell & Co. and HMV, human emotions, social situations and geographical places are associated with sounds, all ordered and categorised in easy-to-retrieve formats for whatever commercial purpose. In this new format, performing artists were rarely, if ever, named or credited on recordings, so the music could be used without paying royalties or musicians' union fees for public broadcasts. Rather than be given titles, these stockpiles of musical product were specifically linked to their intended purpose, the spaces into which they would be broadcast, or with what they were intended to be associated:

> Natural landscapes, the flight of migratory birds, sunsets, rippling shores, orphans running through war-torn streets, a Tokyo massage parlor, cattle roaming through vast fields, lovers ogling on a beach, big city lights, Saturday afternoon at the rodeo, international travel – the infinite tableaux of set designs with their attendant tunes boggle the mind.
>
> (Lanza 1994: 62)

From the 1960s this music became widely used in shopping malls and theme parks, where the intended effects of broadcasting music largely involved consumers remaining in a retail space, and maximising their purchases. In shopping malls different types of music were used to reinforce corporate images created by certain stores, and music formed part of those desired images. Instrumental music, orchestral arrangements and light classical pieces were common in more open spaces and hallways, providing a sense of continuity and uniformity. This music neither interfered with the consumers' conscious interactions with space, nor overpowered the various noises and distractions offered by individual shops, either as 'background' music, played in hallways and other quasi-public spaces in a shopping mall (eateries, toilets, car parks, etc.) and 'foreground' music, generally played inside speciality stores, where retailers encouraged a different mode of listening from the passive light music of the hallways (see Box 9.1). Music is used not simply to reflect certain marketing objectives, or to add a layer of environmental and sensory input over the existing layout; rather music is an integral component of how these corporate images, strategies and built structures were developed in the first place:

> The Mall of America both presumes in its very structure and requires as part of its maintenance a continuous, nuanced, and highly orchestrated flow of music to all its parts. It is as if a sonorial circulation system keeps the Mall alive....Music becomes a form of architecture. Rather than simply filling up an empty space, the music becomes part of the consistency of that space.
>
> (Sterne 1997: 22–3)

In shopping malls, then, music can delineate different parts of the architecture, marking out social and economic boundaries, and effectively 'labelling' zones of consumption intended for particular demographic groups.

Box 9.1 The Mall of America

The Mall of America, in Minneapolis/St Paul, the second largest enclosed shopping centre in the world, attracts millions of visitors per year, and is marketed not simply as a place to purchase needed goods and services, but, because of its sheer scale, is presented as a tourist destination in itself, 'a built space devoted to consumerism' (Sterne 1997: 25). The Mall itself mimicked certain spatial images – hallways inside the centre are called 'avenues', while the Mall as a whole attempted to recreate a sense of national identity, representing uncontested versions of 'American-ness' in non-threatening promotional material. While 'background' music was provided by the 3M Corporation throughout the Mall, different strategies were employed by various shops to draw in consumers, or enhance their particular marketing image. In a Levi's fashion store, a wall of television screens played music videos pre-programmed onto laser discs, clearly visible from outside the store, to distract and draw consumers in. At Victoria's Secret, a lingerie store, selections of romantic, light classical music were played in the store to enhance the impression of refinement and European sophistication that was part of a broader marketing strategy – a particular equation of taste and class. The selections of classical music were available for purchase in the store, so that consumers could re-create the same environment at home. By 1995 Victoria's Secret had sold over 10 million specially packaged tapes and CDs of music played within the store (Sterne 1997: 36). Meanwhile, at Johnny Rocket's, a chain of 1950s style diners, recordings of old rock 'n' roll numbers were integral to the 'experience'. The store attempted to seduce shoppers through music and the promise of a nostalgic journey into a 1950s environment. Very loud music alerted passers-by to this potential experience; décor, menu design and store layout reinforced the retro trip, while, occasionally, employees mimed along with the piped music into ketchup bottles.

The ways in which music was used as part of these marketing strategies not only reinforced, but also could actively create, social divisions and hierarchical structures within the shopping mall:

> Music programmes correspond to the demography of the Mall's desired, rather than actual, visitors. While the Mall desires an affluent (and usually white) adult middle-class population, there is strong evidence to suggest that the real enthusiasts of the Mall are teenagers from a diversity of racial backgrounds...these teens must make use of an environment that is not immediately welcoming to them; or rather, which welcomes them as consumers first, and people second.
>
> (Sterne 1997: 43)

The use of programmed music in the Mall of America was not only integral to commercial strategies to convince consumers to stay longer, spend more and identify with particular taste markets, but could also maintain and perpetuate wider social boundaries and divisions.

In a variety of contexts beyond factories, offices and shopping malls, music shapes and even controls space. Various attempts have been made to alter human behaviour in public spaces through the broadcast of music: at Olympic venues upbeat live jazz was used to distract impatient crowds waiting in line for trains or for hot dogs, while classical music has been used as a deterrent to youths congregating in public spaces. In 1999 a suburban shopping centre in the Australian city of Wollongong began playing Bing Crosby music to stop teenagers hanging out there. Simultaneously an Operation Music was launched there and in various outer Sydney suburbs to play classical music on railway station platforms to soothe potential thugs (Luckman 2001). Music served as both 'welcome mats' and 'keep out' notices (DeNora 2000: 136). Music also shapes tourist spaces, as in New Orleans, where 'music is used to assist in marking spaces where revelry is permitted. Music serves as a signal that a space is open for occupation. Within the French Quarter, where the music stops, tourists hesitate to venture' (Atkinson 1997: 105; Chapter 10). In very different arenas of the private and public sectors music has become an emotional and commercial tool.

'Lounge' and ambient music: domestic dreaming

From the 1950s the recorded music industry, which had a distant relationship to Muzak and pre-programmed music in general, began to adopt the rhetoric, if not the format, of these types of music. Whereas the assumption had always been that consumers bought recordings to deliberately listen to them, many records began to be released that were specifically designed to re-create particular home environments: 'music as wallpaper' played in the background at home to add a mood, to generate a particular atmosphere, or to convey the sounds of exotic, otherworldly places that consumers might dream of visiting. The liner notes to various albums indicated the particular uses to which recorded music could be put (*Music for Dining*, *Music for Dancing*...), encouraging listeners to perceive certain things in the music. Many albums from the 1960s attempted to 'capture' the sounds of far-away places, vicariously transporting the armchair listener to idyllic holiday destinations, mysterious Pacific Islands, Alpine heights and cosmopolitan European streetscapes. *In Love in Paris*, by the Musette of Renaud and Carlini's *World of Strings* (*c.*1968), re-created the archetypal romantic French encounter: 'Montmartre, the Champs Elysee [*sic*], The Seine – all speak of the romance of Paris. Here is a musical journey to Paris by night...Renaud with the World of Strings is your passport to Paris'. Meanwhile, Mantovani's *Continental Encores* (1959) took the format of the 'musical journey' to its extreme – the record's packaging included a ten-page photo album of various exotic European locations: a map of Paris, shots of cafés, Spanish churches, Venetian gondolas at sunset, folk dancers, and bikini-clad bathers on Mediterranean shores:

> For a long time now entertainers have known about the power of music to evoke atmosphere. The circus and the fairground have always used their own special brands of cheerful melody, which are as much a part of those entertainments as the exuberant, closely packed type on their posters. With sound track and disc recording far advanced in technique it is possible now to use this power of association over an ever wider range, and moreover to bring it into our own home. To put it another way, it can now be employed to take you out of your home, and lead you gently by the ear to familiar or imagined places. And so with all the tunes in this album Mantovani produces a sort of heightened impression of the places he takes us to in his music...a musical flip round the continent takes in Italy, France, Switzerland and Germany, each one a natural home of easy, spontaneous melody....Coupled with the photographs in this great Mantovani album, this is as vivid and comfortable a journey in sound as anyone could possibly devise.
>
> (Anon. a, 1959)

Stanley Black and His Orchestra's *Place Pigalle* emphasised these ideas in a particularly essentialised view of Paris, appealing to an Anglo-American fascination with European exotica, and the fears of the outside world that often accompanied overseas travel:

> Why go to all the bother of getting a passport, travelling on boats, trains and planes, buying sea-sick tablets, worrying about the customs, bothering your head by complicated sums with francs and pounds involved. Why not just put on this wonderful record and let Stanley Black take you on a conducted musical tour. This most reliable of guides who has already taken us to South America, the tropics, and the South Sea Islands with the utmost safety, will find no difficulty in such a short flip as Paris. Sure enough, the moment you put this record on the turntable you only need to close your eyes and there you are strolling along the banks of the Seine being shocked by the postcards – the price that is, or nibbling at your croissant accompanied by a French coffee at a table on the sidewalk as the sun rises over Paris, dodging the rush of French taxis...wandering in the Place Pigalle in the lamplight.
>
> (Anon. b, 1958)

The liner notes to this recording speculated on the reasons why music evokes place:

> Even if you have never been nearer to Paris in your life than the stony shores of Folkestone, you can quite easily imagine you *have* been there simply by listening to the music. There seems to be some extra sense, not yet explained by the scientists, that connects every place in the world with its representative themes.
>
> (Anon. b, 1958)

At a time when mass tourism was just getting underway, these recordings became a feature of Western suburban homes, alongside other creature comforts. They were thus bound up in the social and political issues that surrounded the emergence of suburbia as an urban economic form. In liner notes, intended audiences were targeted, and the functions of these recordings were explained, revealing assumptions about the gendered nature of domestic spaces (see Box 9.2). Half a century later, the artefacts of lounge music have ironically become fashionable: from collectors of throwaway vinyl (such as Australia's 'op-shoppers'), to the resurgence in lounge music in both reissues of old recordings (such as Capitol Records' Ultra-lounge series), consumed as parody and kitsch.

Box 9.2 'Lounge' music spaces

Liner notes to albums released in the 1950s and 1960s emphasised the ways in which music was increasingly marketed to consumers as a medium to transform domestic space. Generally associated with certain stylistic features (a predominance of strings, brass and orchestral arrangements of popular tunes, and a lack of vocals, guitars or drums), these 'lounge' releases were promoted as mood-setters, accompaniments to family life or to domestic social occasions (dinners, birthdays, Christmases, etc.), and as aural 'postcards' of far-away places beyond the confines of suburbia. Music was aligned to activities within domestic space, illuminating entrenched gender roles and assumptions about women's position in the home. The BBC accentuated gendered divisions of labour, targeting its 'Music while you work' broadcasts to an assumed female home workforce. Individual releases such as those of Frank Chacksfield offered advice for the 'appropriate' musical accompaniment to a dinner party:

> The musically aware hostess no longer allows the butler, or her husband, to sling records on to the turntable in a haphazard way. She no longer risks the dangers of the soup being spilled by Haydn's 'Surprise' symphony, of Mrs. Alias-Jones choking over the fish because an ill-timed bit of jazz trumpet has frightened her. She now supplies a ready-made background of elegant and suitable music to smooth the evening into one long feat of pleasure and unshattered nerves.
>
> (quoted in Lanza 1994: 77)

For men, the Melachrino Strings offered relaxation and respite after the drudgery of a working day:

> It's time to relax. Let the big automobiles crawl along, bumper-to-bumper, let the airplanes drone overhead, let the trains rattle along the rails. It's time to relax. To go loose all over. Read the titles of the songs in this album. By the Sleepy Lagoon...isn't that restful? Vision d'Amour...doesn't that feel good? Autumn Leaves...feeling drowsy? This is music, too, for soft romance... music for just sitting and thinking about some day soon....In some far-off countries, people relax by standing stiff as ramrods, or by easing themselves onto a bed of nails....Listening to Melachrino's superb essays on familiar themes certainly seems very much more civilized. A farmer may relax by leaning on the handle of his hoe and staring out over a field of softly

> swaying wheat; a plumber may relax by sagging his weight onto the handle…listening to this music is much easier.
>
> (The Melachrino Strings and Orchestra, *Moods in Music: Music for Relaxation*, 1958)

These recordings filled domestic spaces through sound (whether as exotic journeys, sophisticated dinner parties or relaxing breaks from domestic chores), and also helped to define those spaces: particularly through the division of labour, where women worked and men relaxed, against the noisy, crowded world of outside public spaces, beyond suburban utopia.

See: Lanza 1994.

Producing music to create a sense of space, or to re-create the impression of particular environments, boomed with the advent of 'New Age' philosophies, esoteric book stores, herbal remedies, a range of spiritual beliefs and new sorts of mood music: sounds of the rainforest, recordings of whale-songs, babbling brooks and the soothing sound of rainfall. Syntonic Research, in New York, was among the first to release a series of such recordings in the 1970s (simply titled *Environments*) that in themselves were not strictly musical, capturing thunderstorms and other natural sounds. In the case of the Mystic Moods Orchestra, the recording of sound effects included thunderstorms and rainfall to accompany their string arrangements, meant to develop an intimate, sensual environment for the listener, 'music to do whatever you want with whoever you want whenever you want', complete with soft-pornographic inner-sleeve covers and titles such as 'Stormy Weekend', 'Erogenous' and 'Touch'. In many 'New Age' releases, from the sexually explicit nature of the Mystic Moods Orchestra to the platonic spiritualism of Kitaro's Silk Road or the science fiction fantasies of Jean-Michel Jarre, central themes included establishing imaginary spaces for the music consumer, whether this involved passive or active listening, and a sense of journey, of travelling to otherworldly locations.

Ambient music, designed to encourage relaxation and even sleep, emphasised 'special' places both generic and real, which were remote from urban centres, physically attractive – usually involving mountains, falling or flowing water (in streams rather than rivers), rainforests, coasts and oceans, occasionally deserts, and more generally 'wilderness'. Such places were imbued with certain spiritual powers, or associated with particular animals regarded as having special qualities, especially birds, dolphins and whales. Bird sounds were incorporated into many tracks. The liner notes to Andy Holm's *Blue Mountain Tales* (1996), on Primal Harmony Records, stressed the significance of the World Heritage site that is the Blue Mountains National Park, immediately to the west of Sydney:

> During my frequent travels to the Blue Mountains I have always been inspired to pick up my instruments and play out the feelings that arise within me. This album is the product of the precious moments. Whether I played my flute at the bottom of Wentworth Falls or the didgeridoo at Echo Point, the vibration of the surroundings always influenced a strong tribal and spiritual sound, a reflection of my inner self opening up at the Blue Mountains beauty. Although this album is a fairytale story, it is reflecting on my true emotions brought forward by the ambience of the Blue Mountains. Close your eyes while you are listening, let the narrator be your guide and join me on a magnificent voyage through time, sound and space, and a dimension you will find within yourself listening to this work.

Similarly the Australian musician Ken Davis, seeking to enter the North American market, released four albums in 2001: *Tai Chi Music*, *Pan Flutes by the Ocean*, *Dolphin Magic* and *Spirit of Cedona*. On the last of these, Cedona was described as 'a spiritual centre one hundred miles north of Phoenix and is world-renowned for its spiritual energy and geological splendour...a focus for people seeking a spiritual environment. It is known for its healing spirits.' Some performers actually recorded within the natural landscape, but more frequently entitled their albums with the names of specific places (such as Uluru (Ayers Rock) and Kakadu, in Australia) that had particular natural and spiritual significance. Similar themes recurred in ambient techno: DJ and producer Carl Cox encouraged listeners to immerse themselves in certain spatial contexts in order to better experience his music. Accompanying *Two Paintings and a Drum* (1996) was a series of notes for each track, explaining their intended meaning, and the imaginary spaces that consumers should create:

> Close your eyes, on your sofa, on the dancefloor, wherever you are. Let the rhythms and melodies take you deep 'In Your Mind' to an arctic world of freezing tundra and barren mountain ranges. You are in an ice cave high on the steppes, outside a tempest of ice and snow rages. Rubbing two sticks together you make a fire and settle down for the night praying for safety from the maelstrom outside.
>
> (Cox 1996)

Ambient music also stressed the mystical and 'cosmic' life forces that were associated with 'traditional peoples', their links to the earth and their musical instruments. In Australia the didjeridu features in much ambient music, especially that of the Aboriginal performer David Hudson, but panpipes and flutes have also been common (though not necessarily tied to their geographical origins). Beyond generic indigenous peoples, others who had particular spiritual powers might also feature: shamans on some North American music and monks in Ireland. For David and Steve Gordon's *Sacred Spirit Drums* (Planet Earth Music, 2001) the company website exulted in a:

> Melodic, magical, danceable story about a Shaman who journeys into the Spirit world to find a way to heal the Earth and its people....David and Steve both use many types of drums and indigenous rattles, Native American Flutes and Incan Pan Pipe. David adds keyboard orchestrations while Steve plays classical, steel string acoustic and electric guitars, mandolin and various electronic instruments. Finally the copious nature sounds were recorded at various National Monuments and parks.

Irish ambient music often incorporated uilleann pipes, and Scottish music the bagpipes. Natural sounds (an explicit conjunction of indigenous peoples as 'natural') contributed to the primitivist fantasies of tranquillity, timelessness and human interactions with nature. Relaxation and spiritual healing were the anticipated consequences.

Ambient music stressed virtual tourism, both in its content and marketing, which was centred on tourist venues and airports rather than music stores. The notes to Ken Davis's *Early Morning in the Rainforest* (1985), which blended bird song and electronic music, enjoined 'Experience the feeling of being in the rainforest while still in the comfort of your own home.' Rather differently, the website of Australian performer, Tony O'Connor, included a 'guestbook' where listeners could record their thoughts on his music, as Clare Power from the Isle of Man registered in 2001:

> Australian friends introduced me to your music some years ago....I introduced the wonderful flowing sounds to the T'ai Chi Chuan Group, for Shibashi practice. Also your *In Touch, Bushland Dreaming*, *Kakadu* and *Uluru* keep me in touch with my daughter's family and my friends 'down under'....Thank you again, Namaste.

Emigrant Australians, such as Robert, an 'Aussie in Denmark', may 'listen to your music when I am homesick'. Ambient music, with its links to New Age and world music, and its combination of the electronic and the natural, creates a series of generic and particular places, which reflect and induce harmony, in a genre that has become exceptionally popular yet largely distinct from other forms of popular music, through its tourist orientation and the common absence of live performance.

The places and politics of performance

The advent of club cultures, raves and new styles of music such as drum and bass, house, trance and acid house was predicated on the use of certain forms of music, patched together in particular ways to transform spaces symbolically, if not materially, through sound. Raves, clubs and other formats of dance parties became common youth practices in many countries in the West and to some extent elsewhere, including Brazil, Thailand, Indonesia and Israel, in the late 1980s and 1990s. In dance parties, participants deliberately created spaces to suit certain types of music and drugs, while producers of music created sounds to suit particular spaces. Some themes of 'New Age' music – escape, transformation, ambience – were also apparent in these louder, rhythmic, electronic landscapes.

The particular use of space gives credibility to dance music scenes – whether in the transformation of otherwise unused urban spaces, subversively appropriated for the consumption of music, or in the lavish décor of expensive clubs. Dance sub-cultures measure the success of events in terms of the music (and drugs), but also of the specific sites chosen as venues, and the transformation of these sites into imaginative landscapes (Gibson 1999). This could include the physical space – its size, position of DJs, chillout rooms for relaxation, quality of lighting rigs – and attempts through decoration and interior design to construct imaginary play-scapes, reflected in the common practice of mapping the layouts of venues on flyers and advertisements. Physical sites create particular atmospheres; the 'original' rave spaces of the late 1980s and early 1990s relied on the appropriation of locations generally associated with purposes other than musical performance, and their transformation (Thornton 1995). Early raves took place in rural locations, challenging usual urban–rural disjunctures, and destabilising 'the perceived axis between urban location and authenticity' (Gilbert and Pearson 1999: 23), but later returning to the anonymity of the cities.

Such alternative, transformed spaces – albeit temporary – took raves beyond the immediate control of local authorities, emphasising their transgressive nature and suggesting parallels with anarchist concepts of the temporary autonomous zone (TAZ), and its transitory potential for dynamic liberation in the face of commercial culture (Bey 1991). In Blackburn, England, dance parties surrounding the acid house music scene, inspired by New York loft parties, almost all held illegally in empty warehouses and linked to the consumption of illicit drugs (predominantly ecstasy and speed), provided a sense of release, emanating from the collective – almost conspiratorial – use of space, with venues kept secret until the last moment, mainly to avoid the police (Ingham *et al.* 1994; Ingham 1999). In London:

> The derelict locations – warehouses, car parks, railway arches – with their dusty floors and industrial ambience, offer only crumbling walls, a

> loud sound system, and the potential for anything to happen; unpredictable and unrestrained, the parties are imbued with the thrill of their dubious legality.
>
> (Lewin 1997: 90)

As in similar contexts – such as raves in inner Sydney and in rural areas beyond – actually getting to the venues and evading detection became central to the experience. In response to unsanctioned use of space and through moral panics over illegal drug use, various urban and national authorities reacted with legislation to silence the music. In Britain, a conservative government enacted the 1994 *Criminal Justice and Public Order Act*, which made the convergence in public space of large numbers of people, and lengthy broadcasts of music with repetitive beats, punishable offences (Wright 1993; Martin 1998; Sibley 1994). In countries like Australia such regulatory mechanisms as noise restrictions, codes of practice for dance parties, environmental protection legislation, fire and safety laws and alcohol licensing regulations, were used to shut down clubs and turn off disruptive styles (and volumes) of music (Homan 1998). As institutional control over raves intensified, they were effectively forced to give way to more ordered 'corporate clubbing', with less sense of adventure but similar music.

Numerous sub-cultural practices were employed to promote the images of motion and escape essential to rave and club culture. Motion was evident in the physical act of dancing, and metaphorically in the ability of a combination of music, drugs and collective consumption to 'move' dancers to another sensory level; meanwhile the idea of escape was reinforced if an event was held in an innovative space, removed from such 'standardised' spaces as rock venues. Where spaces became more standardised, as in most commercially oriented clubs, decorations enhanced the illusion: images of tribalistic societies or future 'other-worlds', Celtic symbols, mandalas, depictions of Krishna or science fiction imagery (computer-generated designs, abstract shapes and constructions) – all enhancing the feelings of distance from suburban life, and suggesting some forms of altered state. The names of events also similarly projected a sense of other-worldliness: 'Utopia', 'Field of Dreams', 'Bent in Space'. Even perhaps 'through a tapestry of mind-bending music, the DJ is said to take the dancers on an overnight journey, with one finger on the pulse of the adventure and the other on the turntables' (Hutson 2000: 38). Drugs reduced inhibitions and created an atmosphere of goodwill, distinct from the predatory aspects of clubs, and emphasised by more individualised styles of dancing, contributing to feelings of egalitarianism, and a context where, for many women, raves represented freedom from sexual invitation, opening up a 'new space for the exploration of new forms of identity and pleasure', and some challenge to the centrality of heterosexual masculinity (Pini 1997: 154; Malbon 1999: 41–6). Yet such spaces were never wholly removed from gender relations: men continued to dominate as DJs or subcultural gatekeepers, while sexuality resurfaced in less androgynous club scenes.

As dance parties became increasingly associated with licensed clubs, neo-tribal imagery partly made way for more luxurious surroundings – prestigious nightclubs with more attention to interior design, creating the impression of exclusivity and 'class'. In ways reminiscent of the opulence of disco clubs of the 1980s (such as those captured in the film, *The Last Days of Disco*), 'super clubs' such as Ministry of Sound, Gatecrasher, Cream and Home, which originated in British cities such as London and Liverpool, offered weekend retreat from the drudgery of office life, and opportunities for conspicuous consumption, necessitating the latest street or urban fashions, footwear, mobile phones and accessories, including drugs. Clubbers were particularly closely linked to ever-fluctuating fashion trends, and house music magazines devoted many pages to clothing and footwear, while similar clothing styles – and drugs – occasionally created moral panic (Verhagen *et al.* 2000; Gilbert 1997; Gibson and Pagan 2002). Dance music communities were further distinguished through the particular linguistic spaces they created, usually to subtly differentiate a panorama of musical sub-genres that had global resonance, but were impenetrable to those outside these scenes. The press and fanzines are replete with introspective descriptions of musical styles. One CD track, 'Bambaataa', was described in a Sydney club newspaper as:

> rejecting two step mathematical lumber and ice sculpture bass in favour of uncomplicated bottom end and polyrhythmic breaks. The bones of the breaks originally lie in Reprazent's 'Heroes' but Shy skin grafted new ideas of rhythm and injected organic life into the computer beast of Neurofunk.
>
> (*3D World*, 5 July 1995: 27)

Another mix:

> still uses the vocals with lots of trancey, wobbly, under the water noises mixed with some squelchy acid. The very melodic, dubby DJ Pierre remixed grooves are quite slow and repetitive, yet nicely hypnotic. More dubby wild pitch grooves from DJ Pierre on *Closer*: a nice melodic camp garage groove.
>
> (*3D World*, 5 July 1995: 27)

Knowledge of the symbols and language of such niche styles is crucial to participation, yet causes frustration for producers and DJs who often prefer more 'open' symbolism and stylistic descriptions. Thus, more than most other genres and sub-cultures, participants in electronic dance music scenes have been perceived as 'neo-tribes': loosely organised small groups, temporarily but intensely identified with particular styles and attitudes (Maffesoli 1988; Halfacree and Kitchin 1996; Malbon 1999; Gibson 1999; Bennett 2000), where opposition from mainstream culture heightened the sense of collective unity and exclusivity.

Much electronic dance music came to be consumed in ways similar to other previous eras of 'passive' music – from collections of electronica to café soundtracks of 'laid back beats', from the use of dance music in aerobics classes to techno segued to car advertisements or sports shows. However, while dance music may have become commodified, threatening the 'underground', community nature of scenes (Weber 1999), it has also been widely used to give new and sometimes radical meanings to spaces that were otherwise apolitical. Political uses of dance music have included the idea of 'techno as noise'; as with punk, certain sounds may alienate some listeners, or draw attention to loud political campaigns in immediate ways (McKay 1996). This underpinned anti-global free market rallies in major cities around the world, targeted at the World Trade Organization (as with S11 in Melbourne, M1 – May Day – in various locations), and more specifically 'Reclaim the Streets' events in such diverse cities as London, Stockholm, Bogota, Tokyo, Turin, Dublin, Geneva and Toronto (Luckman 2001). In these events, thousands of people gathered at strategic locations to 'reclaim' major intersections and disrupt city traffic, converting streets into dance venues, with quickly erected public address systems, road blocks and a range of distractions (including fun fairs, informal market activities, political stalls, etc.). Such events protested against the increasing privatisation of public space and the dominance of cars in urban transport and planning. In transforming motor-dedicated space, music contributed to challenging assumptions about the use of open spaces. As one participant stated:

> The whole point is to get out on the streets and make some noise. Other sorts of protests involve speeches from people, only those who are interested will hang around and listen to this. By playing really loud techno music, we are able to attract the attention of passers-by, some of them might get into the music and join in the action, others might be appalled by the music, but at least we're attracting their attention and making them think about it a bit.
>
> (Reclaim the Streets organiser, Sydney, 1998)

In other contexts, too, dance music could be seen as oppositional. Dance parties were particularly important among gay men, acting as a means of expressing group identity, meeting partners and, in the 1980s especially, 'permitting expression of psychological coping with the HIV epidemic' (Lewis and Ross 1995: 42).

Dance parties in Sydney, notably the Sleaze Ball, which accompanied the annual Gay and Lesbian Mardi Gras, were to some extent prototypes of subsequent house party scenes in their acceptance of diverse sexualities, of individual performance as creative expression and as one form of differentiation from mainstream society.

There are parallels here in the ability of (usually) Afro-Caribbean carnivals to create a space beyond the reach of racism, where black music has the power 'to disperse and suspend the temporal and spatial order of the dominant culture' (Gilroy 1987: 210), and of events such as gay pride marches (where music has a more muted role compared with spectacle) to challenge the patriarchal, heterosexual order of the city, in a manner that combined the political and the carnival. While all carnivals, often initiated as symbols of political dissent and celebrations of difference, became increasingly commercialised, in that commercialism lay some measure of wider acceptance of what was originally transgression: a transition from 'a temporary spectacle of inversion for a more permanent revolution in the social order'. At the same time their 'subversive power is reabsorbed into dominant structures of power and ordered norms of culture' (Mitchell 2000: 162), as also accompanied the absorption of rave into more commercial dance party scenes. Similarly, the Berlin Love Parade, established in 1989 by underground DJ Dr Motte, when only 150 people took to the streets in affirmations of 'peace, love and friendship', now attracts over a million participants from all over the world, and, somewhat ironically, has become a major tourist attraction and income-generating event in its own right (Richard and Kruger 1998; Borneman and Senders 2000) and is the largest regular mass event in Europe.

Placing gender and sexuality

Gender relations infuse all levels of musical activity, in expression and song, but also in the tangible places associated with the industry. Many such places are characterised along gender lines. The local pub, with its important role as a 'training' ground for aspiring musicians, the mosh pit, a typical guitar shop or the mixing board at any gig are all stereotypically defined as male spaces. Beyond commonplace descriptions of stereotypical male and female spaces in the world of music production and consumption persist subtler geographies of popular music that are equally gendered. Music plays a role of some 'social and political significance...[in] the construction and performance of identities; the empowerment of individuals and communities'. This relationship between music – now recognised as inherently gendered and sexualised – and identity gives it 'not only the power...to articulate sexual identities and communities but also [the] ability to facilitate the production of sexualised space (Valentine 1995: 474). Gendered identities and spaces in popular music remain contested. Indeed geographical space, rather than being simply the 'background' to these relations, is inherently embroiled in gender politics. Space 'is not an innocent backdrop to position, it is filled with politics and ideology' (Keith and Pile, quoted in Bell and Valentine 1995: 18). Space carries within its form the dominant ideologies and politics that, in respect to gender and sexuality in the realm of music, remain overtly conservative, patriarchal and heterosexual.

The earliest music scenes among traditional societies were usually gendered: men were performers and composers, and women were audiences, responsive to male prowess and apparently appreciative of it (Chapter 2). Little has changed despite the spectacular success of such different performers as Umm Kulthúm, Celine Dion or Barbara Streisand. The spaces that have been carved out by those who challenged the existing order have usually been limited in extent, transient and often carnivalesque: the 'old order' inverted but as play rather than future possibility. While the simplistic patriarchal attitudes represented in such 1960s hits as Brian Hyland's 'Itsy Bitsy Teeny Weeny Yellow Polka Dot Bikini' gave way to more subtle marginalisation, the rise of rap brought misogyny to the fore and with it more sexist lyrics. In an emergent era of constructed groups, male, female and mixed, beauty was as important as creativity (where procreativity and domesticity remained norms). From music hall days women were expected to 'grace' the stage with charm and sweetness; those who did not were aberrant, impossible to categorise and consequently perceived as hard to market and thus 'too difficult' for an industry that avoided risk. Gendered identities were particularly difficult to challenge.

Music may nonetheless be a means of challenging dominant paradigms of gendered identity. Many women, marginalised because of gender, and gay men and lesbians, marginalised because of sexuality, have used music both to generate community identity and appropriate space. The riot grrrl movement, readings of kd lang's music (alongside others) and subversive twists on conventionally 'straight' music styles, such as country and western, contest gendered and sexual norms, through generating musical expressions and through the use of real and figurative space. Popular music, as a 'facet of the cultural contest that begins where institutional politics ends' (Smith 1994: 236), is one component of social change and an important means for men and women, whether straight or gay, to appropriate, protect and delineate their 'space'.

Patriarchal gender stereotypes found elsewhere in society are familiar in popular music, and obvious in the lyrics of many songs sung by both female and male artists. Tammy Wynette's famous 'Stand by Your Man' (1968) talks of the need for women to support and care for male partners and Dolly Parton's 'Jolene' (1974) portrays a vulnerable lover concerned with losing 'her man' to a seductive woman. Men, like Ben E. King, demand of their female partners that they 'Stand by Me'. Country music as a genre has been especially conservative on issues of gender: the discourse of country is heterosexual, replete with traditional representations of gender and the 'nuclear family', and imbued with patriotic references (Mockus 1994: 259). Similarly, Asian women have sometimes been labelled as innocent, shy and docile, essentialising female identity in an uncritical manner (Hisama 1993). However, alternative readings of any song lyric are possible; thus within 'Stand by Your Man' the lines 'Sometimes it's hard to be a woman/ Giving all your love to just one man' have been interpreted as an expression of dissatisfaction with women's traditional roles within monogamous relationships (Kruse 1999: 87; Wilson 1995). Gender stereotypes, however, run rather deeper than their depiction in lyrics. They penetrate all facets of the music business and more generally restrict women's access and potential role within it.

Music contributes to the gendering and sexualisation of space through its role in the creation and maintenance of identity. Private meanings and signifiers may be attached to particular styles of music or musicians that are radically different from those intended by the artist. The personal act of consuming music – of listening to it in a specific way – creates a space for listeners where they are free to express a particular identity. Thus kd lang's music can create private, lesbian spaces in public, heterosexual settings via its communication of particular identity markers only relevant to lesbian listeners:

> [kd lang's] music facilitates the fleeting creation of a lesbian space. A 'private' space momentarily shared in 'public'. A space that is dynamic and fluid. It is a private space because, whilst a mutual response to a lesbian anthem allows women to read each other's identities, it does not mark them out as visibly 'different' from those others present.
>
> (Valentine 1995: 480)

Even in the individual act of playing kd lang on a Walkman or Discman: 'At the press of a button, you can summon up lang and all that she signifies, whilst the music, and hence the space, that you have created for yourself remain invisible to others' (Valentine 1995: 481). Many lyrics reiterate particular gender constructions, maintaining the 'normality' of these constructions as well as their corresponding geographies. The contrasting spaces described in the lyrics of Carole King and Bruce Springsteen provide a good example. In Carole King's song 'Home Again' from the album *Tapestry* (1971), she yearns for home as the place where she 'feels right' singing:

> I won't be happy til I see you alone again
> Til I'm home again and feeling right...
> I wanna be home again and feeling right

Later in the album, in 'Where You Lead', King again stresses the virtues of home when she sings: 'I always wanted a real home with flowers on the windowsill.' By contrast, Bruce Springsteen's lyrics, especially during his 'promised-land phase' between 1975 and 1981, reveal that:

> The setting for most of the lyrics is the street in public space, as in 'Born to Run', 'Racing in the Street', 'Meeting across the River', 'Jungleland', and 'Backstreets'. It is in these streets that the masculine dominated working class environment is best portrayed. A *machismo* prevails adding to the oppression of working class women.
>
> (Moss 1992: 175; see Box 2.4)

The contrast between King's depiction of the private home as the source of female happiness, a theme that also pervades country music, and Springsteen's public street as a source of male power is redolent of the spatial gender divisions found in the lyrics of popular music.

Vast numbers of songs have marginalised women, in some case deliberately as in the misogyny of several rap performers such as 2 Live Crew and Eminem. Jamaican dance hall, infused with violence, provided a template for rap. Bujo Banton's 'Boom Bye Bye' was denounced for advocating murdering gays. This, and his misogyny, resulted in a national ban on radio and TV stations playing songs that advocated gun violence (if not misogyny), in a gendered political space where men assumed absolute power in sexual relations with women. Women became bitches and whores, or 'ball-breakers' and 'gold diggers' in gangsta rap (Chude-Sokei 1997: 222), in a macho world of bragging about violence, illicit sex and substance abuse – the antithesis of more progressive lyrics on some form of emancipation and perceptions of popular music's focus on romantic love. The aggressive verbalisation of male dominance in Jamaican dance hall and in gangsta rap may actually be the impotent manifestation of a diminished masculinity, hence the recurring metaphor of the penis as gun in rap and ragga lyrics (Cooper 1998: 165). Other genres, including banda, the Dominican Republic's bachata and contemporary salsa, all with Hispanic origins, have similar themes. Before that, West Indian calypsos often highlighted stereotypes of women that appealed to the prurient interests of men, including one of the best known, 'Rum and Coca-Cola', which featured a mother and daughter 'working for the Yankee dollar', evidently as prostitutes. Lyrics and music emerge in part from difficult economic circumstances: 'one of the ways that men react to a global economy increasingly organized around the low wage labor of women is through affirmations of masculine privilege and denigrations of female independence' (Lipsitz 1999a: 222). Patriarchal ideas infuse many songs, subtly or crudely reinforcing sexist ideas in listeners, as in such songs as Buddy Holly's 'Peggy Sue' (Bradby and Torode 1984; cf. Berger 1999). Modern salsa lyrics are characterised by sexual objectification and increasingly explicit gender hostility, resulting in women 'singing back' to challenge misogynistic stereotypes, and middle-class women selectively distancing themselves from the lyrics and their performative context. Gender ideologies here, just as in urban Thailand, were thus both reinforced and contested (Aparicio 1998; Siriyuvakak 1998), a situation evident in most genres, for example in the rise of riot grrrls, gay transformations of lyrical contexts towards and beyond androgyny (Walser 1994) and the parody of lyrics, even in rap.

Any 'signifier' of cultural meaning in music only becomes significant when interpreted by consumers, who adopt and adapt music, often subverting its meaning. Frith and McRobbie claimed that 'the sexual meaning of rock can't be read independently of the sexual meaning of rock consumption', hence 'sexuality is constituted in the very act of consumption' (1990: 389). Meanwhile depictions of sexuality are ambiguous, being at once progressive and oppressive. On the one hand, music is a form of sexual expression. Much popular music is 'sexy' because it is inherently erotic, with pulsating rhythms and overtly sexual lyrics (Dyer 1990). Here, in its use of eroticism, music can be most radical. Using eroticism, artists campaign for sexual liberation – for both sexes – even positioning sexual freedom as 'one of rock's motivating forces' (McDonnell and Powers 1995: 1). At another level the personal is the political, evident in Wise's discussion of her own days as an Elvis fan. As a confused and isolated lesbian adolescent, being an Elvis fan 'filled a yawning gap in [her] life....He was a way of being acceptably "different" because it simply wasn't fashionable to be an Elvis fan when [she] was one.' For her, Elvis was 'a private and special friend', hardly the 'butch god' or 'phallic hero' proclaimed by his industry image (Wise 1990: 395, 397). This subtle, and personal, encoding of issues of gender and sexuality makes popular music, and more generally popular culture, pervasive tools for reiterating particular gender norms and conditioning society's acceptance of this as 'natural'.

Men dominate every facet of the music industry, monopolising both the creative and production sides, and replicating gender relations in wider society. In production, men have power over many of the crucial positions of influence; for example, as music technicians they have particular control over the manufacture of various 'sounds'. Similarly, men have monopolised 'gatekeeping' positions in the industry, as rock journalists, publicists and talent scouts, so filtering via a male lens what is and is not appropriate (see Frith and McRobbie 1990; Cohen 1997b; Gay 1998). Institutional barriers define areas where women can legitimately negotiate a role in the music industry, categorising these, demarcating their boundaries and, in the process, limiting any perceived threat to male power. Even where men's role has been limited they are likely to be attributed or claim credit for women's success. Janis Joplin found herself doubly disadvantaged, in terms of gender and sexuality, when she first tried to enter the industry. She was initially denied access: 'record companies...did not want another lesbian on their label: to quote one "we've got enough lesbians already"' (cited in O'Brien 1995: 4). Both Joni Mitchell and Deborah Harry suffered humiliation at the hands of male publicists. Joni Mitchell was described in an advertisement as '99% virgin', while Deborah Harry's single 'Rip Her to Shreds' was marketed with the slogan 'Wouldn't you like her to rip you to shreds?' (Raphael 1994: xiv, xvi). Understandably both women were infuriated by this representation of themselves and their music.

On the creative side of the industry, male musicians outnumber female, especially in the ranks of instrumentalists. Women in bands are conventionally the singers. When a 1996 *Mojo* article featured a ranking of the hundred best guitarists of all time only three were women: Sister Rosetta Tharpe, Joni Mitchell and Bonnie Raitt (cited in Bayton 1997: 37). While there may have been gender bias in judging this ranking, there remain few female guitarists. Women are relatively over-represented in one instrumental speciality – the electric bass – and in one genre – indie rock music. The bass is relatively easy to learn and of lesser attractiveness to men, but legitimises women's presence in a male site of artistic production; women argue that bassists have rhythm, hence this is emotionally supportive creativity, enabling self-portrayal as caring and group-oriented, but also intuitive and primal: resistance and reproduction combined (Clawson 1999). The electric guitar, the most symbolic instrument in rock music, is explicitly gendered as male and 'connotes phallic power' via its shape, volume, playing style and intensity (Bayton 1997: 43). Guitars are coded as both male and heterosexual. When Charlie Benante, drummer in heavy metal band Anthrax, was asked if he'd ever thought of using keyboards, he answered: 'That is gay. The only band to ever use keyboards that was good was UFO. This is a guitar band' (cited in Walser 1993: 130). The drums are likewise gendered. Karen Carpenter was barred from taking her drums with her on tour because her agents believed it inappropriate to the image of the fragile woman that they wished to convey (Raphael 1994: xxvii). However, the short-lived success of the Honeycombs in the 1960s was primarily due to the novelty of the band having a female drummer. Guitar and drum shops are commonly gendered as male spaces. Women are alienated from this space, blocking their entry into the rock world at one of the very first points (Bayton 1997: 41–2). Not surprisingly 'gender constraints operate most strongly in the early stages of women's musical careers' (Bayton 1993: 192; Goh 1996; Cohen 1997b). In Liverpool the 'indie' scene is thus largely 'created apart from women', resulting in structures at the grassroots level of the local music scene that restrict women's access and perpetuate it as a male domain. Yet the scene 'is not naturally male....[r]ather actively produced as male through social practice and ideology, contributing to the process through which patterns of male and female behaviour...are established' (Cohen 1997b: 34), and which continue to pervade most music scenes.

Women's role in the industry is further compartmentalised and restricted by stereotyping particular styles of music, such as folk music, as 'women's music'. The folk singer/songwriter genre is supposedly characterised by women who are 'long-haired, pure voiced, [and] self-accompanied on the acoustic guitar' (Frith and McRobbie 1990: 377), enhancing 'the notion that women should play quieter, gentler music...based on a sexist stereotype of conventional femininity' (Bayton 1993: 185–6). Country and western music occupies broadly the same position, and both genres have provided contexts where women's involvement is particularly acceptable. These versions of femininity are nearly always heterosexual. Lesbian musicians have been reluctant to 'come out' largely because of fear for their musical careers if they break this stereotype. Though many women have defied being stereotyped and marginalised, few have taken their rightful places in the history of popular music (Gaar 1992; Hirshey 1994), though certain moments in popular music history have expanded the space occupied by women in the industry. Performers such as Sarah Vaughan, Ella Fitzgerald and Billie Holiday were foregrounded in jazz and some were able to evade the image-moulding producers of the time to explore new territory, as with Ella Fitzgerald's renowned 'scat' singing, Alice Coltrane's piano ambience or Nina Simone's frank and politicised lyrics. The 1960s saw a proliferation of all-girl groups, such as the Supremes, the Ronettes, the Shirelles and the Shangri-Las: 'not since the blueswomen of the 1920s was there a decade that saw so many female artists at the fore' (Hirshey 1994: 521; Smith 1999). Movements in popular music such as punk with its 'fuck anyone who doesn't like it' attitude (Raphael 1994: xxi), and riot grrrl, which propagated a staunchly feminist agenda, contributed to greater freedom for women in the industry, by broadening the types of music that women could acceptably play and subjects they could sing about. In the late 1970s punk gave increased access to women, through a 'DIY ethos and anti-muso attitude [which] allowed women a much-needed space to perform without fear of ridicule' (Raphael 1994: xiv), while 'the traditional association of love and romance with popular music (as well as the association of sex with pleasure) came apart' (Gottlieb and Wald 1994: 258). Categorising 'masculine' and 'feminine' in music became less relevant despite debate about moral and political issues, including feminism. Punk challenged traditional notions of sexuality; in contrast to the 'free love' ideal that had preceded it, for some 'punk sexuality was angry and aggressive, implicitly feminist' (Steward and Garratt 1984: 158), while Patti Smith's androgyny challenged notions of femininity. Riot grrrl, associated with bands such as Bikini Kill and Bratmobile, both from Olympia, Washington, was an underground, feminist music movement in the early 1990s that aimed both to increase women's space in the popular music industry and challenge gender relations in the broader society, 'punk with politics' (Raphael 1994: xxiii):

> Riot grrrl is foremost about *process*, not product; it's about the empowerment that comes from 'getting up and doing it', and the inspiration audience members draw from witnessing this spectacle of self-liberation.
> (Reynolds and Press 1995: 327–8)

All-grrrl gigs were essential, increasing women's access to live venues and giving women greater confidence to become involved themselves in music production. Even at mixed riot grrrl gigs, the area directly in front of the stage, usually defined as male at hardcore gigs because of the sheer physicality of moshing (Gottlieb and Wald 1994: 257), was cordoned off as female-only. More recently, Ani di Franco mixed feminist lyrical sensibilities with a more political-economic agenda, establishing her own label, Righteous Babe Records, in Buffalo, New York, in a deliberate attempt to work outside the male-dominated music industry. Frustration with persistently patriarchal structures in music led to new expressions, movements and enterprises, including developing women's touring circuits, such as Lilith Fair.

Outside the West more subtle changes emphasised the emerging place of women in society and in the music industry, though gains were relatively limited. In Muslim societies singers such as Umm Kulthúm (Box 7.2) were able to achieve success, though she was the exception to the rule. In Mali Oumou Sangare became exceptionally popular and used her popularity to emphasise lyrics that sought the greater liberation of women in Muslim and traditional African societies, but the transition was complex. Malian popular music effectively encouraged women to overcome the difficulties and corrupting nature of everyday urban life, but through expressing conservative ideas about relations between the sexes and generations and emphasizing the 'contented disposition' of rural women – providing moral lessons that both praised women for upholding 'traditional' values while absolving them of complicity in undermining such values (Schulz 2001: 366). Similarly women in the city of Kinshasa, Congo (Zaire), achieved some degree of autonomy through using music as a terrain for gender struggle. In the colonial city women were identified with tradition and thus 'uncivilised'; they accepted their status, rejected the 'modernity' of European civilisation and chose African music, a cultural form treated with condescension by the colonisers. By doing this, they questioned the European definition of the city itself and, in so doing, elaborated a new model of gender relations that was crucial to social life in the city (Gondola 1997: 81–2). Challenges to existing patriarchal orders took quite different forms.

Challenging heterosexual space

Even in 'liberal' societies, gay men and lesbians are rarely free to openly express their sexuality in public places. Most public spaces are coded as heterosexual, and in many cases the atmosphere (if no longer the legislation) is outwardly hostile. Gays are only able to freely express their sexuality in certain localities, whether private houses, clubs or particular neighbourhoods. Within such 'common action spaces', particularly the club scene and related social networks, an overt gay community can exist (Bell 1991: 324). The greater liberalism of the 1970s enabled movement 'out of the closet', and from conservative suburbia to the non-conformist and transitional inner city, producing 'gay ghettos' or 'pink triangles' in such places as Castro (San Francisco) and West Hollywood (Los Angeles). The 'flight' of Boy George, from suburban Bromley (South London) into the city, exemplified the search for safe city space:

> George's undisguised homosexuality ensured his unhappiness in the suburbs, but pop offered a way out. In a carnivalesque inversion of the millions of respectable commuting journeys made along the railways of south-east London, George and his cronies fled at night to the centre of the city, seeking out punk clubs, gay clubs, and subcultural space where the respectabilities of suburbia could be temporarily forgotten by immersion in the delights of loud music, cheap drugs and easily available sex.
>
> (Medhurst 1997: 265–6)

In the largest cities of the rich world, new affluence stimulated a gay inner-city renaissance partly centred around entertainment and cultural economies.

The earliest geographies of gay communities focused on the location of gay bars and clubs, along with some other sites, which evoked a sense of place (Ashworth *et al.* 1988). From disco in the 1970s through to house and its variant styles in the 1990s, dance contributed to transforming the sexuality of these spaces, attracting many gay men because of its mixture of 'whole body eroticism', utopianism and romanticism; glamour and rhythm created an eroticism of disco that differentiated it from other genres of popular music. Unlike what is said to be the phallocentric nature of most rock music, disco was said to 'restore eroticism to the whole body and for both sexes, not just confining it to the penis' (Dyer 1990: 415). Hence, in the club scene:

> a moment of community can be achieved, often in circle dances or simply knowing people as people, not anonymous bodies.... This can be achieved [at discos]...which, when not just grotty monuments to self-

> oppression, can function as supportive expressions of something like a gay community.
>
> (Dyer 1990: 417)

Disco provided a soundtrack that was relevant in style and content, and enabled some resistance to heterosexual hegemony over private and public space.

Glamour also created gay icons, from Liberace to Dusty Springfield, Kylie Minogue, Madonna – whose music was routinely remixed for the gay club scene – and Diana Ross. The latter's catchy lyrics and rhythms, her drag-like dress-changes and the dominant theme of her music – the ecstasy and fleetingness of the modern love affair (notably 'I'm Still Waiting') were 'important in the gay male scene culture, for she both reflects what that culture takes to be an inevitable reality (that relationships don't last) and at the same time celebrates it, validates it' (Dyer 1990: 416). While gay icons were relatively numerous, and even more performers, such as David Bowie, Marc Bolan, P.J. Harvey and Patti Smith, were androgynous or played with camp (Geyrhalter 1996), few – after Janis Joplin – declared their homosexuality, until kd lang led the way for Janis Ian and Melissa Etheridge, to become role models and means for overt identification with individuals and lyrics.

Concerts by out or suspected gay musicians, such as the Indigo Girls and kd lang, the Pet Shop Boys and Jimmy Somerville, created temporary or transient gay spaces. Audiences were provided with rare opportunities to redefine the sexuality of public space, creating a sense of community empowerment. kd lang concert goers described this experience: 'I just think it's great to be among a whole load of dykes. It's a real sense of community knowing that everyone in the audience is gay' (quoted in Valentine 1995: 478–9). Beyond such tangible places as bars, concerts and dance clubs, gay and lesbian communities use popular music in more subtle ways to create a private, gay space within the predominantly heterosexual public sphere. Where homophobia is considerable, as in several developing countries, opportunities to create such spaces through music (or in other ways) are rare (Skelton 1995) and gay spaces are only domestic.

All forms of music create personal spaces, and, where sexuality is repressed, the role of 'gay-friendly' performers is particularly important, hence, as one music journalist described:

> The Indigo Girls have provided the soundtrack to which I, and many of my lesbian contemporaries, came out. . . .Much of their work has entered the proverbial lesbian collective unconsciousness via the back door; you'll hear them from a market stall, over a loudspeaker during breaks at a music festival, soft in the background at a dinner party.
>
> (Catt 1998: 31)

Hearing the Indigo Girls, Paul Oscar, Frankie Goes to Hollywood or Bronski Beat alters the definition of sexuality in the spaces where it is played. Particular songs such as Dolly Parton and Kenny Rogers's 'Islands in the Stream' hold similar meanings (Kruse 1999). While dance music and club scenes epitomised the creation of gay spaces, even such seemingly heterosexual genres as country music could be transformed in appropriate contexts (Box 9.3). More frequently, even where lyrics were gender-free, as in much of the music of Joan Armatrading or George Michael, performers dismissed discussions of their sexual identities. Nonetheless, mainstream music, without gender-specific lyrics, might be appropriated in a range of circumstances, especially where performers emphasised control of their own destiny (Bradby 1993; Wise 1990). Hearing music that can be both a soundtrack for the gay community and acceptable in mainstream society aids in both the creation of identity, and the transgression of everyday space.

Box 9.3 Queer country

Country music is widely acknowledged as being one of the more heterosexual styles of popular music (Mockus 1994; Valentine 1995). Mockus argues that any reference to homosexuality is completely banished: 'If other genres of mainstream popular music allowed for a minimal queer presence, country music made no such allowances whatsoever. Guys are guys, gals are gals, and anything queer is entirely exscripted'. Consequently country music intensified homophobia because of its ideologies of traditionalism, patriotism and the nuclear family (Mockus 1994: 259). Nevertheless, gay and lesbian communities have adopted country music, resulting in its artists and dancing styles becoming an integral part of gay and lesbian culture. The most obvious gay and lesbian icon to come out of country music is kd lang. While lang's music is no longer exclusively country in style, her roots were firmly planted in its traditions. Her androgynous appearance, her vocal and performance styles and acoustic sound, contribute most to her lesbian appeal (Valentine 1995: 476). So too do her lyrics in which she very rarely uses the male pronoun, preferring the ambiguous pronouns of 'I–you' instead (Mockus 1994: 261). Not all lesbian country music icons are female. Country singer, Johnny Cash, has also been adopted as a lesbian icon, through his 'troubled and suffering masculinity', his 'cozy us-against-them relationship with his audience', his 'legendary path towards butchness', which is a 'perfect correlative to many lesbian experiences of becoming gendered as "mannish"', his contribution to lesbian fashion (Ortega 1995: 260, 262) and, not least, his rendition of 'A Boy Named Sue'. Country music bars, theme nights and dancing styles have also become common in gay and lesbian culture since the mid-1980s. One of the events in the 1998 Sydney Gay and Lesbian Mardi Gras Festival was the Harbour City Hoedown, 'a bootscooting country dance party' (Sydney Gay and Lesbian Mardi Gras Festival Guide 1998: 83). Mockus attributes part of this interest in country music themes to the sentimentality of country music songs, as well as its dance styles being attractive in the age of AIDS through 'being sensual without being sexual' (Anderson, cited in Mockus 1995: 259), the antithesis of disco. Appropriation of such a heterosexual, and even homophobic, style of music is a powerful example of consumption being an active rather than a passive act (Valentine 1995: 475), even to the extent that destinations such as Dollywood in Pigeon Forge, Tennessee, the theme park created by and around Dolly Parton, have become significant lesbian destinations (Chapter 10). Country music is actively consumed in a deliberate effort to highlight and destroy the homophobia inherent within it: 'to queerify country is, in a sense, to expose and even undo its homophobic deeds', a form of consumption seen as 'turning country into camp' because it sees 'through its ridiculous falseness by delighting in the dorkiness of country music' (Mockus 1994: 260). Thus 'lesbian country and western can be seen as a parody, or reworking, of the "traditional" masculine cowboy of the western film genre and a reappropriation of the "butch" identity' (Allen, cited in Valentine 1995: 477). Gay and lesbian communities have adopted and adapted a style of music, originally from a hostile section of society, and used its adapted form to define sexuality within the walls in which it is played. These communities are then able to challenge the heterosexual paradigm that traditionally exists in space by consuming country music in a new way, adding a layer of meaning to its sound and soundscape, and then projecting this interpretation of the music back to its traditional listeners.

Into the music

Music is a means of recreation and pleasure. In domestic spaces, sounds are used and have been marketed directly as 'aural wallpaper', as background music to a range of social functions, or as 'armchair journeys' into exotic places, via essentialised representations of place, culture and ethnicity in 'lounge' and ambient music. The 'virtual tourism' that this stimulated, despite its caricatures, eventually gave way in a more affluent age, to new forms of mobility and actual tourism. That also enabled the sites of music production, and the places of inspiration for diverse lyrics and styles, to become transformed into places of pilgrimage and pleasure (Chapter 10). Just as frequently music has sought to change factories, train stations and shopping malls into places where mundane tasks might be less irksome, and old abandoned warehouses into places of consumption and recreation. Yet many transformations are transitory and carnivalesque, hemmed in by institutions, regulations and moral panics, which slow the processes of social and political change.

Sounds in themselves can occupy the site of considerable political tension. Political musics, from the acoustic folk of Woodie Guthrie, Bob Dylan or Billy Bragg, or the harsh sounds of punk, hip hop, to the penetrating bass thump of industrial techno, have all been associated with certain movements and ideological concerns, and employed in ways that define particular spaces and events. These same disruptive sounds have triggered reactions from the state, and sometimes are seen as reflecting the general state of popular culture in society, and, by inference, a critique on contemporary society itself – as in one account of rap:

> The malaise of modern life has become embedded not just in lyrics...but in the very form and fabric of the music. A music that has surrendered melody to beats is a music that trusts the body more than the mind....A music that speaks in fractured, elliptical gasps instead of telling a story with a beginning, a middle and an end is a music that implies that there is no future – or at least no future that could be imagined with any certainty.
>
> (Holden 1994: 8A)

To others, control over sounds, and the contexts into which they are broadcast, offers broad new possibilities. Sounds, and the spaces they inhabit and transform, open up room to critique, to subvert and to reconstruct social realities. Ironically, despite transnationalism, and for all the diverse transformations of music in place, and the social changes that marked the last half century, gender relations and the structure of public and private space have changed little. The material power of the industry has been a remarkably powerful conservative force, as in other cultural industries. Place, space and social relations have exhibited continuity rather than change.

10

MARKETING PLACE

Music and tourism

Music is bound up in place and in transformations of material spaces; increasingly this occurs through tourism and its promotion. Some tourist destinations have developed because music (or the performers themselves) had some connection with those places, but music is more subtly connected with tourism in other, more diverse ways. Music is a cultural resource bound up in how places are perceived, and how they are promoted. It is one means by which places can be represented in wider mediascapes, shaping local or regional identities; and, by design or default, music influences the images that attract tourists. Few holiday brochures are without some reference to music, whether in bars and clubs, through shows of various kinds, in references to 'folkloric' local musical customs and performances, or through the depiction of buskers on street corners. Whether actively, through attendance at events, or visiting 'heritage' musical sites, or merely passively, music plays some part in many holidays, as in Bali, where one travel brochure notes 'Asia's first Hard Rock Hotel covers a 3-hectare site...418 tribute rooms and suites are situated in 6 blocks named after different musical styles: rock 'n' roll, blues, reggae, psychedelic, alternative and rock'. Music, like other elements of culture, from literature to sport, has increasingly stimulated new tourist niches, which have become critical sources of income in several places.

Music can also be used as a cultural resource to enhance or define tourist activities and material spaces – it fills commercial sites, creates atmosphere in theme parks and results in dedicated spaces for music performances aimed at tourists. Visible 'reminders' of the presence of musical scenes and sounds that have since declined or disappeared, or of artists who have died or whose careers are finished, have generated some successful year-round tourism economies, yet they are difficult to establish, given the relative 'invisibility' of music and sound. Because music is aural in the first instance, experienced in the transient moments of listening, this conflicts with the tourism industry's reliance on stable, visible phenomena and links to the 'tourist gaze' (Urry 1990) that structure networks of sites and tours.

Tourism relies on evidence of cultural activities, incidents from the past, tangible artefacts that can be photographed and attract tour buses and backpackers, a process that 'reflects the central importance of such cultural forms as museums, art galleries and historic houses in tourist consumption patterns' (Quinn 1996: 383). Without these visible reminders of a musical past or an annual music festival, tourists are likely to be absent, and benefits to local economies accrue unevenly.

Musical expressions and events may only occupy spaces in ephemeral ways, as particular scenes are surpassed by new trends elsewhere. Hence, numerous attempts have been made both to revamp 'musical landscapes', through restoring heritage sites, and to construct buildings that represent and capitalise on cultural traditions, as in Nashville, Tamworth and Liverpool. Music tourism involves a range of practices where sites of music production and expression (whether in past or present 'scenes'),become the points of attraction for tourists, central to strategies employed by the local state, tourist promotion boards and companies to market musical heritage. Music sites have emerged in the last decade of the twentieth century, particularly in the West, drawing visitors to experience, or visualise, the contexts within which particular musical styles were developed, recordings were produced, where 'stars' were born and died, or were mythologised in songs. Some tourists might take in a musical performance in a city famous for its music, particularly if that place was a source of an internationally successful style or movement.

Nostalgia and pilgrimage

The growth of music tourism (and its attendant cultural and economic significance) involves a sense of historical accident. With more people being able to afford to travel, coupled with faster travel times and cheaper fares, places have diversified in order to promote different kinds of tourism, and music has been one aspect of 'culture' drawn into tourism, a historical legacy that helps to define and differentiate places. Many of those with the time and resources (retirees, settled couples with secure incomes, etc.), grew up with the popular culture of the 1950s and 1960s, the period that has spawned the most visited, and the most profitable, music tourist sites (such as those surrounding the life and times of Elvis Presley and the Beatles). Affordability and access, alongside nostalgia, are key explanations for the structure and character of music tourism.

Music is heard in different ways by individuals, and may be emblematic of a particular period, generation or experience; for those people who shared and enjoyed those same recordings, 'Music has the ability to evoke personal memory, to place something in one's life in a personalised period context' (Wheeller 1996: 336). These 'affective investments' in music cannot be revealed simply through lyrics or particularly memorable tunes – they are bound up in the ways in which audiences construct meanings for and around songs, artists or regional sounds.

Music can evoke memories of youth and act as a reminder of earlier freedoms, attitudes, events; its emotive power (in the music itself, and through visual stimuli such as album covers) serves to intensify feelings of nostalgia, regret or reminiscence. Tourism taps into this, providing opportunities to relive emotional attachments to music; to 'return to roots'; to acknowledge the past, visiting sites that are associated with favourite performers; or to see the buildings from which influential sounds emerged (for some an implicit acceptance of the determinist notion that places somehow intrinsically shape musical expression). Music tourism is bound up in wider issues of fandom, how audiences generate 'superstars' and 'idols' out of performers, and how lifetimes can be dedicated to following the careers of particular artists. Music tourism may form part of living out 'retro' fantasies, of 'being young again', at once celebrating the music's obsolescence and yet keeping its cultural meanings alive; it might also constitute the final stage of a process of commodity fetishism ('I've collected all the Beatles albums, the biographies, the fanzines, now I have to visit the place they came from and that made them famous...') that becomes place fetishisation.

Inherent in some forms of music tourism is the notion of pilgrimage. Indeed:

> [popular music] stars are the closest thing that contemporary U.S. culture has to living gods and goddesses: they're highly charismatic, larger than life figures; they're deeply embedded in cultural myths and legends...the behavior that fans exhibit towards them is often nothing less than worshipful in its adulatory and awestruck quality.
>
> (Rodman 1996: 111)

This sense of quasi-religious faith in the power of music icons also renders the range of physical spaces that they occupied or passed through sites of crucial importance, imbued with sacred meaning. Such faiths operate without organised churches, so seemingly meaningless, ordinary spaces are transformed, as for Elvis Presley: 'what makes these locations sacred, is not so much the central roles that they played in Elvis' life as the practices of the Elvis faithful with respect to these sites' (Rodman 1996: 116). Similar pilgrimages are directed towards sites associated with dead performers, such as Jim Morrison's grave at Père Lachaise Cemetery, Paris; John Lennon's Imagine memorial at Strawberry Fields (Central Park, New York), and the nearby apartment block where he died; Jimi Hendrix's resting place at Greenwood Memorial Park, Seattle; the site of Kurt Cobain's suicide death in 1994 (much to the annoyance of his family who still live there); but, above all, Elvis Presley's Graceland, Memphis (see Box 10.1 and Figure 10.2). Even the late English folk singer Nick Drake inspired pilgrims to visit his grave in the tiny village of Tanworth-in-Arden. Testimonies in church guest books, and graffiti at such sites, suggest the depth of tourists' nostalgia, attachments to the particular performers, and claims that their lives have been influenced, even transformed, by their music pilgrimage (see Figure 10.1). Thus at Abbey Road Studios, London, the site of Beatles recording sessions (see Box 10.2): 'I have been waiting 20 years for this moment. John, Paul, George and Ringo you have touched my life forever. From New York to London, I have finally made it.' At each of these sites, and many others, the diverse languages of both tourists and graffiti indicate the global extent of music pilgrimage.

Box 10.1 'I'm going to Graceland…'

Graceland is arguably the most famous music tourism site in the world, the home of the 'King of rock 'n' roll' and a focal point for the quasi-religious cult that has surrounded Elvis since his death in 1977. Since then it has become part of a tourism empire, consisting of tours of the house itself, and other related enterprises: the Graceland Plaza, the Sincerely Elvis museum and other museums dedicated to his cars and jet planes, among the ubiquitous souvenir shops and re-created 1950s-style diners. Graceland operates as an archetypal music tourist site that was central to constructing Elvis as a star in the first place. Even while alive, Elvis's seclusion behind the mansion's gates, and his often bizarre activities (including six month binges on meatloaf), served to separate him from fans' everyday lived experiences. Graceland intensified the Elvis myth because it located all that Elvis encompassed in an easily identifiable, specific site: 'this heightened sense of permanence helps not only to render Elvis's stardom unique, but also to make his posthumous career possible' (Rodman 1996: 123). The processes through which this site has been transformed into a music tourism centre are multifaceted: Graceland serves as the authentic site of Elvis, providing a focal point for a large community of fans that continually reproduces the mythology and sanctity of Elvis: 'an environment where the type of "extreme" adulation that fans of other stars can't openly express is able to flourish comfortably with relatively little fear of public censure' (Rodman 1996: 127). Some commentators have read further into the mythology that has surrounded Elvis in his afterlife, and the pilgrimages conducted to Graceland:

> It's a way of coming to terms with our own sense of loss, with what's become of us as a nation – the transition America has made from the young, vital, innocent pioneer nation we once were (the young, vital Elvis we put on our stamps) to the bloated colossus we feel we've become: the fat Elvis of nations.
>
> (Rosenbaum 1995, quoted in Rodman 1996: 114)

At one level, as one graffito said, 'Elvis, it's been 20 years and we still can't fight our tears….I made it here, love always.' At another, accounts of Graceland by travel writers have both captured this sense of religious awe, and highlighted the contradictions of the star's excesses:

> A small bus carried us up the drive to Graceland itself. We entered with the hushed reverence of visitors to the Sistine Chapel, except that here was not symmetry, beauty and the spirit in flight, but heinous taste, berserk ostentatiousness, the evidence of a soul unhinged.
>
> (Brown 1993: 55)

At Graceland, fandom, and the successes and excesses of a star performer, have come to represent much more than either pilgrimage or crude commercialism, by entering into popular national mythology. Indeed, Graceland has not just become a site of pilgrimage for Elvis fans but, unlike most music tourism sites, has become part of an American tourist itinerary.

See: Rodman 1996; Alderman 2002; Marling 1996.

Figure 10.1 Graffiti at Abbey Road
Photo: J. Connell, 2000.

Figure 10.2 Elvis Presley's grave at Graceland
Photo: C. Gibson, 1998.

Box 10.2 Magical mystery tours

The lives, itineraries, songs and stories surrounding John, Paul, George and Ringo have become a multi-million dollar industry in the United Kingdom, in particular at Liverpool, their home town and mythologised source of the 'Mersey sound' of the 1960s. Liverpool has undergone significant changes as deindustrialisation has changed the role of heavy industry in England's north during the last thirty years, in terms of a marked decline in the role of ports and reduced levels of manufacturing. Music tourism, related to the city's famous past, and in particular the Beatles, has been offered as one possible solution to these complex problems (Cohen 1997a). Mike Wilkinson, Head of Liverpool Tourism Arts and Heritage emphasised this strategy:

> The importance of the Beatles, and indeed of the whole of the Mersey sound, cannot be overstated where tourism is concerned.... When you ask foreign visitors what they know about the city before they

> came here, it boils down to football teams and pop groups. Pop related tourism has developed a lot already but there is clearly scope for a great deal more. New Orleans has jazz, we have The Beatles, it's definitely an important way forward.
>
> (quoted in Wheeller 1996: 339)

Previously abandoned sites on Merseyside have become tourist destinations, with the Albert Docks being transformed into a variety of tourist-oriented uses: souvenir shops, a Museum of Liverpool Life (featuring many Beatles artefacts) and *The Beatles Story*, a museum and interactive display featuring depictions of 'authentic' scenes – such as a replica *Cavern Club* – the legendary venue where the Beatles began their English career. This venue itself has also been reconstructed and plays host to tourists. As an acknowledgement of the importance of this tourist market, Liverpool eventually renamed its airport John Lennon Airport, and tourist authorities produced a number of publications to link the various Beatles-related sites (such as Strawberry Fields and Penny Lane), thus integrating music tourism infrastructure in the city. These have included a Beatles map of Liverpool, social histories and organised Beatles tours – including the *Magical Mystery Tour*. Local music traditions formed the basis of plans to reshape the city's image to the outside world, generating more 'cultured' representations of place to counter the myths of 'rust' and 'roughness' that accompany industrial decline.

Yet Beatles tourism has expanded in many other places in which the band spent its time, with the band's tours, recording sessions and reputation providing the catalyst for many tourist initiatives. In London, Beatles 'sites' are incorporated into an international tourism market that has already become a major income-earner in the capital, thus being absorbed as part of the unofficial 'circuit' traversed by tourists, along with the Tower of London, Big Ben and Piccadilly Circus. Abbey Road studios, made famous through the Beatles' album of the same name (and the famous image of the four band members using a nearby zebra crossing), is now celebrated on postcards, T-shirts and replica street signs.

In other less obvious places, the music of the Beatles has also been incorporated into tourism strategies, despite the absence of evident links to the band's career or the members' personal lives. In Prague, for example, a large public portrait of John Lennon, which under communist rule formed part of a wall covered in protest graffiti, has since become a standard attraction (now known as *Lennonova Zed*); while Beatles-inspired stage shows including

Rock Therapy and *Yellow Submarine* (running in different theatres simultaneously) attracted large crowds. These shows, set to famous tunes of the Beatles, displayed on-stage 'film clips' of images, alongside, in the latter case, traditional Czech Marionette puppet interpretations of the Beatles film. Here, the Beatles music formed a part of strategies to attract tourists, not through actual connections to the Beatles' career, but through appropriating their music as a sort of 'nostalgic language' of protest and pleasure (recognised by all those exposed to some popular culture over the last thirty years). Not only have the Beatles stimulated tourism in such places as Liverpool and Abbey Road (and also Hamburg) but their music has been used to boost tourism far beyond their own itineraries and particular personal influence.

See: Cohen 1997a; Wheeller 1996; Brabazon 1993.

Music, tourism and sub-cultures

Music tourism sub-cultures have emerged around the tours of particular artists, with groups of highly committed fans (even 'groupies'), who follow performers around from concert to concert, even generating a sense of 'communitas' (McCray Pattacini 2000) through shared experiences, fan clubs and traditions maintained on-tour. Famous examples of this rather particular form of music tourism include 'Deadheads', fans of seminal San Francisco rock group the Grateful Dead, a sub-set of which, known as 'tourheads', traversed the United States with the band (until its demise in 1995), developing new sub-cultural practices, words and symbolic devices of their own. Original Deadheads kept in touch via newsletters (called Grateful Dead Almanacs), and established a utopian subculture connected to the group, exchanging tapes of Grateful Dead performances (sanctioned by the band) and travelling from gig to gig as part of an 'alternative', drop-out experience. Deadheads could 'spot each other on the road via the semiotics of window decals and bumper stickers, or on the streets via tie-dyed uniforms' (Rheingold 1993: 49). Audiences established zones of exchange and experimentation outside gigs; venue parking lots became sub-cultural spaces where fans camped out, 'colourful street fairs' where some attempted to subsidise their trip by selling T-shirts, tapes, hash cookies and necklaces, 'the culture outside the concerts became a ritual, a homecoming, almost a religion to some' (McCray Pattacini 2000: 2). The Haight-Ashbury precinct in San Francisco, California (the original hang-out of the band in the late 1960s), was subsequently transformed into a key tourist site, largely because of its significance in the career of the Grateful Dead and the counter-culture 'summer of love'. Deadheads thus had a physical site for pilgrimage, which became filled with cafés, souvenir shops, record stores, iconic graffiti and even an ice cream store advertising flavours such as 'Cherry Garcia'. After Jerry Garcia's death in 1995, many Deadheads turned to the Internet to keep the ideas and sub-cultural traditions alive; others followed newer groups such as Phish (whose fans were known, in a deliberate comparison with Grateful Dead fans, as Phishheads). Phish-devoted Internet sites acted as important points for information flow, enabling Phishheads to plan journeys around the country in ways reminiscent of nascent 1960s counter-cultures. Thus a different generation of wandering music fans had revisited a distinctly American alternative tradition, although many ex-Deadheads complained that Phish attracted a 'jocky, punky element', creating 'a haven for overindulged white kids who were looking to rebel against their parents, but not rebel so far as to never be able to return at the first sign of trouble' (McCray Pattacini 2000: 3–4). Thus, even among Deadheads and Phishheads, sub-cultures often characterised as sharing, peaceful and generous, 'old' and 'new school' distinctions sought to separate the credible from the inauthentic.

In very different settings, an international network of 'dance music' sites emerged during the 1990s, closely associated with routes of backpacker culture, places where electronic, and overtly drug-related, psychedelic music could be experienced as part of overseas holidays by mostly European backpackers. These places included Ibiza (Spain), Amsterdam (Netherlands); Goa (India), Byron Bay (Australia) and Koh Samui (Thailand), all part of an international backpacking, budget trail that provided dance parties and drug cultures as features of this style of tourist package (see Figure 10.3). Tourists directly influenced domestic dance music scenes. Many of both Stockholm and Sydney's major DJs and local music producers originally travelled as backpackers, subsequently deciding to stay. Similarly, large-scale providers of dance music (such as Ministry of Sound, Cream and Renaissance – labels that have gone on to promote album and single releases as well as fashion items) now operate within this tourist network, staging events in nightclubs in Australia each summer, alongside venues in London, Liverpool and Ibiza. Tour operators have sprung up to cater for this specialist market, capitalising on the commercial success of dance music, particularly in Britain, where one company sent over 65,000 people per year to Ibiza alone (Sellars 1998), and in Australia (Figure 10.3).

Explanations for such patterns of sub-cultural travel relate to the emotive pull of music, its capacity to enhance 'self-actualization, self-enrichment, self-expression and regeneration or renewal of self, feelings of accomplishment, enhancement of self-image, and interaction and belongingness' (Stebbins 1996: 949). Increasingly large-scale movements of tourists have also formed the basis for the creation of entirely new transnational flows of investment (organised around networks originally part of 'alternative' culture, but now overly capitalist), and unforeseen transformations of place. Tourism connected dance cultures with a new politics of place (Saldanha 2000, 2001), creating a variant of 'neo-colonialism', as Western

Figure 10.3 STA flyer

Source: © STA Travel Australia, 2001; design Rob Banks.

Note: The rise of dance music in the 1980s and 1990s was closely linked to place. Many places associated with dance styles, or particular sites of consumption, have spawned tourism economies. Here, the Student Travel Association company encourages international travel to locations known for electronic dance music. The reverse of the flyer states: 'Tech-trance, drum 'n' bass, funky break beats, happy house, acid tech, deep psych trance, deep house, french house or just plain and simple loud hard thumpin' techno...whatever your style the world has it to offer – get out there.'

travellers (often from middle-class backgrounds) mythologised sites of music consumption with 'mystical resonance' (as with beach parties at Ibiza, Goa, and full-moon parties at Koh Samui). Tourism changed the material configurations of such places, engaging rural communities in tourism development cycles that encouraged mass tourism and increased external ownership of property and resources (Westerhausen 2002). At the same time, music tourism spurred new kinds of informal-sector economic activity, encouraged controversial cultural practices that sometimes offended local sensibilities (such as drug-taking and certain modes of dress and behaviour), sparked moral panics and affected new waves of complex cultural change. Local communities variously resisted or negotiated this sub-cultural colonialism, yet such practices remain problematic (Box 10.3).

To complete the circle between music and tourism, towards the end of the 1990s dance music labels began releasing compilation CDs that capitalised on connections between electronic styles and places of tourism: the Café Del Mar series sought to recreate the atmosphere of the famous club of the same name in Ibiza (the island also spawned a British television series), Goa became a key word to include in the titles of countless trance CDs, while Ministry of Sound released a series of CDs more deliberately aimed at tourists (such as *Clubbers Guide to Australia* (1999), complete with guides to clubs in each capital city), and tracks mixed to suit the tourist experience – 'you have a global sound that perfectly captures the cosmopolitan vibe of Australia'. Such releases enhanced the appeal of tourism in clubbing locations.

Box 10.3 Goa trance: backpackers and local development

Backpacker tourism is often thought of as an alternative form of travel, which brings with it quite distinct and unique economic and cultural impacts. Where backpacker tourism has transformed places (such as Bali and Kathmandu) there have been, in some cases, new opportunities for local participation in economic development, and opportunities to earn incomes through commercialising or appropriating music, art and craft for tourist consumption (Hampton 1998; Westerhausen 2002). Elsewhere, backpacker tourism has merely prefigured mass tourism, encouraging foreign investment, increasing the rate of external ownership of resources and changing the way that local communities engage with development agendas.

Dance music has driven the development of backpacker tourism in particular locations, including Ibiza, which, like Bali, was transformed into a mass-market holiday location for European backpackers, centred on club and dance party scenes. Koh Samui, an island on Thailand's south coast famous for its full-moon parties, and Goa, on the Indian west coast, were other places early transformed by alternative tourist sub-cultures: 'pioneer' low-budget travellers who sought places to visit and move through on an extended basis. Such locations were particularly known as sites of hedonism within international traveller networks – places where music and drugs combined to attract young people. While acid (LSD) and marijuana remained a constant, psychedelic rock music in the 1970s eventually gave way to psychedelic electronic music in the 1980s and 1990s (Saldanha 2001), coalescing in a distinct style of techno that combined deeper and more powerful 4/4 beats with swirling synth arpeggios and samples of exotic Indian (and other 'tribal') music: now known as Goa trance (Cole and Hannan 1997; Chan 1998). The music was produced by English, Israeli, Dutch, German and Belgian artists, often back in their home cities and towns – to some extent a style made in the memories of past trips, or in anticipation of future travels. Images and ideas of Goa sparked off new scenes in those cities and towns, leading to the commercialisation of Goa trance and encouraging new generations to travel to India.

In Goa itself, 'proper' clubs in tourist areas were increasingly taken over by British capital, an experience similar to Ibiza (Saldanha 2001), yet relatively disconnected and more localised informal activities continued in places such as Anjuna – where events were connected more fully to a drug-trading economy and 'alternative' sub-culture. Such activities, more harshly looked upon by critics such as the Hindu-nationalist BJP party and Catholic activists, enticed backpackers to track down secretive locations for open, free events put on with little marketing or promotion. Enchanted by the 'charm of the scene' and the exclusivity of such events, travellers sought an 'intertwining with nature…the presence of sand, red rocks, ocean…makes it all the more exciting and authentic' (Saldanha 2001: 8). Trance parties would be structured in particular ways, early sets were dominated by most backpackers and local Goan men, while later, the regular visitors to Anjuna, the core of the Goa trance scene, would take over. Parties supported a 'mat economy' in chillout zones; local women on ground stalls sold cigarettes, chai tea, food and coffee to tourists in fierce competition. Such activities took place on highly unequal terms: 'the space in which music is enjoyed has to be constructed on other people's suffering…the locals and seasonal workers in Anjuna earn less in a month than the foreigners spend in one night' (Saldanha 2001: 19). Goa trance parties, organised under the guise of the original rave slogan 'peace, love, unity, respect', revealed more complex and problematic politics, as local communities variously engaged and reacted against changing cultures of tourism, resisting or participating in uneven ways.

See: Saldanha 2000, 2001.

Backpackers are not only audiences for dance music styles such as house and trance. Young budget travellers are also consumers of a diverse range of musical styles, influencing the creation of a range of specialist sites for the consumption of music, or the transformation of inner cityscapes into musical spaces, through clubs, Hard Rock Cafés, buskers, Irish pubs – each stimulating new areas of activity for local (or travelling) musicians, creating venues and improving chances for performers to earn a stable income from their music. One frequent element of traveller sub-cultures – an extended trip away to 'find oneself' – has a parallel manifestation in music, particularly when certain locations become known for 'neo-tribal' or 'new primitivist' ideologies, activities and sounds. In various backpacker destinations, drumming and percussion especially have been associated with attempts to 'regain one's inner self', as has the didjeridu, whether in Eilat in southern Israel or Byron Bay, Australia (Gibson and Connell 2003). This has included the consumption of artefacts, images, and cultural experiences such as music, in a range of locations that accrue value due to their association with 'alternative' economies, pre-capitalist, non-Western lifestyles and spirituality. Traveller activities in such key nodes rely on an amalgam of exotic images and discourses that are constructed in a similar fashion within traveller cultures. These symbols of exoticism commonly involve Hindu art, Africanesque artefacts, neo-tribal jewellery and particular styles of dress, as well as certain taste cultures in music.

World music is said to have 'some discernible connection to the timeless, the ancient, the primal, the pure, the chthonic; that is what they want to buy, since their own world is often conceived as ephemeral, new, artificial, and corrupt' (Taylor 1997: 26). While Taylor argues that world music represents a form of sonic tourism for consumers in the metropoles, through their position as global backpacker nodes, locations such as Byron Bay have also become centres of world music consumption for younger audiences, through tourism itself. The marketing of world music relies on the association between popular music and bodily pleasures, yet it does so in reaction to the perceived loss (or absence) of bodily focus in Western musical forms, as they are commodified, studied academically and institutionalised. In contrast, backpacker or traveller cultures involve the body in the tourist experience. Music enthusiasts, and the paths people travel through in search of embodied musical experiences, constitute a new network of subcultural tourism, as backpackers seek the places of world music, and connections 'back' to more 'primal' cultures, through drumming tours of Senegal or trips to remote northern Australia to find 'proper' didjeridus (see Figure 10.4). Travellers seek authenticity by engaging with music aimed at bodily transformation, whether as witness to performances or when taking part in drumming or didjeridu workshops and seminars. Thus, African drum workshops in Byron Bay were described as 'a dynamic and powerful way to access your inner joy and celebrate your being using African movement to make you move, groove, sweat and

African Drum & Dance Tour
Senegal 2001
4 June to 1 July

Pape Mbaye
Griot & head drummer of 25 years
Will lead you into his world & share his culture

Located
In Yoff, a fishing village off the beautiful
coast of Senegal

Learn
Djembe, Sabar & Tama
African Dance

Participate
Traditional Ceremonies
Senegalese Meals
Village Life

Ask Margaret
0411 718 032
antaki@mpx.com.au
http://keur_afrique.tripod.com

THE ULTIMATE AFRICAN EXPERIENCE

MOHATA 2002

Percussion and dance cultural study tour to Guinee - West Africa with Mohamed Bangoura

Each year Mohamed Bangoura runs a very successful music and dance cultural study tour to Guinee. Next years tour will run from 18 February to 15 March 2002.

Mohata 2002 aims to provide students with the ultimate African experience. Music and dance is a very important part of every day life in Guinee and students from all over the world are drawn there to study percussion and dance with the many outstanding artists still practicing their musical heritage.

Guinee has a lot more to offer than just its music. During your time in Guinee you will also experience life in Africa. You will attend traditional ceremonies, eat traditional food and visit villages and some of the islands off the coast of Guinee but perhaps the best part of visiting Guinee is experiencing the warmth and hospitality of the Guinean people.

The study tour will be focussed on music and dance and aims to provide all students with high quality tuition and the opportunity to learn djembe rhythms and dance in a traditional African environment.

Figure 10.4 Advertisements for West African music tourism
Source: © Margaret Antaki, 2001 and Karen Buchan, 2001 (k.buchan@unsw.edu.au).

smile all over'. Engagements with world music have been said to 'wake up' the body, 'using ritual, rhythm and dances from all over Africa to excite and inspire you and let your wild side out to play' (quoted in Gibson and Connell 2002).

Outside the West only Asia, Latin America and the Caribbean – the home of Bob Marley (Box 10.4), reggae and carnival (Sampath 1997) – have stimulated significant music tourism. Rumba and the revival of Cuban music, through the Buena Vista Social Club, has drawn tourists to experience an isolated country and its musical traditions, while the tango has provided an image for Argentina (Goertzen and Azzi 1999), just as gamelan music has for Bali, and each has been a stimulus to tourism. Otherwise buskers, nightclubs and shows are part of most tourist scenes, but as backdrop rather than centrepoint. With rare exceptions, such as minority tourism to experience African drumming in West Africa (Figure 10.4), tourists have not been drawn to the 'homes' of world music, and music tourism is highly concentrated in the most developed countries.

Backpackers have not been the only ones to use music as a means to self-actualisation or self-discovery as part of tourist experiences. Music enthusiasts of a variety of ages and backgrounds may shape their own form of tourism and performance. The construction of personal music tours has become increasingly possible for rock, blues, jazz and country, taking in a series of particular sites, most popularly following part or all of Route 66, whose towns were mapped out in the song made famous by Chuck Berry and the Rolling Stones. According to the Lonely Planet guide:

> it is really worth taking the time to follow the old road for a while and visit the 66 towns....It really felt like we had entered a time warp...our music education was addressed in Memphis, the Sun Records studio tour was terrific and captured the embryonic sound and atmosphere of rock 'n' roll.
>
> (Wheeler 1994: 3)

Since being superseded by a series of interstate freeways in 1984, Route 66 has been preserved and designated as a 'national historic corridor' through special federal legislation (Krim 1992, 1998), and became subject to numerous acts of mythologising in American popular culture (Alvey 1997; Brown 1993; Bull 1993). Specific guide books now map out places associated with particular artists (such as Farren's *The Hitchhiker's Guide to Elvis* (1996) and Hazen and Freeman's *Memphis: Elvis Style* (1997)), key music locations (see Scanlon (1997) on Camden Town, London; McKee and Chisenhall (1993) on Beale Street, Memphis) or for whole nations, as with Fodor's *Rock & Roll Traveler USA* (1996), and its British equivalent. The former provides a comprehensive, highly detailed guide to every state in the United States, filled with information regarding famous sites (such as the Apollo

Theatre, Harlem) and much more obscure locations such as the New York college that the Beastie Boys attended; the Boston clubs where the Pixies and Lemonheads first played; or directions to the tiny Roy Orbison museum in Wink, Texas.

In the United Kingdom, the British Tourist Authority published its own music tourism guide, the *Rock and Pop Map: One Nation Under a Groove* (1998), which directed tourists towards a plethora of rock sites across the country: from the Luton home of Jethro Tull and the Hertfordshire school at which Wham members George Michael and Andrew Ridgely met, to UB40's recording studios in Birmingham (an ex-abattoir). The various sites were located on a map of the UK, shaped not by topography, but in the image of an electric guitar, with nearby Ireland an amplifier. Thus, the UK itself was portrayed, even defined, as an island of rock and roll locations, an island of music:

> THIS ISLAND ROCKS: If you don't own anything by a British band in your record collection, you ain't into music.... This map gives you a guided tour of the cities, towns and places that have inspired and shaped British pop makers. Whether you're into The Beatles, Bowie or Blur, The Stones, The Sex Pistols or The Spice Girls – Britain is where the beat is.

More specific publications have guided visitors through localities renowned for musical expression. New York City, Liverpool and Dublin, Ireland, developed city walking tours of important music sites (the 'Music Trail', 'Liverpool Beat Route' and 'Rock 'n' Stroll' tours, respectively), through a series of plaques on the walls of prominent studios, venues and company headquarters. At an even more localised scale, Scanlon's (1997) esoteric account of rock sites in Camden Town included the Roundhouse venue (famous as the home of early psychedelic 'raves' and performances by bands such as Pink Floyd and Soft Machine in the 1960s); the stalls once run by Annie Lennox and Dave Stewart of the Eurythmics at Camden's weekend markets; and even the fish and chips shop frequented by members of Camden's ska band, Madness. Thus, rock tourism began to encompass not only the mass-consumed sites of Elvis or the Beatles, but a variety of offbeat places tenuously connected to other lesser performers.

Music, tourism and commerce: constructing musical heritage

The extent and commercial value of music tourism has gradually been recognised. In New Orleans, visitor spending attributable to music totals around US$600 million per annum, almost 20 per cent of all tourist expenditure, while the New Orleans Jazz and Heritage Festival alone attracted over 350,000 visitors, producing a revenue of approximately US$70 million (Atkinson 1997). In Ireland, music has been a major part of the spectacular growth in tourism in the 1990s. In 1993–4, 69 per cent of all visitors to the Republic of Ireland rated traditional Irish music as 'a very important' or 'a fairly important' determinant of their visit, while over 45 per cent of overseas visitors witnessed musical performances in Irish pubs (B. Quinn 1996:386–7). The growth in tourist numbers seeking out 'authentic' Irish music as part of their holidays has increasingly been translated into specific tourist promotions, including the Irish Music Hall of Fame and the Dublin 'Rock and Stroll' tour, marketed in a specially produced brochure (Dublin Tourism, 1998), which consists of a series of discs attached to the walls of prominent 'sites' around town (such as the café where Sinéad O'Connor once worked before stardom, the mall that the Fureys once busked in, or the site of a Thin Lizzy film clip).

Where visible evidence of musical significance is absent (even when 'scenes' themselves are still quite active), many places have attempted to physically 'capture' sounds, in a deliberate attempt to counter the transient, ephemeral quality of music scenes. These include music museums (such as the Beatles Story, Liverpool, and Bob Marley Museum, Jamaica – Box 10.4); historical precincts (such as Bourbon Street, New Orleans; Beale Street, Memphis); statues and monuments (such as the Buddy Holly statue in Lubbock, Texas); guided tours (the *Sound of Music* tours in Salzburg, Austria); supporting publications (guide books, maps, tourist drives); and places of production (Sun Studios, Memphis; Motown-Hitsville USA, Detroit). Split Enz's frequent references to their tiny but grandly named New Zealand home town of Te Awamutu eventually led to the town museum devoting an exhibition to them. In a range of cities, politicians, local planners and entrepreneurs were quick to capitalise on connections between place and music. Larger-scale projects included the Experience Music Project (EMP) in Seattle – an interactive music museum, exhibition and performance space, in a town already known as the hometown to such artists as Ray Charles and Jimi Hendrix, and the grunge sound of the early 1990s. The project was funded by Paul Allen, co-founder of the Microsoft Corporation (which also has its head offices in Seattle), and grew from its original plan as the Jimi Hendrix museum to include:

> interactive exhibits and hands-on facilities...such Hendrix icons as the guitars he used at the Monterey and Woodstock festivals...a smashed-up Kurt Cobain guitar, the original sheet music to 'Louie Louie'...[and] a massive 'Electric Library' stocked with all sorts of Northwest recordings, common and rare, including early stuff by Ray Charles and Quincy Jones.
>
> (Perry and Glinert 1996: 298)

Sheffield, in the north of England, opened the Museum of Popular Music, but it fared badly.

Box 10.4 The Bob Marley Museum, Jamaica

The most important musical site outside the West is the Bob Marley Museum in Kingston, Jamaica. Not only is it internationally famous but it is also probably the only musical site to be the most important tourist destination within a country. Bob Marley, the major figure in reggae music (Chapter 8), was born in 1945 in Nine Mile, a poor mountain village near the north coast of Jamaica, but moved to Kingston in his teens. In the 1970s Marley became a figure of national and international renown, courted by Jamaican politicians (and shot in 1976 in political violence) and a strong voice against racism and injustice. He died in 1981, of brain cancer, at the age of thirty-six. What is now the Museum was his home from 1975 until his death, while a theatre at the back of the house housed his Tuff Gong recording studio. The inside of the compound wall, which has an Ethiopian flag over the gate, displays a six-panel mural, 'The Journey of Bob Marley Superstar', and, in the garden, the Rastafarian artist Jah Bobby's statue of Marley portrays him with guitar, and a football and portrait of Haile Selassie at his feet. The tree under which Marley practised guitar playing, and smoked ganja, is also a prominent feature. The house itself displays bullet holes in the wall (from Marley's near-assassination), Marley's favourite denim shirt, stage dresses of his backing vocal group, the I-Threes (who included his wife Rita), his Order of Merit – the highest decoration of the Jamaican government – and a rod of correction from the Ethiopian Orthodox Church, gold and platinum records, Rasta cloaks and a chart of his world tours. His bedroom has been retained as it was and there is a recreation of Wail 'n' Soul, the tiny Trench Town record store where he hung out with Peter Tosh and Bunny Wailer. Marley was buried at Nine Mile, and the Bob Marley Centre and Mausoleum there have also become a major tourist destination. There is a prayer space for orthodox Rastafarians, an 'inspiration stone', painted in the green, yellow and red Rasta colours, where Marley is said to have meditated, slept and learnt to play the guitar. Marley's body, and his guitar, are buried in a marble tomb inside a church of Ethiopian design. Package tours that combine Mausoleum and Museum, with other musical sites such as the present Tuff Gong recording studios, have ensured that Bob Marley and reggae music remain critical elements in Jamaican identity and tourism.

Music was used in a different way as the basis of Hard Rock Cafés, a syndicated network of themed restaurants based around collections of rock 'n' roll memorabilia, which 'were popular with tourists because [they] allowed them to return home with the evidence that, not only had they gone to Europe, but they had gone somewhere fashionable'. Hard Rock Cafés became 'outposts for American rock 'n' roll culture abroad, complete with hamburgers, ribs and apple pie' (Hannigan 1998: 94). Memorabilia were regularly cycled through these 'outposts', from a central distribution warehouse in Orlando, ensuring that each restaurant could claim to house a suite of original gems. Once again, as discussed in Chapter 5, rarity would endow music, and in this case a chain of eateries, with credibility and economic value.

Almost accidentally, Woodstock, New York State, whose name was central to the 1969 festival, became a haven for a particular type of visitor in a rather less orchestrated – and sometimes resisted – context:

> Japanese tourists and straggly-looking backpackers...turn up in town on the anniversary, asking for directions to the Festival site.... The shops along Tinker Street were full of [souvenirs]...with its shops selling tie-dye shirts, scented candles, joss sticks, Tibetan wind-chimes and Indian prayer-mats.
>
> (Brown 1993: 13)

While the town has been the retreat of many famous musicians (including Bob Dylan, Van Morrison and John Sebastian of the Lovin' Spoonful), the actual Woodstock festival took place at Bethel, some 100 km south-west of Woodstock. Indeed, the town of Woodstock actively opposed the promotion of '3 days of peace and music' (as it was billed), forcing organisers to shift the festival to Bethel. Yet Woodstock has been the recipient of an accidental heritage, highly reliant on the tourism that has followed the festival even after thirty years, despite initial resistance:

> People complained it was a bore having to keep directing them 60 miles south-east [to the Festival site at Bethel], but the inconvenience could be partly mitigated by selling them a Woodstock Festival T-shirt, poster or badge.... [Woodstock] has become as dependent on tourism as an addict to methadone: in summer, it was the only part of the local economy that counted.
>
> (Brown 1993: 13, 28)

While Woodstock is integrated into the international tourist industry, the absence of visual reminders of musical heritage at the actual festival site taints the tourist experience, rendering it empty of meaning:

> The mythology of Woodstock seemed to evaporate with every mile that I travelled. Bethel was too insignificant to be marked on my road map and I stopped several times to ask for directions. The replies were polite,

> bemused and ultimately pitying. 'You can git there, but I don't know *why* you wanna git there. Ain't nothin' there but a field.'
>
> (Brown 1993: 31)

Nashville, Tennessee, the self-appointed 'home' of country and western music, more than any other city, has attempted to capture and preserve the artefacts and material spaces of a 'sound' – constituting a diversity of commercial celebrations of music in the South (see Box 10.5). From theme parks to hotels, performance spaces and precincts to tacky souvenir shops-cum-museums, Nashville has cemented its reputation in the music world as the centre of country music production, and simultaneously marketed this reputation in the tourist world through constructions of 'spectacle'. Central to the city's success as a tourist destination has been its ability to locate the sounds of country music in physical, photogenic space. Meanwhile, Branson, Missouri, made famous by country performers such as Roy Clark and Boxcar Willie, became a location of choice for investment in new theatres, facilities and tourist attractions, particularly after Roy Clark opened a major theatre complex in 1983 (Carney 1992). Branson too sought to make concrete associations with music, through targeted marketing campaigns, dedicated streetscapes (Highway 76, the main street, became known as '76 Country Music Boulevard), and new larger-capacity venues. The success of Nashville and Branson in attracting large numbers of tourists, in particular retirees, group tours and families, led others to copy the formula: towns such as Myrtle Beach, South Carolina, and Pigeon Forge, Tennessee (the birthplace of Dolly Parton), encouraged country music stars to build theatres (as Alabama, Alan Jackson and Louise Mandrell did) and embraced Dollywood, a theme park dedicated to the town's most famous citizen. Country music became much more than a style connected to a place of origin (Nashville), and was a catalyst for tourism in a range of towns throughout the South, a trail along which particular demographic groups travelled.

Box 10.5 Nashville, Tennessee, and country music tourism

Despite its reputation as the capital of the country music world, Nashville has been central to the growth of music across many genre boundaries, serving as a focal point for blues, R&B, rockabilly and hillbilly, since George Hay announced the beginning of the Grand Ole Opry on WSM radio in 1927. (At this stage the term 'country music' had not become widespread.) This show, originally broadcast from the Ryman Auditorium in downtown Nashville, is the longest running show in American music history (although the Opry was later moved to its present location within the Opryland theme park). By the 1930s Nashville had become known as a country music centre, a reputation solidified in the 1950s with the emergence of a Nashville 'sound' (see Chapter 5). Since then the city has attracted millions of visitors, keen to see the home of this now highly internationalised genre, to bootscoot and line-dance – or to visit some of the various museums, performers' homes and souvenir shops. Nashville has been shaped by music, with recruiting and publishing offices built for the activities of music companies, and whole districts specifically catering for music tourism. In addition to the Ryman Auditorium, open for daily tours and occasional concerts, the famous 'Music Row' includes a 'serious' industry-focused business park and the commercial glitz of the Country Music Hall of Fame; the Opryland complex, as well as hosting the Grand Ole Opry, has its own museums, souvenir shops and cafés. While Nashville's music infrastructure includes a range of smaller businesses (shops, cafés, etc.), Gaylord Entertainment Inc. owns many of Nashville's major sites; comparisons to other world famous entertainment empires are inevitable: 'Where Disney wears Mouse ears, Gaylord wears a Stetson' (Jameson 1995: 45). Gaylord relocated the Grand Ole Opry to the city's outskirts, built a country music-based theme park (which has since become another shopping precinct), and diversified into other areas of broadcasting, production and tourism marketing, including the enormous Opryland Hotel, which features:

> the Delta, a 4.5 acre interiorscape that rises 15 storeys high....[including a] flowing river more than a quarter-mile in length complete with four 25-passenger flatboats...an 85-foot high fountain...a 400-seat restaurant...[and] a spectacular 4.5.acre indoor garden filled with southern live oaks, plants, flowers, a 110-foot-wide waterfall.
>
> (quoted in Jameson 1995: 45)

Such commercialism has led many to react in harsh terms: 'Nashville is hell. I've never been to such a crass place....Nashville is America at its shallowest, most schmaltzy, money-grubbing and exploitative' (Bull 1993: 79). Yet as travel writer Julietta Jameson was keen to assert, 'That's the world of Gaylord. Then there's the real Nashville.' She invoked a sense of artificiality about the Nashville tourist infrastructure, constructed around its country music past, yet stresses a hidden, 'authentic' Nashville found in alternative sites: 'Venture outside the world of Opryland and you'll find the Nashville they write songs about – tiny bars that just serve Budweiser beer and coke and which feature the greatest musicians you'll ever hear' (Jameson 1995: 45).

Nashville's tourism partly capitalised on the country music revival that propelled artists such as Garth Brooks, Trisha Yearwood, Shania Twain and LeAnne Rimes into mainstream charts, yet it also involved wider cultural issues, as the aural 'codes' of country and western were captured in physical space. Conservative themes of family, religion, wholesome lifestyles and 'home', which are common in country music, were transplanted onto the built landscape, prioritised and celebrated as 'quintessential' elements of an American lifestyle. In particular, the notion of 'home', central to country and western songs, was idealised and captured in physical structures of tourism, as performers opened up their homes to the public as tourist attractions. Many country song titles play on the idea of a utopian 'home' as a site that contains a set of solid values, prescribed domestic roles, and a permanent 'haven' to return to from journeys into chaotic, urban landscapes. As Blair and Hyatt (1992: 80) argue, 'the home is a universal symbol of love, security and family. It is the one thing we would all like to go back to, somehow expecting that it will still be there, just as we remember it.' Such themes are (re)presented in the homes of country music stars such as Conway Twitty, in Patsy Cline's complete, re-created lounge room in the Country Music Hall of Fame and personally sponsored displays such as the Willie Nelson and Friends Showcase, which are designed, built and decorated in ways that reflect particular ways of life, reaffirming morals and values that the music itself parades:

> Between the rows of cheap souvenirs and the trophies of success, one could read a deeper story. At a time when America was fragmented, seemingly spinning inexorably out of control, country music was a touchstone of more reassuring times...a buffer, and a consolation, against the giddying uncertainties of modern life, embodying a vision of America of unchanging and dependable values. There was a reassuring simplicity in the songs about loving, cheating and drinking too much – everyday human frailties and emotional dilemmas whose very familiarity offered a kind of comfort in itself.
>
> (Brown 1993: 218)

These ideological stances, values and moral positions, captured in the museums, memorials and malls of Nashville, are now part of a multimillion dollar industry. In addition to catering for tourist desires to experience 'authentic' country music, tourism in Nashville involves larger issues: representations of idealised domestic space, contested notions of American national identity and morality, and specific questions of race, class and gender.

Many attempts to create more permanent and lasting images of musical significance are tenuously connected to the original performers, recordings and expressions that are being marketed – musical associations that are 'invented' to various extents, through whatever resources and structures are available. Where access to lengthy music histories is more marginal, or where little-known reputations need greater exposure, and places are overshadowed by more obvious centres of music fame, invented tradition is necessary. Tamworth has successfully promoted itself as 'Australia's Country Music Capital', modelling itself on Nashville's experience. (Many of Tamworth's tourism planners have travelled to Nashville to compare campaigns and event organisation.) Rather than being the outcome of some 'organic' process, Tamworth's transformation into a country music capital developed over time as part of a deliberate attempt to increase exposure across the country, first through radio and then though a major festival, inscribing markers of musical heritage on a largely non-musical built environment. Tamworth was also able to access and mobilise particularly Australian versions of rurality that appealed more generally to tourists from capital cities. Rural images, of the white pioneer, the conqueror of nature, and pastoral expansion, bushmen and drovers, are central to an idealised vision of Australia's national identity, glorified in stories of the 'outback' and in movies such as Paul Hogan's *Crocodile Dundee*. Tamworth mobilised images of rurality in marketing itself as the home of Australian country music (Box 10.6). While country music has been heard around the world as a quintessentially American sound, it took on new meanings in other countries where rural ideologies played a large part in myths of national heritage.

Box 10.6 Tamworth: country music capital of Australia

Tamworth has had an association with country and western music at least since the 1940s, when local radio station 2TM began to broadcast its daily hillbilly hour. This later became the famous 'Hoedown' show that broadcast on an unusually clear signal to many parts of Eastern Australia. In the 1960s Tamworth featured pub venues that hosted country music performances as part of a larger rural touring network. A regular country music festival from the 1970s staged a number of performances and a talent quest over the

January holiday surrounding 'Australia Day'. In 1969, Max Ellis, manager of the 2TM radio station, was the first to officially denote Tamworth as 'Australia's Country Music Capital', a catchphrase to describe a place that became central to the promotion not only of the festival itself, but also of year-round tourist strategies for the town as a whole. In the 1980s, the festival was extended to a nine-day programme, attracts over 50,000 visitors to the town, generating an estimated A$40 million for the local economy, with over 650 official individual performers and bands, let alone scores of unofficial performers and buskers. The Tamworth Country Music Festival is also now part of a wider geography of music festivals that span several states and includes the Gympie Muster, the South Australian Riverland Country Music Festival and the Emerald Country Music Spectacular.

While Tamworth has had associations with country music in the past, music tourism has solidified these connections, particularly in the built spaces of the town. The local council observed: 'very early country music became an excellent marketing tool for Tamworth...the catalyst for other "country music landmarks" to be established thus reinforcing the city's image as Country Music Capital throughout the year'. Tamworth attempts to promote tourism all year round, having built an entertainment centre and country music theatre for regular performances, and by the 'Roll of Renown' plaques, initiated in 1976 (where country music 'stars' are added on an annual basis to a granite memorial); the 'Handprints of Fame' (where cement renderings of famous performers such as Slim Dusty are entrenched in the footpath); a Gallery of Stars Waxworks; and, most notably, the 'Golden Guitar' – a 12 metre fibreglass replica of the trophy that is presented to winners at the Australasian Country Music Awards each year. In addition to these features, Tamworth council has invested in a guitar-shaped tourist information complex on the outskirts of town, local hotels feature guitar-shaped swimming pools, and images of bush lifestyles and music abound. Country music placed the town on a national cultural map, guaranteeing widespread media coverage at festival time each year, but also providing the raw materials for it to position itself in relation to a wider commodification of rural life. Country music, with its attendant links to the rural, and ideas of 'honesty', 'traditional values' and country attitudes, was reworked through Antipodean rural myths of working-class mateship, rural solidarity (against distant, uncaring urbanites) and triumphs over nature. These associations with rural ideologies allowed Tamworth to market itself as the archetypal rural Australian location, the 'heart of the country', as billboards at the entrance to the small town proudly announce.

Tamworth has largely invented its musical heritage, mobilising what resources were available and transforming the built landscape of the town into a marketable, permanent centre for rural tourism mediated through music: to some extent a 'simulation' of Nashville, Tennessee – mimicking its icons and mobilising the rural ideologies inherent in country music – yet in vernacular ways.

Music festivals have become common features of music tourism industries – particularly since Woodstock, Monterey and the Isle of Wight festivals in the late 1960s (Hinton 1995). Festivals provide places with 'spectacle' and a sense of 'uniqueness '– associating spatial locations with one-off performances, collective gatherings associated with a style, a sound, a genre of music, as with the Brighton Festival, '[helping] to project the town's individuality and validity' (Meethan 1996: 188). Festivals have also formed wider musical networks through which performers migrate, connected to particular musical niches: bluegrass, country and western, and hillbilly festivals in the United States; folk festivals in Israel (Waterman 1998), Sweden (Aldskogius 1993), Germany and Britain; 'alternative' festivals such as Lollapalooza and Lilith Fair in North America, Glastonbury in the UK, Japan's Fuji Rock Festival and Denmark's Roskilde, all of which attract major international performers and tourists. Many smaller festivals are aimed at specific audiences from a limited, domestic tourist market, or are primarily aimed at enhancing the cultural awareness and experiences of local populations, and are thus less explicitly concerned with generating tourist income or catering for tourists' tastes and needs (e.g. Duffy 2000). Others attempt to cater to narrowly defined and otherwise neglected niches of music, often supported by the state or local cultural non-profit organisations, such as the Irish Wexford Opera Festival, which aims to 'be the recognised world leader in the production of rare or unjustly neglected opera and to continue to win for Ireland a reputation as a centre of cultural excellence' (quoted in B. Quinn 1996: 391). It draws over 85 per cent of festival-goers from overseas, a similar situation to the Kfar Blum festival in Galilee, Israel, at which 'a particular version of cultural tradition and identity is performed or paraded…at the cutting-edge of conflicts over the definition of "culture" and over what it is to have culture in Israel' (Waterman 1998: 264). Even more dramatically, Womad – the global network of 'world music' festivals – features a diverse mixture of 'exotic' musicians as sonic tourism, with tourists experiencing the festival both as an event and as vicarious tourism.

Most festivals are explicitly commercial, either from the point of view of tour promoters seeking out fruitful markets in which to stage festivals, or of local planners seeking ways to boost local economies. Thus Tonga 2000 sought to draw dance music tourists to the central Pacific, on the edge of the International Date Line, to mark the millennium (though this venture failed). In some cases, such attempts to appeal to a wider commercial tourist market have ironically ignored or erased signs of local musical cultures at the same time as they seek to musically define the locality at an international level. In Ireland, the Clifden Country Blues Festival exemplified this, as organisers attempted to 'package' the festival in particular ways, appealing to a particular type of overseas tourist with very specific tastes, constructing 'uniqueness' with reference to place, yet ironically defining this uniqueness through repressing local culture:

> the choice of music and of musicians was based at least in part on an expectation of what was most likely to attract visitors to the area. In the case of the Clifden Country Blues Festival, there was a deliberate strategy to exclude local bands on the grounds that the quality of musicianship would not be sufficiently high to enhance the event's reputation and to attract audiences to the event.
>
> (B. Quinn 1996: 391–2)

In many older industrial centres, where tourism has been promoted as one means of generating new economic activities linked to heritage, music-based ventures, including festivals, are among a series of activities promoted by the local state as panaceas for urban decline and unemployment. Detroit tourism campaigns have revolved around its musical heritage, using pictures of groups such as the Supremes; Memphis has remodelled its image as a centre for blues and rock 'n' roll; New Orleans public officials have offered tax breaks to clubs scheduling jazz performances in an attempt to boost the city's music-based tourism economy (Wade 1994), while Liverpool sought to generate income and employment through its musical past. Such place marketing relies on 'branding' locations; music is a cultural resource, hence trademarked labels such as 'Nashville: Music City USA', 'Memphis: the City of Kings [i.e. BB King, Elvis Presley]', 'Liverpool: Birthplace of The Beatles', 'New Orleans: America's Cultural Capital/The Birthplace of Jazz' and, less famously, 'Tamworth: Australia's Country Music Capital'. Despite the incorporation of musical events into tourist schedules, local reactions are not always positive. Festivals have produced local antagonism towards crowds, noise, drugs and environmental degradation. Music tourism, where it has transformed places, raises questions over how local communities are involved. The impact of Elvis-based tourism on Memphis brought disjunctures between tourism, with the associated tacky souvenir shops, and local concerns:

> The contrast between the increasingly antiseptic glitter of the Graceland giftshops and the nitty-gritty, slightly run-down, working-class flavour of much of the rest of Memphis…is striking: wandering around town, one

> gets the distinct impression that the average Memphian would be quite happy to see all the 'Elvis Zombies' wither up and blow away so that the city could get over Elvis already and go about more interesting business.
> (Rodman 1996: 108)

In line with these reactions, local record store Shangri-La published an alternative music guide to Memphis, which actively avoided references to Elvis sites, and constructed a radically different perspective on the musical heritage of Memphis, beyond the 'line of 800 Elvis Zombies waiting to shell $15 out to smell Elvis' bicycle seat' (quoted in Rodman 1996: 108). Others resented the neglect that other important, and arguably more formative, sites of musical history experienced, as tourism resources were channelled into the 'Elvis-town'. Other sites were actually destroyed: Stax recording studios, the source of famous soul and funk records by Otis Redding, Wilson Pickett, Isaac Hayes and others, was demolished in 1989, 'to make way for a soup kitchen' (Rodman 1996: 109), while Beale Street was almost completely destroyed as part of urban regeneration programmes during the 1970s (McKee and Chisenhall 1993: 96). Only later did the heritage in this district become part of larger tourism schemes, attracting visitors away from the Graceland complex to hear blues in reconstructed and highly entrepreneurial venues – including, ironically, Elvis Presley's Memphis Blues Bar.

The commercialisation of music for tourists affects local musicians both positively and negatively. In Ireland and other locations, avenues for local musicians performing for tourists 'formalised' what previously might have been precarious touring schedules. Local 'scenes', while often vibrant at the height of activity, are always vulnerable, reliant on sympathetic venue owners, dependent on a critical mass of musicians and constant innovation to 'keep the scene alive' (Chapter 5), but tourism in some instances has provided a mechanism to enable local musicians to earn steady incomes and to conserve local musical traditions, however transformed (Daniel 1996; Goertzen and Azzi 1999). The demand for 'traditional' Irish music has meant a formalisation of regular performances in pubs, but reducing the scope for musical innovation. In Bali musical events have been greatly truncated from their original durations to respond to tourists' limited attention spans. Similarly, Tjapukai Aboriginal Dance Theatre in Kuranda, Australia, became a new source of income and employment for indigenous youth, and a national success in creating employment and in the cultural industries. This too involved compromises. Regularity of tourism performances stifled spontaneity, and limited the styles and range of music that could be performed. At Alice Springs in the centre of Australia, tourists perceived traditional Aboriginal music as authentic – an example of new primitivism – being unaware or uninterested in Aboriginal appropriations of country music, reggae, rap and hardcore metal. Hence within Alice Springs there has been a proliferation of didjeridu shops (an instrument that actually originated from different tribal groups 1500 km north, and not played in Australia's central desert regions). Aboriginal musicians and producers have to some extent responded to this, with didjeridu CDs, artefacts and so on, yet non-Aboriginal labels, musicians and distributors dominate the market, simultaneously reinforcing notions of 'primitive' Aboriginal culture through their marketing efforts, visual images and sounds. Other Aboriginal people were unhappy with essentialised representations of their diverse cultures in tourism promotions, which denied regional variation and contemporary realities in contrast to a mythologised loin-cloth-wearing, spear-throwing past. Tensions between Alice Springs as an Aboriginalised tourism destination (and celebrated as such), and the economic situation of the Aboriginal population, in terms of their benefiting from their music, have increased. More generally, material aspects of tourism development, including labour relations, wages, a lack of common-sense cultural sensitivity, local backlashes and divisions, have occurred at many music tourism sites.

In the large post-industrial city of Manchester, cultural tourism (including music and other forms of 'lifestyle' consumption) was part of urban regeneration initiatives centred on the city's reputation for dance music and club cultures (growing out of the famous 'Madchester' sound of the late 1980s). Cultural industries and tourism were incorporated into the official strategies of local authorities, who relaxed licensing regulations in an effort to encourage more clubs, cafés and restaurants to open, catering for the large influx of visitors to the city each week for music (Halfacree and Kitchin 1996; O'Connor 1998; Haslam 2000). In peak season anywhere between 25,000 and 85,000 revellers were estimated to come into Manchester on Friday nights, to visit famous clubs and take part in sub-cultural dance activities. During the late 1980s, at the height of the scene's international prominence, over 40 per cent of young American tourists to Britain listed Manchester as their number one destination, largely due to its musical reputation. Yet, while local councils have loosened licensing regulations and encouraged the growth of music venues over the last ten years, to cope with and capitalise on this surge in popularity, such moves were not without drawbacks. The decline in licensing regulations, coupled with difficulties surrounding drugs and crime, eventually damaged the original scene that made Manchester a well-known music site. The Hacienda club (a site once known within club cultures as 'the Eiffel Tower of Manchester', where artists such as New Order, the Stone Roses and Happy Mondays began to perform), was abandoned, a victim of rising prices demanded by monopolistic door-security cartels and pressure from local crime figures: an ironic emptiness, given the 'opening-up' of Manchester to the outside world encouraged through strategies to market musical heritage. Furthermore, urban regeneration strategies in deindustrialised cities such as Manchester mobilised cultural industries that promoted a particular mix of 'procapitalist' philosophies and deregulation, potentially alienating those very populations that had suffered most from deindustrialisation:

> the resultant development, while based on images of leisure and consumption and aestheticisation taken up by urban boosterists and sociologists alike...had limited cultural resonance, especially among those whose labour would be crucial to the transformation of the centre into a cultural landscape.
>
> (O'Connor 1998: 231)

Renewed cultural zones and precincts, the sites of cultural tourism (and gentrification), often benefited visitors and upwardly mobile professionals more than the workers for whom economic regeneration strategies were intended. Here, too, cultural tourism, development and reconstruction were contested.

Music tourism strategies can be understood simply as a way of extracting profits from relatively gullible tourists, convincing visitors of the authentic nature of a place's musical heritage. However, at another level pleasure is derived by tourists at performances and festivals, and through music sites as spectacle – the artificiality of the places themselves. In these instances, tourist experiences on offer at Nashville, or Tamworth, or in Salzburg's *Sound of Music* tour, involve a sense of parody invoked by the participants – read into the places by tourists as the consumers. The concern with play, parody, a lack of meaning and the celebration of 'surface' in leisure and entertainment economies (e.g. Baudrillard 1981) is reflected in many music tourism industries. Through the emphasis on leisure and play, a sense of artificiality, superficiality and simulation of mythical 'original' artefacts can be celebrated. Several sites qualify for such parodic, almost absurdist pleasures to be read into the tourist experience: from Elvis Presley's gold Cadillac in the Country Music Hall of Fame; to sites such as Dolly Parton's theme park Dollywood or Conway Twitty's Twitty City. Similarly, the vast network of Hard Rock Cafés, that present standardised combinations of rock 'n' roll memorabilia, menus and T-shirts (whether in Amsterdam, Bali or Toronto) offers only surface placelessness; Madame Tussaud's 'Rock Circus' of wax figures in London celebrates simulation and artificiality in the extreme. Tourists are often quite aware of the artificiality of these encounters with music. While overt commercialism and a lack of meaning may appal some visitors, for others artificiality can be consumed as a pleasure in itself. Hence, tourism sites where musical significance is to various extents 'invented' can represent mimicked simulations of an original concept, scene, sound or event enjoyed by visitors in often unintended ways.

Yet music tourism is much more than just the tourists; it cannot escape the social, economic and complex cultural politics of other examples of travel industries – the seasonal nature of most tourist activities, issues of cultural representation and local participation, effects on local musical cultures and musicians, and the role of music and tourism in regeneration strategies (alongside leisure industries and magnets of cultural capital such as film, fashion and design). Music tourism plays both economic and cultural roles in reshaping particular geographies, boosting development, and incorporating – or excluding – the local musical industry, and is therefore to be both welcomed and rejected by the communities themselves.

11

TERRA DIGITALIA?

Music, copyright and territory in the information age

As globalisation unfolds, more places are drawn into the reach of companies distributing music beyond its country of origin, though this process has been very selective (Chapter 3). This has involved the increasing penetration of American and European record companies into new territories (Sadler 1997), global distribution of both relatively 'universal' and more specific 'national market' recordings (such as Latin music), and some measure of standardisation in the music product, with new formats (CD, MiniDisc) and digital transmission. The shifts towards digital recording and global music distribution, and their impacts, are analysed in this chapter. Music as a 'digital' product is now one niche within much larger information and media empires while, at the same time, the emergence of digital music recording, storage and distribution has presented challenges to established ways of producing, buying and listening to music. As the speed of flows of culture increases (with ever more places subsumed into the networks of influence of media corporations), digital technologies have an increasingly important role to play in how we hear music. Technological shifts link people in diverse places, increasing the range of music available and, in turn, allowing various cities, towns and regions to contribute to global cultural economies, promoting local music much more effectively to outside audiences. Such decentralising effects, associated with the expansion of computer-mediated communication (Internet, email, intranets), present challenges for music economies, although they too are mediated by issues of uneven access to technology and telecommunications infrastructures. This chapter examines how the corporate domination of cultural industries has influenced music, its marketing and its sounds; whether the emergence of digital music, and new forms of distribution such as the Internet, 'democratise' the world of music; how Internet cultures have changed the way music is thought of as information and whether major music companies will be able to 'control' how people access, reproduce and share music in the computer age. An important starting point is to consider the ways in which music has become part of a much wider 'information' economy.

Information economies

A number of authors have discussed the emergence of new industries, networks and institutions based on the storage, sale, dissemination and use of information, as a key feature of late twentieth-century capitalist development (Scott 1986; Castells 1989; Lury 1993; Borja *et al.* 1997). While information has always been a key element of the production of any given good, a series of events has pointed to the emergence of distinct information economies. These have included developments in computer technology, the growth of a more globally integrated financial system, convergence between corporate interests in the telecommunications and high-technology industries; state deregulation of media and communications sectors; and the appearance of new forms of dissemination (cable, Internet, intranet). Music has been described as an 'information industry', or even a 'content-providing' creative industry (alongside film, multimedia, publishing, TV, newspapers, graphic design). The ways in which music has been absorbed within these larger umbrella industries can be thought of in two ways: first, in relation to the ownership of music companies by larger corporations that themselves primarily deal in information, and, second, in terms of the impact of Information Technology (IT) itself on music production, distribution and consumption.

Publishing, production and distribution have been caught in a cycle of corporate purchases and amalgamations. The five companies that dominated music recording in the 1960s were largely replaced by the 1990s by an oligopoly of entertainment companies, usually referred to as the 'majors'. Independent labels such as Motown, Stax and Def Jam (which heralded new music styles and social movements) were all bought by the majors after the 'boom' period of their label, style or 'scene'. By the 1980s and 1990s, despite the continued emergence of local independent labels, interests in the recording industry had become consolidated into a few major companies (Peterson and Berger 1996; Scott 1999). More recently, entertainment companies have in turn been drawn into the domain of other larger corporate empires, as part of a 'convergence' variously labelled an 'infotainment industry', 'info-communications sector' (Herman and McChesney 1997) or 'Siliwood', the conjunction of 'Silicon Valley' and 'Hollywood', which suggests the continued importance of geographical locations within increasingly global oligopolistic corporate control (Hozic 1999). 'Infotainment' implies the merging of the worlds of IT and entertainment, both in terms of how cultural products like music are received and consumed, but also in terms of the companies that own and distribute these products. Corporations created their own versions of 'infotainment': Sony, for example, began as a technology company producing domestic electronic products, before purchasing CBS in 1988, diversifying its interests in music. Seagram, with headquarters in Montreal, Canada, began as a wine and spirits company, then bought interests in film and music (as a point of entry into the 'infotainment' world), before being purchased by Vivendi, a French cable television company. Conversely News Corporation, controlled by expatriate Australian Rupert Murdoch, specialised initially as a news media company before becoming an 'infotainment' giant. News Corporation's interests in music include the Festival/Mushroom Group, Australia's 'major' label, structured alongside digital publishing firms, sports teams and competitions (the Los Angeles Dodgers, Australian National Rugby League), radio stations (Sky Radio), e-commerce investment (eVentures) and polling research (Newspoll). News Corporation has therefore been able to secure control not only of the means of information flow (such as through film, newspapers, TV stations and Internet sites), but also of several areas of 'content', including music.

The increasing dominance of all-encompassing corporations in the fields of entertainment, media, and now IT and telecommunications, calls into question the existence of a distinct 'music industry' separate from other sectors of the cultural economy. By using the different outlets available to them, companies such as TimeWarner (who control CNN cable TV networks, and merged in 2000 with the world's largest Internet service provider, AOL), Sony (who also own the CBS television network and Columbia Pictures) and News Corporation can market music across media: in film soundtracks, commercials or television programmes, or in the new field of digital distribution – the Internet. Music plays a vital contemporary role in such conglomerates, worth 26 per cent of Bertelsmann's annual turnover (US$3.5 billion), 24 per cent of Time Warner (US$4.2 billion) and 11 per cent of Sony's combined hardware and content businesses (Shuker 1998). Collectively, the information–communications–entertainment sector accounted for a global economy worth approximately US$1.5 trillion in 1997 (Herman and McChesney 1997: 108). Music has been bound up in processes of expansion, consolidation and monopolisation of a global magnitude.

These shifts have had a major impact on the business of selling music. The convergence of media forms 'has taken the historical relationship between music and other media into an altogether different realm by marrying the intensified entertainment economy to the global-telecommunications-computer infrastructure' (Breen 1995: 498). One effect on music is the increased emphasis on copyright – the rights to reproduce and distribute an actual recording of music – which also accompanied a change in status of the music commodity itself (see Chapter 3). Whereas the actual material good (the CD, LP or cassette) has long been the 'product', in 'infotainment' industries these physical goods have been somewhat de-emphasised. The physical music product has become a conduit for transmitting copyright; the real commodity is the intellectual property.

The image machine: manufacturing and marketing music

One debate concerning the growth of 'infotainment' corporations and their control over music production and distribution is the extent to which there has been a shift in the styles of music offered for general consumption. A vast amount of popular music distributed by media/information corporations has been characteristically dismissed or ignored by music critics (if not by sociologists or the popular media). Major artists associated with each of the media/information/entertainment corporations – Madonna and Michael Jackson, who had transcended their pop star status to become global icons, and a pantheon of others such as Elton John, Mariah Carey, Celine Dion, Britney Spears and Whitney Houston – rarely gained critical acclaim, yet were exceptionally successful in their dominance of popular music and their emergence as global superstars. Major corporations sought to maximise sales in particular niches, linked to perceptions of tastes within age groups: teenagers, but also 'middle of the road' and 'adult contemporary' genres, with the preponderance of 'golden oldies', the rise of film soundtrack albums and the incursion of music into new arenas (lifts, shopping malls and telephone holding patterns) all contributing to a widespread notion that contemporary popular music has little that is innovative or interesting. The resurgence of corporate-marketed country music in the 1990s was also perceived to be part of the malaise. When Garth Brooks (Capitol/EMI) displaced Michael Jackson at the top of the American charts in 1992 the surge in sales was attributed to an increasing Republican emphasis on 'traditional values' and an adult backlash against rap, but also to the successful fusion of country themes with rock: 'a hybrid music with enough of a twang to qualify as country but without the deep southern inflections that scare away rock fans' (Tannenbaum 1992: 21). Brooks, originally an advertising executive, went on to combine country music with stadium rock, and sentimentality with patriotism, to sell over 100 million units in the United States alone, in November 1999 superseding Elvis Presley as the most successful performer of all time. As music marketing became increasingly linked to other forms of entertainment, 'synergies' were sought between music, film, television and promotions through store displays, merchandising and product placements. Michael Jackson starred in Pepsi commercials; Coca Cola introduced a new Coke; Pepsi retaliated by sponsoring Lionel Richie while Coca Cola brought back 'Classic' Coke and signed Whitney Houston for the commercials (Ennis 1992: 377), all of which flooded not just American television screens, but those of a vast number of countries, in a comprehensible form. Indeed, if music was 'the universal language' then 'Cola was the universal solvent' (Savan 1993: 89). Artists and their output became 'brands', while new albums became multimedia events.

Such convergence maximised profits, not simply by selling a wider array of products, but also by licensing the rights over a recording to other users, as material is used and transmitted through a number of media. Infotainment companies licensed more tracks than ever for advertisements as artist backlash (pioneered by Neil Young) receded: Nissan used the Breeders' 'Cannonball', Apple used Elvis's 'Blue Suede Shoes' to market iMac computers, Apollo 440's 'Stop the Rock' became the Mars Bar theme, while The Kinks' 'You Really Got Me' was used to sell kidswear by fashion label Gap. Virtually all of Moby's album *Play* has been used in different advertising promotions, making it the most licensed album ever. These strategies also served to boost marketing, with music linked to particular concepts, cultural signifiers and visual meanings. Soundtrack albums from films and television series (alongside other multimedia visual representations of cultures and places) were highly successful, with a percentage of the takings of each film eventually filtering back to the music copyright owners: the media/information corporations (who in the case of *Moulin Rouge* also owned the film studios and distribution channels). Music companies were sometimes able to use both film and advertising, as with Lenny Kravitz's 'American Woman' (1999), a song that was 'durable and open to any number of interpretive uses' (Henderson 2000: 6), and which featured in the soundtracks to both *Austin Powers: The Spy Who Shagged Me* and *American Beauty*, and has since been used to entice consumers to purchase Castrol Oil, Tommy Hilfiger designer products and Gatorade drinks. Further examples of 'convergence' include video games, segued to licensed recordings (such as Chinese singer Faye Wang's 'Eyes On Me', the theme song for the video game *Final Fantasy VIII*) and multimedia products (CD-ROMs, websites), which featured film clips as well as interviews, live footage and the music. In terms of corporate strategy, the 'aural' and 'visual' became inseparable, evident in the parallel rise of DVDs.

One outcome of the incursion of 'infotainment' corporations into popular music has been the increasingly unapologetic 'manufacture' of popular music performers for local and international audiences, rather than the promotion of already existing bands. Various bands have been created in the hope of market access, from the Monkees in the 1960s to Milli Vanilli and then the Spice Girls in the 1990s. Neil Tennant, lead singer of the highly successful Pet Shop Boys, once proudly stated 'We can't cut it live', at once a challenge to authenticity (Goodwin 1992: 36; Heath 1990) and a recognition that image was more important than performance. By the end of the millennium this form of 'creativity' was much less likely to be considered a threat to authenticity or commercial success, or even an issue to be discussed. Overtly 'manufactured' artists dominated singles charts, with an abundance of boy and girl groups (Take That, Human Nature, Destiny's Child, B*Witched, Bardot), solo 'stars' (Christina Aguilera, Vitamin C, Mandy Moore) and Australian daytime soap actors (Kylie Minogue, Natalie Imbruglia), mostly marketed to teenagers. Emphasis returned to short-run profits from 'hit singles' rather than building careers through album sales. Stars were recruited for television and music, as with S Club 7 (stars of their own Miami-based family television show), or fabricated to recipe, as with the Backstreet Boys and 'N Sync. Even members of the Backstreet Boys themselves lamented the formulaic construction of such acts; at the apex of their 2000 promotional world tour one complained 'we're like a good car design, [but] they just keep stamping them out. Making copies! Copy-rama! It's kind of frustrating...how can we ever make a move or differentiate ourselves?' (quoted in *Sydney Morning Herald*, 23 November 2000: 14). Image, fashion and teen appeal, always a presence in music capitalism, had come full circle.

These trends indicate that Adorno and others might have been right about the 'culture industry'; terms such as 'convergence', 'synergies' and 'cross-promotions' all suggest standardisation and homogeneity. However, although a very small number of companies control and make profits from most of the world's English-language music (a classic oligopoly), at the local level, a whole range of different actors, geographies of production, distribution and consumption are present for diverse genres. No single 'music industry' dominates global music production. At both international and regional scales, music economies, like wider forms of capitalism, have been as much disorganised, fragmented, uneven and variable as they have been governed by constant and predictable patterns. Media and technology giants like TimeWarner and Sony aim to wield global power, yet the dynamics of South and South-East Asian cassette cultures, or even the Christian and New Age music industries (largely independent from major companies, and simultaneously ignored by academics) demonstrate that their influence is not universal.

Given the enormous power of infotainment companies, it is no surprise they have drawn much criticism. Insecure conditions for employees and artists signed to infotainment labels is common practice; as one commentator argued after the merger of PolyGram with Universal (which itself resulted in major job losses across the world): 'there are no good guys in this, it is rich people wanting more and record companies liking the existing deal' (Alloca 2001: 22). After the purchase of CBS by Sony, one of its biggest recording stars, George Michael, began a process that would eventually end in litigation. The singer, who claimed he worked under conditions of 'professional slavery' and was being treated by Sony as 'no more than a piece of software' (quoted in Charlesworth 1994: 1172; Coulthard 1995: 733), sought to extract himself from an eight-album recording contract but eventually lost the case. Numerous musicians including Ani di Franco, Hole and Public Enemy either entered into legal battles over record contracts, severed relationships with infotainment conglomerates or established their own independent labels. Meanwhile the Beastie Boys, They Might Be Giants, David Bowie, Roger McGuinn (of the Byrds) and others circumvented the interests of the corporations to which they were signed by releasing songs in the freely distributable MP3 Internet format (see below). Monopolisation was not without resistance.

Digital technology: turning music into 1s and 0s

One key element of change in the way music is produced, distributed and enjoyed by audiences (linked to the growth of infotainment) has been the digitisation of sound. This involved a system of recording, transmitting and reproducing information with a level of hitherto unparalleled accuracy, durability and universality of application, allowing 'any signal, whether a sound or an image [to] be transmitted or manipulated in similar ways' (Lury 1993: 157). At their most basic, digital technologies convert audio sound waves into binary code by sampling the audio signal for volume (amplitude) and pitch (frequency) at a given rate. Digital technology improved on older analogue equipment, offering extremely low noise levels and considerable flexibility. Digital information could be stored in computer files, with less danger of deterioration over time, and be converted into a range of formats (pressed into a CD, segued with film or television footage, converted into a video game soundtrack) more quickly and with more reliable results compared to analogue technology with its range of recording media, formats and incompatible equipment. This flexibility was a crucial element of the support for digital technology by the major infotainment companies who continued to govern and administer extremely large copyright catalogues and thousands of master recordings.

Digital recording techniques emerged in the 1970s for music that was intended to be distributed on conventional vinyl albums, yet the main push towards embracing digital technology in distribution formats only occurred after the widespread slump in the recording industry in the late 1970s and early 1980s. Digital technology and the promotion of the new CD format of music distribution kick-started a lagging market for popular music, as music already consumed by 'baby-boomer' generations on vinyl albums was resold in the new format. Parallel to previous inventions in the music industry (discussed in Chapter 3), the widespread adoption of digital technology and the push for convergence triggered new kinds of legal and territorial battles over 'culture' and intellectual property that filtered through music's diverse geographies of production, distribution and consumption.

Digital decentralisation?

The impacts of new digital recording technology included improved formats and better storage and transmission of copyright material for major labels, but also involved unintended effects, with the potential to change the ways in which musicians were able to distribute their sounds to wider audiences. Technological development in the digital realm, aimed at making recording gear more powerful for studio applications, also made equipment cheaper for the home recording and small studio market. Various companies specialised in producing equipment that recorded professional-quality digital information at much lower costs than previous analogue technology. In addition to these specialist units, sophisticated software for hard-disk recording and sequencing meant that all recording and sound-processing activities could be completed through a personal computer.

Digital technology meant that producing high-quality recordings became possible for musicians without recording deals and for artists scattered across greater distances. Record contracts previously required major labels to advance production costs to the musicians, who then repaid these costs as an advance against future royalties (a system that meant that production remained capital-intensive). This further evolved:

> Record companies have been the providers of capital, giving them de facto control over the promotion of new talent. With the cost of creating a compact disc now within a performer's reach, record companies will have to rely more on marketing or distribution to exert market control.
>
> (James 1995: 62)

In turn, software and specialist digital recording equipment manufacturers targeted amateur musicians, bedroom enthusiasts and home studios with emphasis on the ability to produce high-quality recordings without relying on corporate investment. The rapid emergence of music editing software, Musical Instrument Digital Interface (MIDI) capabilities and sequencers, along with advances in digital recording gear, meant that professional recordings were much cheaper, and within closer reach of amateur musicians and localised scenes. The whole notion of 'making music' shifted (particularly with the growth of techno and other dance music styles), from mechanical skills associated with playing instruments (with years of tuition and practice) to expertise with the interface of a computer screen, mouse and keypad. Synthesisers – from Robert Moog's early models to on-screen oscillators – enabled seemingly endless combinations of sounds; drum machines allowed for unprecedented mechanical precision and repetition. Sequencing, mixing, sampling and looping became part of the creative process.

The rise of home recording cultures in many non-metropolitan areas has suggested the potential for decentralisation, through cheaper and more accessible technology, Internet resources and capabilities for global distribution and marketing for unsigned bands. Durant argued that technological change could bring about 'democracy as something which results from the cheapness of the equipment' (1990: 193), and which could be attained by a wider range of people with fewer skills in musicianship, a parallel to the earlier rise of cassette culture. In music scenes as diverse as house music production in bedrooms of Japanese teenagers to country musicians in inland Australia, digital technology was taken up, allowing home 'demos' to reach new heights of quality. Sometimes whole 'digital cultures' emerged, as with indie 'DIY' movements and networks of distribution for DAT tape recordings among DJs in the global trance music scene (Box 11.1). Other notions of 'decentralisation' implied that the various technological changes could widen the access of people from more diverse backgrounds and places to production. Redefining the means and geography of production seemed possible.

Box 11.1 DIY music in Japan

Japan has always been a unique setting for popular music production and consumption. While a select range of Western genres and artists have been successful – rock, hip hop, R&B, teen pop and big-ballad adult contemporary (such as Celine Dion, who won Japan's Best International Artist for three consecutive years) – the domestic market has been dominated by local styles as well as J-pop, Japanese language variations on Western sounds such as rock act L'Arc En Ciel and solo star Utada Hikaru, whose album *First Love* (1999) became the biggest selling Japanese release of all time (selling over 8.4. million copies). Against this background of corporate-dominated popular music (particularly by Sony Music and Toshiba–EMI), Japan has recently experienced a substantial growth in home recording and 'indie' label activities, reminiscent of early stages of cycles of innovation and consolidation in the American and British recording industries during the 1950s and 1960s. Parallel to Western 'indie' and punk scenes, fanzines have become crucial, as with *Beikoku Ongaku* and *Indies*, but the low costs of production equipment allowed experimentation and a wider range of releases across genres, from the 'melo-core' (melodious hardcore) of Hi Standard, who sold over 700,000 copies of their hit single 'Making the Road' (1999), to the 'dark ambient musings' (McClure 2000b: 65) of Kyoto's Mana. Electronic dance music and the use of samplers provided new possibilities, as with *taku-roku*, the Japanese home recording scene that produced the electronica of artists such as DJ Krush, Nobukazu Takemura and Tatsuya Oe (Captain Funk). Such independent releases now account for nearly 10 per cent of all music sales in Japan, a major slice of one of the world's largest markets. As one industry source put it:

> the indies are more focused than the majors...they're able to concentrate on a relatively small number of releases...[they're] not blowing their money on hostess bars in Ginza; they're working in a relatively simple world where they know their medium and have low budgets.

The incursion of 'indies' into the Japanese market forced Warner Music to set up its own low-cost production subsidiary, Warner Indies Network, in an attempt to tap into this growth. Traditional impediments to commercial success (limited finance, expensive means of production) have been overcome through accessible technology, vibrant sub-cultural production and determined DIY attitudes.

See: McClure 2000b.

The emergence of 'dance music', as a primarily computer-generated genre, is particularly illuminative of the new geographies of production opened up through the digitisation of music. House music, and Detroit techno, like other genres of popular music, led to sub-styles, such as acid house, happy house, garage, jungle, electro and so on, which were principally produced through the sampling of existing records (repeating elements and changing the speed), or through the use of synthesised beats and basslines. As in other genres, certain locations were established as points of origin (Chicago, Detroit, Manchester) yet, unlike many other sub-cultural styles, dance music has a decentralised geography of scenes with more continental European and Asian influence and participation in production. The relative placelessness of dance music (a function of its lack of lyrics) has allowed more people to contribute to successful distribution beyond domestic markets. Hence, dub techniques (such as the use of deep basslines, 'sweeping' sounds from filters, echoes and effects) have become popular for French and German producers; cities such as Brussels, Tokyo, Hamburg and Vienna have contributed to global sub-cultural production; trance producers have cropped up in places as diverse as Japan, Sweden, Israel and Croatia. Flows of independently produced dance music reflect sub-cultural linkages so that musical fragments from each of these locations could be consumed in any other site without a necessary awareness of the tracks' origins. Yet ironically, given this relative placelessness, dance music genres often coalesce in unique ways in different locations, giving rise to new sub-cultural blends and further identifications with place: furious gabber techno from Rotterdam, dark dub in Hamburg, Belgian New Beat, Viennese dub, Goa trance, New York illbient and so on (see Chapter 5).

While some measure of decentralisation has occurred, complicating factors temper claims of genuine democratisation in music economies. Such claims often assume that many people will take an active role in cultural production. While these technologies certainly improve the situation for amateur musicians, bands and artists chasing a record deal or attempting to 'get a CD out', there is a 'danger of abstracting these technologies' intrinsic "capacities" from the social contexts of their actual use' (Morley 1995: 309). The 'decentralising' effects of allowing access to distribution for thousands of musicians resulted in a profusion of unsigned, unpromoted artists on websites, and of demo CDs sent to radio stations, record companies and publishers. One unintended effect was a rise in demand for new ways of differentiating products, of sorting through, categorising and finding music genres. 'Branding' of artists and releases, a central element of corporate strategies, became as crucial on the Internet as the physical manufacture of compact discs, simply due to the vast range of choices for customers. Local artists engaging in DIY music production competed with thousands of other similar artists, all seeking success as small-scale entrepreneurs. As Binas has succinctly put it, 'technology marks and maps relations of power and hierarchies as well as processes of potential democratisation' (1999: 7). Unless musicians could generate significant links from other websites, or could mobilise audiences for self-produced material, their sounds were likely to be lost in a 'sea' of digital noise (see Box 11.2).

Box 11.2 DIY production in Byron Bay, Australia

Byron Bay, on the north coast of New South Wales, Australia, would seem a location ripe to benefit from the digital production of music. The town (with a population of only 5,000) has become mythical in Australia as a site of 'alternative' culture, supplemented by recent waves of counter-urbanisation that have brought many former urban residents to the area, searching for natural beauty, sub-tropical lifestyles and the 'creative' arts scene. The town has sustained a considerable amount of music production, far more than other Australian rural centres, and many individuals within local scenes seized upon the digitisation of music for regional music production. Until the mid-1990s, home recording primarily consisted of the production of low-cost cassette 'demo' recordings; however, with the advent of more affordable technology, it became common for bands to press their own CDs, either as self-funded projects, as independent releases signed to local labels or through labels established by the artists/scenes themselves. Digital recording provided good-quality sound at a more accessible and cheaper price. As one long-term local producer commented:

> It's a great thing democratically. It's empowered people who previously couldn't afford studios. You couldn't afford to do it. It was an

> elitist thing, and it was part of the framework that enabled the multinational labels to maintain a stranglehold on the industry, so in that sense it's great. In terms of quality and accessibility for people of lesser means, it's a really positive thing.
>
> (quoted in Gibson 2000: 230)

Home studios in Byron Bay were used across a variety of styles of music, from 'New Age', funk and Christian music through to techno, with CDs sold at gigs and distributed through local retail outlets, social connections, independent labels and websites. One musician, Tarshito, even established a solar-powered recording studio to record his blend of New Age, folk and world music. The advent of home recording enabled artists to control more aspects of the production process, a trend that troubled many studio engineers in the area: 'The days of going into a big recording studio with a huge budget are pretty much over. Bands are recording at home and may be going into a studio to mix it' (quoted in Gibson 2000: 234). New opportunities for performers constituted threats to producers.

Internet promotion and distribution became extremely popular within the region, with sites for solo artists, groups, indie labels, festivals and record shops. The Internet was strongly identified by musicians, promoters and labels as a means of overcoming corporate distribution; as one manager argued: 'geography doesn't matter any more, and why should it?' (quoted in Gibson 2000: 245). Yet despite all this, new technology has not yet constituted a panacea to dilemmas associated with a rural music scene. Considerable problems remain with production standards, promotion and distribution. Music commodities produced by local musicians were rarely purchased without consumers already having seen the live performance, and often formed an adjunct to the event. Furthermore, most Byron Bay musicians were unlikely to sell enough units to make a reliable living, even with the supposedly limitless space of the World Wide Web. Bands were able to record a CD and create websites, but found it difficult to be able to convince other sites to provide links, in order for people to access, enjoy and eventually purchase their music. As one musician put it 'there are lots of bands around with boxes of unsold CDs still sitting under their beds'. Digital technology enhanced amateur recording, and improved the means of dissemination, yet technology alone could not guarantee success.

At the same time as digital music technology spawned more grassroots CD production, the Internet centralised other activities – as very large music shopping sites (such as CDNow) operate from large warehouses, depriving local retail outlets of customers. Decentralisation in music production and consumption could only occur where access to digital production gear and the Internet itself was apparent. Digital decentralisation re-created a selective geography at the global scale, with most of the world's population unable to afford or access computers, and hence participate in new musical networks.

Copyright, capital and 'geographies of piracy'

> The barbarians are at the gate...they're in the moats, and they're climbing up the sides of the castle.
>
> (Robert Goodale, CEO of Ultrastar, a New York Internet firm promoting links for unsigned artists, quoted in Mardesich 1999: 96)

Since the emergence of the 'music industry' from its early origins in music publishing, copyrights – the rights to a master recording, including mass reproduction – have become central. Developed in the United Kingdom in 1710, 'copyright' was intended to protect the rights of authors or composers to benefit from their own work, yet copyright law was never implemented in that way: 'it was always a publisher's right rather than that of an author, since it vested the right of reproduction in the copyright holder rather than the author' (Lury 1993: 165). Copyright became a major industry in its own right: in 1996 the total value of United States copyright industries (film, TV, music, software, advertising, publishing) was over US$278 billion; they employed over 6.5 million people and earned the United States over US$60 billion in export earnings, more than agriculture, automobiles or chemicals (Huck 1999). Copyrights are products of the legal systems unique to each country, whereas Internet technologies, the new means of distribution for music, operate on a transnational basis largely beyond the operation of intellectual property structures. This variability of legal systems from country to country is one reason why global integration of music industries has been uneven. The absence of legal protection for copyright has produced 'geographies of piracy', which have long characterised bootleg cassette and CD production, and, more recently, the illegal digital transmission of copyright material. Counterfeit cassettes and CDs are part of legal, economic and geographical struggles over copyright shared with video games, software, film and design.

A significant amount of unsanctioned production emerged in a range of countries, including several in Asia and East and Central Europe, where production 'overcapacity' of the CD manufacturing industries was substantially greater than 'legitimate' domestic demand, as counterfeit trade became a high proportion of total market value. 'Piracy' also thrived in locations such as Bali (Indonesia) and Hong Kong, where tourists constituted the bulk of the retail market. Counterfeit trade reached as high as 85 per cent in Estonia, and other nations such as Paraguay and Israel were deemed 'priority cases' by the Record Industry Association of America (RIAA), which represented the interests of music companies, for urgent intellectual property reform. This geography of illegal production influenced the structure and pace of globalisation. PolyGram, for example, expanded its global activities significantly during the 1990s, buying labels across continents but establishing a real presence only in those markets where reforms in intellectual property and copyright laws had been undertaken. Thus:

> India's rich culture, large population and improving economy make it a very exciting entertainment market. With compact disc penetration still at an early stage there, and following recent amendments to the Copyright Act which will help curb music piracy, we believe India represents great potential for PolyGram.
> (N. Cheng, President, PolyGram Asia, quoted in PolyGram N.V. 1995: 1)

The protection of their catalogues – privatised databases of sorts – was of utmost importance to the continued survival of recording companies, particularly as they became very large, partly due to the incessant logic of expanding a copyright repertoire (in order to cover a vast range of artists, styles and potential opportunities for broadcast and reproduction). Universal, for example, holds over 650,000 copyrights. In many cases music companies were involved in direct pressure tactics, threatening United States economic sanctions (as in the dispute with China in the 1990s over intellectual property laws). RIAA actively pressured nations to reform intellectual property laws and provide conducive environments for the entry of infotainment corporations into overseas markets, and reported international copyright breaches to the United States Trade Representative:

> Just as the RIAA is actively involved in working with our elected officials to construct workable and fair legislation for U.S. copyright owners, so too are we dedicated to protecting and enforcing the copyrights of our members in other countries.... Countries identified as Priority Foreign Countries under Special 301 face investigation and the potential imposition of trade sanctions. Internationally, we have two main objectives: the facilitation of commerce where our members are doing business, which includes copyright reform and anti-piracy efforts; and the opening of markets presently closed or restricted to our members.
> (Record Industry Association of America 1999: 1)

International borders thus both discourage trade and enable piracy. Only when legal systems are standardised across the world, 'setting up the framework through which globalisation is furthered' (Sassen 1999: 151), will those in control of music copyrights have a real global market.

The geographical patterns of Internet 'piracy' are thus somewhat different from the trade in illegally reproduced CDs and cassettes, reflecting the selective global reach of computer technology. Most illegal trade (in this case downloads rather than physical products) has been linked to Internet Service Providers (ISPs) in Western nations, with 'about 60 per cent of these sites...hosted in the U.S. and 40 per cent elsewhere in the world, in countries like Canada, Korea, Australia, Sweden, the Netherlands, Belgium, Germany and Italy' (Sherman 1999a). While illegal industries in the developing world rely on lax domestic laws to reproduce CDs and cassettes, Internet 'piracy' has emerged as a phenomenon of the West, with computer-savvy consumers accessing shared electronic files. Counterfeit CDs and cassettes have dominated countries where infotainment companies only recently started fledgling operations, whereas Internet downloading cultures, such as MP3, threaten the stability of long-established markets where intellectual property protection was previously assured. This new pattern of piracy in part explains the ferocity with which music companies have fought the battle for copyright over Internet music formats – a ferocity that attracted strong criticisms and parody (see Figure 11.1).

RIAA and the International Federation of Phonographic Industries (IFPI) maintained a confrontational stance on the development of MP3 technology as part of their wider anti-piracy campaign, viewing it as a threat to the integrity of copyright properties held by member corporations. RIAA Senior Executive Vice President, Cary Sherman, in an address to the 1999 MIDEM Music Industry conference, articulated the political position taken by the Association with regard to MP3 technology:

> To the mostly young people caught up in the MP3 phenomenon, it means: Cool technology! Free stuff! What it also means, of course, is breaking the law. What it means is knowingly or unknowingly, the cyberspace equivalent of lifting a palette of CDs from the loading dock at Tower Records or FNAC and handing them out on a street corner. What it means is taking money out of the hat of an up and coming musician.
>
> (Sherman 1999a: 3)

and, more accurately for RIAA, from the coffers of its corporate music publishing members. In a speech presented to another gathering of computer and music industry executives in the following month, Sherman equated the MP3 format with music piracy, clearly identifying the format as the target of its legal and political strategies for intellectual property reform: 'Our copyrights are as important to us as your patents and trade secrets are to you. They are the lifeblood of

Figure 11.1 Pro-MP3 'propaganda' cartoon
Source: © ModernHumorist.com.
Note: A parody of RIAA's global anti-piracy campaign, from modernhumorist.com, illuminates the extent to which record companies engaged in an ideological 'war' with both quickly established formats (MP3) and consumers.

our industry' (1999b). 'Unofficial' fan pages, photo archives and other sites established by consumers were also threatened with legal action over copyright breaches (as was the case with one Australian 1980s group INXS), prompting vehement reactions from critics and consumers:

> A press release at the time of the INXS site launch…insinuated that the band was taking back – or protecting – what belonged to them, completely missing the point that since the death of Michael Hutchence the only point of unification for INXS fans has been the unofficial sites…some of the sites in question represent tens of thousands of hours work and contain extraordinary archives of articles, extensive discographies….E-commerce by itself is no more than that: e-commerce. It is not looking after fans. It is raking through their pockets and wallets to see how much money you can make.
>
> (Gee 1999: 60)

While Internet sites and distribution of copied files caused concern, all major corporate interests in infotainment simultaneously scrambled for deals and market share in the digital distribution of music, after United States court decisions forced the most famous download site, Napster, to remove access to copyright-protected material in 2001. Corporations purchased stakes in websites and established their own software and formats. BMG purchased Internet music retailer CDNow and signed a multi-million dollar deal with Napster; EMI purchased a stake in Liquid Audio and linked their business to that of Microsoft; Universal launched Duet and purchased MP3.com in an effort to become 'the primary supplier of on-line music and related services' (*Sydney Morning Herald*, 30 August 2001: 28), while AOL TimeWarner introduced MusicNet platforms as their attempts to control digital music. Such activities sent Internet stocks into a rapid cycle of booms and busts with each acquisition, merger and news on technological developments, yet no single format, or stable platform for commercialising digital music on the Internet, came to dominate. In the wake of Napster a suite of other software companies emerged to keep music downloads free, including Gnutella and FastTrack (which obviated the need for any centralised database of music, thus the music 'pirates' were completely 'placeless'). Copyright, a very abstract concept in one sense, continues to inform sites of struggle such as the Internet, governing (and yet never wholly controlling) the geographies of distribution and consumption of music.

Towards a digital world?

Landscapes of music have gradually become borderless, as have the activities of increasingly large companies that own music as one division alongside media, entertainment, technology and telecommunications. Music is now recorded and stored digitally, and is one form of 'content' within infotainment, marketed and cross-promoted over a range of media, from film soundtracks to the Internet. Ideas of 'authenticity', which valued live performance over simulation, or musical training in local contexts, have remained important (as evident throughout this book), yet are now accompanied by a simultaneous acceptance of highly 'manufactured' performers aimed at mass-market success.

Despite the general trend of contemporary cultural industries towards 'control' strategies (of markets, brands and targeted campaigns), in this recent phase of globalisation, musical activities have also become more fragmented, dispersed and localised. Digital technology was developed as a means to extend the expansionary tendencies of infotainment companies, by being able to record and store all acoustic signals as copyright. This effectively created a perfect format for recordings, which could then be owned, collected, governed and reproduced for consumption. Digital technology allowed the ultimate conversion of music into property: copyrights became the commodity. This technology was also more accessible and produced better results than old analogue recordings, particularly for those with small budgets, enabling more grassroots production (regardless of the quality of the music), redefining amateur music making and allowing new independent producers and labels to emerge. Such DIY production, coupled with new geographies of sub-cultures (as with the global connections in trance scenes), provided networks of musical creativity, exchange and consumption that paid little heed to the structures of distribution controlled by the 'majors'.

In contrast to images of Internet technology as 'borderless', the influence of physical distances remained crucial, but producing new geographies of regulation and governance in a world of digital content and surveillance (Leyshon 2001). While electronic media and 'virtual' platforms enhanced 'time–space compression', in dance sub-cultures, where vinyl has remained the preferred format, the 'time-lag' experienced between a release in one scene, and its distribution to others across continents, has actually increased. In contrast to the simultaneous worldwide release of major products from infotainment corporations (such as the co-ordinated release of albums by U2, Mariah Carey and so on), global distribution of independent 12-inch dance singles is haphazard, unpredictable and slow. Small print-runs ensured the scarcity of vinyl mixes, while independent labels struggled to sustain transcontinental linkages and accounts. Hence, DJs in more remote places battled for the 'latest' imports even if they originated in London or New York more than six months earlier. Digitisation has been neither utterly pervasive, nor has it completely erased the 'friction of distance' for the movement of real goods across physical space.

Transformations have occurred in the consumption of music. Audiences have remained active players in determining how new technologies are applied, whether through personal desires to create their own music or to copy and distribute copyright material, and in making decisions to 'absorb' the marketing ideals of infotainment corporations. The relative 'openness' of digital formats (for those with computer access) allowed for unintended uses of copyright material (as with music 'piracy', sampling techniques, digital distribution and cultures of informal exchange of a mixture of legal and illegal material). Debates surrounding copyright were a product of a crisis in the legitimacy of copyright ownership that had become particularly apparent in the age of the digitisation of all forms of audio and visual products.

While technological change, corporate strategy and marketing campaigns imply homogeneity, music production and consumption has become increasingly diffuse, connected to sub-cultures, amateur musical practices and consumer demands to hear, make and buy the music of choice. Musicians also returned to simpler production: dance music DJs retained the use of turntables and reliance on vinyl, and composers, as with French group Air, put aside drum machines and sequencers for acoustic drums. Both producers of dance and 'indie pop' sought out old analogue synthesisers in charity shops, sparking a return of Robert Moog's famous machines, alongside other electronically generated instruments such as the Theremin (obligatory in 1960s horror soundtracks) and Hammond organ (on records by Jamiroquai, Plone and others). As the range of synthesised sounds of contemporary keyboards became too familiar, the constant search for something new in music turned to embrace the old. Globalisation, rather than signalling the 'end of geography' (O'Brien 1992), stimulated both homogeneity and diversity.

12

THE LONG AND WINDING ROAD...

Popular music, like other popular media, has the ability to mediate social knowledge, reinforce (or challenge) ideological constructions of contemporary (or past) life and be an agent of hegemony. Beyond that, and central to its importance, it is a part of everyday life and a key element of recreation for many. It accompanies and transforms daily life for vast numbers of the world's population. Even so it has received scant attention in the world of geography primarily because its spatial impact is supposedly trivial. Though Guattari once described music as 'the most non-signifying and de-territorializing of all' cultural forms (1984: 41) it is now evident that music can be linked to images of place, and to the places where music has been created and performed, and that it is one element of the transformative power of globalisation.

This book has sought to provide a multidisciplinary approach to spatial, social and cultural themes in music, drawing from various social sciences, including the small but slowly growing number of geographical studies, to highlight the complex relationships between place and identity reflected and created through music. It has focused on different approaches to understanding the relationship between fixity and the fluidity, and hence the diversity of globalisation in its many manifestations. Some would argue that technology has heralded the 'end of geography' and that commercialism has killed the lingering authenticity and integrity of popular music; others have launched pessimistic readings of contemporary music. Texan singer, Terry Allen, responding to a question on whether his music was country music responded:

> Which country? There really is no more country music; it's all urban music now. Because in a sense there's no more country – even more so with the Web and all that. It's all corporate music now, because it's a corporate culture – that accounts for the deadness you sense in the music.
>
> (quoted in Gill 1999:15)

We have sought to show here that, while in some respects Terry Allen is right, music is far from dead and place is of enormous significance. Indeed Feld's plea for an 'ethnomusecology', with reference to such places as rural New Guinea (Feld 1996), has much wider reverberations, whether in terms of fixidity or fluidity.

Music is necessarily part of an expanding commercial world, diffused through a handful of constantly restructuring transnational corporations, by means of rapidly changing technology. At once global, it is also local in its ability to provide personal – if technological – links between performers and audiences – new variants on the virtual communities that have always been a crucial element of music scenes. While all this implies global homogeneity, music can and does enable resistance to globalising trends; rap and the ever-evolving hip hop scenes of particular inner-city areas are manifestations of this. Local musics remain vital. Indeed within the context of globalisation, localities may be revalued: 'under these circumstances of global integration, local identities and affiliations do not disappear. On the contrary, the transnational economy often makes itself most powerfully felt through the reorganisation of spaces and the transformation of local experience' (Lipsitz 1994: 4). Consequently 'Music's deep connection to social identities has been distinctively intensified by globalisation...[and] musical identities and styles are more visibly transient, more audibly in a state of constant fission and fusion than ever before' (Feld 2000a: 145), and simultaneously both local and transnational.

Music is sometimes a vital part of transnational networks of affiliation, and of material and symbolic interdependence, well evident in the diasporic ties across the 'Black Atlantic', and across all other oceans. Just as national boundaries are irrelevant to the location of cultures, almost all music is diasporic – evident in the multiplicities of reggae, or in the lyrics of *fado* or *isicathamiya* – products of dispersal and fusion, the stories of migration itself, as loss and nostalgia. The rise of mix music, a 'sonic montage [of] abruptly juxtaposed musical styles heard in rapid succession' (Greene 2001: 169) is a metaphor for the complexity of movement, and the 'unsettled cosmopolitanisms' (Greene 2001: 169) that follow. Hence, in this case in Nepal, listeners are 'dancing, thinking through, and even finding pleasure in the identity issues that arise...and in the many contradictory and ever shifting cultural worlds in which they now live' (Greene 2001: 184). Hybridity and syncretism are inescapable. Music, like other facets of globalisation, draws on: 'a host of references which are fused, rearticulated, played back' (Massey 1998: 122). Simultaneously transnationalism, flux and uncertainty emphasise seemingly lost tradition and authenticity.

Music nourishes imagined communities, traces links to distant and past places, and emphasises that all human cultures have musical traditions, however differently these have been valued. Music and nostalgia combine within 'an ideological field of conflicting interests, institutions and memories' (Walser 1993: xiii).

Authenticity hangs on nostalgia, hence 'there is a nostalgic dimension to any argument for authenticity' (Zanes 1999: 69). Moreover 'Nostalgia can be seen as a new way of imagining communities, harnessed in and by the post nation-state, an attempt at a connivance of a recovery of a lost childhood, a return to the m(other)land' (Sant Cassia 2000: 299). Modernity demands its converse, tradition and even invented tradition, which becomes embedded within modernity. It is as true of popular music as of most other facets of globalisation that it 'reinforces certain kinds of re-localisation, as seen in the strikingly pronounced rediscovery of many ethnic traditions and identities. Thus globalisation in this sense does not produce either global integration or a homogenised universal global culture' (Urry 1996: 1981), though that is almost always the superficial appearance. Likewise 'openness to foreign cultural influences need not involve only an impoverishment of local and national culture. It may give people access to technological and symbolic resources for dealing with their own ideas, managing their own culture, in new ways' (Hannerz 1987: 555; see Bilby 1999: 284–5). Music charts the imperial endeavour, in its legacy of anthems, hymns and popular songs, measures resistance to it in the 'liberation' songs of underground performers, marks the regional and ethnic tensions that have accompanied the creation of new nations and the fragmentation of others, and demarcates the cultural policies of national broadcasting stations.

Music as emancipation?

Critics of Adorno's arguments about the 'culture industry' (Chapter 1) have pointed to the emancipatory potential in music and musical sub-cultures – that popular music can provide opportunities for individuals or groups to assert human agency, to avert cultural homogeneity, to resist symbolically the wider social order and capitalist modes of production, and negotiate hegemonic ideologies. Sub-cultural style represented a contested realm in which youth rebellion, subversion and meaning remained in tension with the 'moral panics', or alternatively the commodification of styles from the 'mainstream':

> the tensions between dominant and subordinate groups can be found reflected in the surfaces of subculture – in the styles made up of mundane objects which have a double meaning. On the one hand, they warn the 'straight' world in advance of a sinister presence – the presence of difference – and draw down upon themselves vague suspicions, uneasy laughter, 'white and dumb rages'. On the other hand, for those who erect them into icons, or use them as words or as curses, these objects become signs of forbidden identity, sources of value.
>
> (Hebdige 1979: 2–3)

The 'signifiers' of deviance and nihilism identified by others as evidence of popular music's transience and 'empty' politics are imbued by the marginalised with subversive, oppositional meanings, the 'safety pin' fashion of punk becoming an iconic sign of deviance and refusal: 'the subordinate class [in Sartre's words] "make something of what is made of them" – to embellish, decorate, parody and wherever possible to recognize and rise above a subordinate position which was never of their choosing' (Hebdige 1979: 139). Ultimately all musical materials are capable of being mobilised subversively: religious music expressed solidarity under extreme conditions of oppression in the cotton fields and plantations of the American South, maintaining folk traditions and constructing fields of meaning (through language, subtle codes of song) that signalled defiance and dignity. Even the most commercial music is consumed in different ways by various social groups: a Kylie Minogue record means something different to a 12-year-old schoolgirl than to inner-city crowds at gay clubs; Dolly Parton records are not necessarily listened to in the same way in country bars in Tucson or in Zimbabwe.

In certain contexts a great deal has been attributed to the power of music. In Haiti:

> a carnival song helped bring down a government, a populist priest and future president borrowed the title of a peasant song as the name for his political movement, progressive musicians served the president and held assembly seats, and military coup leaders implicated a musician as a precipitating factor in the coup.
>
> (Averill 1997a: 2089)

In some lyrics and genres, such as hip hop, challenges are particularly confrontational, and music may function as a 'site of struggle between dominant and dissenting interests...[exhibiting] an essentially dialogic nature' (Lockard 1998: 21). Music was capable of inspiring resistance, and might be even more than that:

> Banda bears the burden of making sense out of terrifying changes. It arbitrates the tensions created by the disruptions of life in rural Mexico, brought on by peso devaluation, NAFTA and structural adjustment policies. It speaks to the stress in relations across genders and generations created by the imperatives of migration, resettlement and low-wage labor. It offers symbolic solutions to real problems and provides rituals for repairing ruptures in the social fabric...banda creates a site for solidarity.
>
> (Lipsitz 1999b: 204–5)

Even more grandly, 'popular music genres and musicians, like their counterparts in literature and film, have sometimes reflected and articulated themes of cultural

Figure 12.1 Interpreting music

decolonization, empowerment, neo-colonialism, cultural confrontation, disillusionment, despair, widespread inequalities, subcultural nationalism or identity, nation-building and anti-imperialism' (Lockard 1998: xii). Yet, while it may be the case that banda, and other particular musics, have extraordinary emancipatory potential, most music is not intended in this way, nor does it come close to transforming lives. In most cases, genres and songs carry a multiplicity of meanings; consumers are 'entangled' within the geographies of consumption for commodities (Crang 1996) as in Tanzania, where both reggae and rap are variously seen as 'a bumpy road to riches...fun dance music, a serious form of communication, a skillful art, or all of the above' (Remes 1999:5). Indeed most consumption of music constitutes passive reception without critical reflection, at most what Bourdieu described as a situation where 'the people chiefly expect representations and the conventions which govern them to allow them to believe "naively" in the things represented' (1984: 5). Some academic writing about music, a form of entangled consumption and reflection in itself, can attribute too much to the emancipatory potential of music. Yet contexts where and when music is resistant or liberating generate new senses of authenticity and thus contribute to commercial credibility. No music is objectively authentic or inauthentic. There is an element of futility in trying to assert absolute authenticity, but there is also every reason to search for genuine emotion and subversiveness in music. Music cannot be fully explained through functionalist explanations, or solely through 'decoding' music as communicative texts (Figure 12.1). Because music is bound up in emotional worlds it is always psychological as well as geographical.

Songs in the key of life

Music pervades everyday life through its production and consumption, through passive recognition and active participation. It circulates through shopping centres, blares from cars, cajoles us into dropping coins in buskers' hats; it supposedly appeases us in elevators, soothes us into purchasing commodities, remaining placid or going back to sleep, puts us on hold and frustrates us when telephone connections fail. Music is about nostalgia, producing tribute and cover bands, allowing golden oldies to dominate radio channels and compilation albums, letting even the Sex Pistols resurrect themselves for 'final' tours, and stimulating the new music tourism, and not only in the most obviously Western societies. In Colombia, Sunday afternoon dances – *viejotecas* (old-theques) – emphasise music's role in the formation of local identity, popular memory and also individual subjectivities (Waxer 2001). It is the background to films (and film music compilations have become best-sellers and instant nostalgia) and advertisements, and it accompanies aerobics and grander sporting events, often as national anthems. It is increasingly connected to other sub-cultural activities, such as sport (snowboarding, surfing, skateboarding and soccer – with the Jamaican team known as the Reggae Boyz), which are both local and global.

Popular music has been accredited with subliminal power. In stores music has become a 'device of social ordering, an aesthetic means through which consumer agency may be articulated, changed and sustained' (DeNora 2000: 132). Music shapes consumer agency, provides an interface between material culture, social action and subjectivity, and enables retailers to target clientele and brand image, and even to structure the temporal dimensions of retailing. As the manager of one British clothing chain, targeted at young women, observed: music was used to 'encapsulate the corporate identity of young, modern females and let the customers know they are in tune with what's going on in the music industry'. Where emotion, uncertainty and spontaneity are critical retail elements, music overwhelms the 'sound of silence' of traditional outfitters catering to a different generation (DeNora and Belcher 2000: 85). On what are left of the terraces of British football grounds, supporters chant local dreams and aspirations, perhaps nowhere better than Sheffield United's 'Greasy Chip Buttie Song' (You fill up my senses like a gallon of Magnet/ Like a packet of Woodbines, like a good pinch of snuff/ Like a night out in Sheffield, like a greasy chip buttie/ Oh Sheffield United, come thrill me again) (Thrills 1999: 229). It has even been used in the case of Deborah Harry's song 'One Way or Another (I'm Gonna Get You)' by the United States Internal Revenue Service. Most dramatically, rock music was used by American forces to remove Panama's leader from his sanctuary in the Vatican Embassy in Panama City by turning up the volume until the Vatican delegates could bear it no longer. Finally popular music, and particularly Frank Sinatra's 'My Way' and Bette Midler's 'Wind Beneath My Wings', accompanies us to the grave.

Music is more than nostalgia, politics, entertainment or marketing tool. This book began by examining the significance of music in societies that had experienced little change. Amongst the Suya, as in so many societies, music 'is said to come from beyond the mind and beyond the body – from the natural order as it is differently conceived by different peoples. This gives music a pre-ordained, transcendent, and often unquestionable reality' (Seeger 1987: 64). Modernity and post-modernity took music and other cultural forms down different paths, away from this kind of centrality towards a recreational sphere and what seemed a more peripheral and superficial position in life. Yet music never entirely lost this centrality; in most religions it shapes consciousness and demonstrates and demarcates belief. Moreover the role of music as a remedy for both physical and psychological imbalances, harmonising forces of the visible and invisible worlds (Diallo and Hall 1989, Roseman 1991), has taken on a new lease of life. Music has become therapy; as one therapist has observed, music:

> can put you into a relaxed state where you are open to things – old pains, mental blocks, feelings of peace or flashes of inspiration...listening to certain notes or harmonies can act as a kind of emotional massage before a counselling session, making them more powerful and in touch with their feelings.
>
> (quoted in the *Sydney Morning Herald*, 14 July 2000: 13)

Drumming has become a new means of restoring life forces and getting in touch with the 'inner self' and thus both a major component of tourism, from Byron Bay to Senegal, and a new form of circulation as West African drummers conduct participatory sessions throughout the Western world. Raves, where ecstatic states may be created through repetitive electronic music (and drugs), raw personal emotions and out-of-body experiences, combined with a sense of returning to roots, are said to offer parallels with spiritual healing and even evangelical conversion (Hutson 2000). Music as cure, solace, inspiration and element of wellbeing has gradually come full circle.

The more things change

Popular music has constantly evolved. Music is dynamic: fashions change, instruments evolve, new technologies provide new means of reception, performance, fusion and diffusion. New genres, such as dance music, emerge and appear to create different senses of space and place, while 'world music' seems to transcend global diversity. A mix of dynamism and continuity makes music exciting and liberating. Music still creates communities of interest, though such communities are more likely to be virtual, links people to places – even if those places have never been experienced – and threatens the status quo of established ways of life. In reception too there is continuity in change; when the discovery of the New World brought a new style of dance music (*ciaccona*) to Europe it inspired what now seems a familiar set of reactions: 'on the one hand it was celebrated as liberating bodies that had been stifled by the constraints of Western civilization; on the other it was condemned as obscene, as a threat to Christian mores' and in Spain in

1615 it was temporarily banned (McClary 1994: 36–7). It was perhaps the first moral panic associated with the introduction of new forms of popular music, and a clear indication of how music was able to both unite and divide communities.

Over time music has increasingly been associated with nostalgia and memory. In Vietnam, for the United States military stationed there, in the words of one:

> You'd hear a certain song and you could remember what kind of car you had, who you were dating....Those memories were important, because most of us were convinced we were going to die, and anything reminding us of normal life was great.
>
> (quoted in Klein 1991: 87)

Popular music betokened normal life. More generally music offered memories of country and countries, remembered places much distorted in the remembering; the folk music revival, country music and the music of vast numbers of migrants often represented 'a time when folk did not feel fragmented...when life was wholehearted...a past that was unified and comprehensible, unlike the incoherent divided present' (Lowenthal 1989: 29). Here too were virtual communities, the places of memory that evoked community, affection and identity, even where they represented 'fakelore' or an invented tradition. Popular music scenes, especially those surrounding youth cultures, invoked their authenticity from inner-city streets; hip hop and punk, reggae and rave all triumphed or challenged inner-city life. Here, as elsewhere, music was situated in place but any hint of success and thus diffusion shattered such simple relationships.

Music transformed some places; the 'Mersey sound' uplifted and romanticised Liverpool, and Liverpool (like many other places) promoted itself as a site of musical heritage and therefore tourism. That too is part of the new wave of nostalgia that has given history the ability to stimulate notions of identity and authenticity (and also commercial development), and mythologise local authentic spaces. In the face of the continued commercialism of popular music, and in a world of uncertainty, the quest for identity and meaning has been rekindled. Place became a critical musical resource, as a symbol in music (requiring strategic inauthenticity for adequate commodification), a means of national identity and the inspiration for ever-evolving local scenes. Invented traditions required an invented geography.

From fixity to fluidity and back again

The twentieth century brought unprecedented changes in technology, including that of transport and communications (and music itself), which seemed to mark both the fruition of the modernist project and the arrival of a global economic and social world. However, as cassette culture in Egypt and India (Box 3.3) demonstrated, new technology did not overwhelm local commercial production, but often created space for it. This has remained so, even at the end of a millennium when CDs had long replaced records and cassettes as the principal commercial form. Though the cost of CD production initially contributed to the centralisation of the music recording industry (particularly since, unlike cassettes, they could not easily be pirated), the declining cost of high-quality sound-engineering technology quickly enabled home recording by individuals or bands: 'direct control over creativity is shifting back into the hands of the musicians themselves, at least for those who can grapple with the technical challenges this presents' (O'Duibhir 1998: 10). The same was true of the Internet, which enabled consumers to have direct access to the music of small, obscure record companies that previously struggled for air time. In every kind of technological change, from paper sheet music to the World Wide Web, there were two phases: the first of mobility, and, second, a range of possible responses and reactions to these processes, from opposition and resistance to standardisation, against the background of frequent revivals of the local. In the case of the Internet, a more anarchic communications context became an intense site of struggle and eventually a cause of greater attempts to survey and control, as in the imposition of global intellectual property mechanisms. Music is a commodity and an agent of mobility and change, affecting the terms and conditions for globalisation.

Even at the Western core of contemporary popular music, in major world cities, spaces for the local remained. Bands and individuals sang about local issues and places (though the sounds might have been internationally recognisable), evoked a sense of place (such as 'the Seattle sound') and claimed that space had an impact on their music (Chapter 5). Certain musical genres – such as cajun or swamp rock – remained prominent in particular regions, a situation partly mythologised through music tourism, in terms of invocations to visit Nashville 'the home of country music' (Chapter 10) and distinct regional variations occurred in the popularity of most forms of music, within countries or within cities. Moreover local musical sub-cultures, such as 'northern soul' (Chapter 5), have continued to appear and be created in particular cultural, institutional and economic contexts. Independent record labels promote the music of tiny local bands seeking to emerge from pub and club circuits – or sometimes avowing their intent to remain there since success might compromise their style and authenticity (Goh 1996). Free street newspapers (particularly in Australia) promote 'alternative' local bands, venues and festivals: the spirit of performance. The notion of community has been invoked by numerous groups, mainly in the inner city, where its existence was threatened by gentrification, while particular sub-cultures – such as ravers – claimed alternative spaces. The resounding success of films such as *Brassed Off* emphasised the deep sense of nostalgia for a local world –

largely lost – of working-class community (and struggle) and the spirit of performance and participation. Whilst we may no longer inhabit a world where the 'local' stands for community, security and truth, but share instead experiences of rootlessness and constant movements of capital and labour (Chambers 1990), the memory and mythologising of that world remain vibrant. At a very different scale, part of the success of Bruce Springsteen was due to his apparent identification with local values and, despite his mass audience, being able 'to confront the inertia and dilution of spirit and energy that accompanies such notoriety' (Marsh and Swenson 1993: 483; Box 2.4). The more common invisibility of bands and performers that celebrate local and regional themes (invariably more evident before commercial success) should not preclude the manner in which a claim to the local is also a claim on authenticity.

Though local spaces survived – however the 'local' is defined – they were mediated by regional, national and international influences – technological, commercial and creative. Musical instruments diffused more rapidly in the twentieth century, tending to add to or replace 'local' instruments with those from the West. This transition was most evident in the growing dominance of the guitar, readily purchased in most countries and easy to repair and replace (compared with sometimes complicated techniques for the manufacture of local instruments), and later of keyboards, especially as these lent themselves more easily to electronic amplification. The absorption and incorporation of certain 'standard' musical instruments throughout the world has pointed to the increasingly syncretic and hybrid nature of much popular music – in the adoption of instruments, melodies and lyrics from a variety of sources, but primarily from the West. At certain times instruments from developing countries, notably the banjo and the sitar, were incorporated in Western musical forms. Such 'experiments' were often short lived, either because the sounds were unwelcome, the playing techniques too complex, the instruments too cumbersome for live performance or the costs too great. They were, however, visible and audible elements of the incorporation of non-Western music in the West and the rise of 'world music'. Musical diversity in most places now offers a range of options open to people through radio stations, clubs and stores, what Edward Said described more broadly as 'atonal ensembles' of cultural expressions (1990: 16). Walking through any Egyptian town for example 'in a single block one might hear Western rock music, young Egyptian and other Arab stars, and the recitation of the Quran' (Danielson 1997: 8), let alone classical music or jazz, and the hybrid versions of all of these. On another night each might emanate from a quite different window. It would be elegant if the 'local' could be equated with the 'authentic' and globalisation with the inauthentic, but it is abundantly clear that such simplistic distinctions have no bearing on the complexity of the migration, the fusion and hybridity of contemporary musical forms, their credibility or their enjoyability.

Journeys through music

Music is by nature geographical. Musical phrases have movement and direction, as though there are places in the music: quiet places and noisy places, places that offer familiarity, nostalgia or a sense of difference, while the dynamism of music reflects changing lives. Sound is a crucial element in the world we construct for ourselves, and the world that others construct and impose on us. Music has been used by governments to stir people into action; the music 'designed to fit certain formal and ceremonial spaces, at once defines and reinforces the disposition of power within those spaces' (Leyshon *et al*. 1997: 8). Alternatively live music creates local scenes and a sense of identity for those who are there: a social link between performer and audience, which reinforces the link between music and place.

Nothing is more redolent of the link between music and place than the music of Aborigines in central Australia, where 'songlines' are the means of navigating desert landscapes, creating identity and relating to other tribal groups, 'the song and the land are one' (Chatwin 1987: x). Songlines were created when the first ancestors emerged from the earth, crossing the land and singing it into being as they went, hence 'an unsung land is a dead land, since if the songs are forgotten the land itself will die' (Chatwin 1987: 58). Contemporary ambient and New Age music echoes such sentiments, while every society, less dramatically and directly, claims some link between people and place, and invariably some means of expressing this through music, while so many imaginative geographies may be remembered through music, whether as tourism or nostalgia:

> Noise is a vital element of travel, with some sounds instantly associated with certain cities or countries. In Shanghai it's bicycle bells; in Muslim countries the call of the muezzin. In Manila the incessant horn-blowing by jeepney drivers provides a mad mobile orchestra. In New Orleans it's the wail of a sly trombone from behind a shuttered window in the French Quarter. In many Asian countries it's the yowl of karaoke from basement dives. In Paris there's always a suggestion of Piaf singing in the shadows. In Vienna, it's Strauss – the whole city moves to the tempo of a waltz – while in Salzburg it's Mozart (unless you're on a *Sound of Music* bus tour, in which case it's Julie Andrews).
>
> (Kurosawa 2000: 47)

Directly, but more importantly metaphorically:

> the best of popular music acts as a tour guide, visiting not only significant places but pertinent times from the past....Rhythm is a glue that adheres the jagged spaces of memory...the familiar is validated and the nostalgic promoted....Between the fluidity of melody and a fixity of

> lyric, a place can be located and remembered.... The craving for rhythm feeds a desire for meaning and memory.
>
> (Brabazon 2000: 112)

Identities are multidimensional, constantly being renegotiated, but never divorced from place. This book has discussed some of the complicated ways in which music occupies sites and travels across space, but in the end we can only hint at the mixture of order and frenzy apparent in the dynamic nature of musical expressions and expansions.

BIBLIOGRAPHY

Abercrombie, P. (1933) *Town and Country Planning*, Thornton Butterworth, London.

Ackland-Snow, N., Brett, N. and Williams, S. (1996) *The Art of the Club Flyer*, Thames & Hudson, London.

Adams, V. (1996) Karaoke as Modern Lhasa, Tibet: Western encounters with cultural politics, *Cultural Anthropology*, 11, 510–46.

Adorno, T. (1988) *Introduction to the Sociology of Music*, Continuum, New York.

Adorno, T. and Horkheimer, M. (1977) The culture industry: enlightenment as mass deception, in J. Curran, M. Gurevitch and J. Woollacott, eds, *Mass Communication and Society*, Edward Arnold and Open University Press, London, 349–83.

Alan, C. (1992) *U2 Wide Awake in America*, Boxtree, London.

Alderman, D.H. (2002) Writing on the Graceland wall: on the importance of authorship in pilgrimage landscapes, *Tourism Recreation Research*, 27, 3 [in press].

Aldskogius, H. (1993) Festivals and meets: the place of music in 'summer Sweden', *Geografiska Annaler*, 75B, 55–72.

Alleyne, M. (1994) Positive vibration? Capitalist textual hegemony and Bob Marley, *Bulletin of Eastern Caribbean Affairs*, 19(3), 76–84.

—— (2000) White reggae: cultural dilution in the record industry, *Popular Music and Society*, 24, 15–30.

Allinson, E. (1992) Music and the politics of race, *Cultural Studies*, 8, 438–56.

Alloca, D. (2001) Is everything alright with Courtney Love?, *Q*, 176, 22.

Alvey, M. (1997) Wanderlust and wire wheels: the existential search of Route 66, in S. Cohan and I.R. Hark, eds, *The Road Movie Book*, Routledge, London and New York, 143–65.

Anderson, B (1983) *Imagined Communities: Reflections on the Origin and Spread of Nationalism*, Verso, London.

Anon, a. (1959) Liner notes, Mantovani, *Continental Encores*.

Anon, b. (1958) Liner notes, Stanley Black and his Orchestra, Place Pigalle.

Anon (1994) Liner notes, *Electric and Acoustic Mali*, EMI, London.

Aparicio, F. (1998) *Listening to Salsa. Gender, Popular Music and Puerto Rican Cultures*, Wesleyan University Press, Hanover.

Appadurai, A. (1990) Disjuncture and difference in the global cultural economy, *Public Culture* 2(2), 1–24.

—— (1991) Afterword, in A. Appadurai, F. Korom and M. Mills, eds, *Gender, Genre and Power*, Philadelphia, University of Pennsylvania Press, 467–76.

—— (1996) *Modernity at Large. Cultural Dimensions of Globalization*, University of Minnesota Press, Minneapolis.

Ashworth, G., White, P. and Winchester, H. (1988) The red light district in the west European city: a neglected aspect of urban landscape, *Geoforum*, 19, 201–12.

Aston, M. (2001) Album review: the Avalanches *Since I Left You*, *Play*, 14 April, 20.

Atkinson, C.Z. (1997) 'Whose New Orleans? Music's place in the packaging of New Orleans for tourism', in S. Abram, J.D. Waldren and D.V.L. MacLeod, eds, *Tourists and Tourism: Identifying People with Place*, Berg, Oxford, 91–106.

Attali, J. (1985) *Noise: The Political Economy of Music*, Manchester University Press, Manchester.

Averill, G. (1994) 'Mezanmi, kouman nou ye? My friends, how are you?': musical constructions of the Haitian transnation, *Diaspora*, 3, 253–72.

—— (1997a) *A Day for the Hunter, A Day for the Prey. Music and Power in Haiti*, Chicago University Press, Chicago.

—— (1997b) 'Pan is we ting': West Indian steelbands in Brooklyn, in K. Lornell and A. Rasmussen, eds, *Musics of Multicultural America*, Schirmer, New York, 101–29.

Baily, J (1994) The role of music in the creation of an Afghan national identity, 1923–73, in M. Stokes, ed., *Ethnicity, Identity and Music: The Musical Construction of Place*, Berg, Oxford, 45–60.

Baker, G.A. (1995) *Faces, Places and Barely Human Races*, Random House, Sydney.

Ballantine, C. (1999) Looking to the USA: the politics of male close-harmony song style in South Africa during the 1940s and 1950s, *Popular Music*, 18, 1–17.

—— (2000) Gender, migrancy and South African popular music in the late 1940s and the 1950s, *Ethnomusicology*, 44, 376–407.

Banerjea, K (2000) Sounds of whose underground? The fine tuning of diaspora in an age of mechanical reproduction, *Theory, Culture and Society*, 17, 64–79.

Banerji, S. and Baumann, G. (1990) Bhangra 1984–88: fusion and professionalisation in a genre of South Asian dance music, in P. Oliver, ed., *Black Music in Britain*, Open University Press, Milton Keynes, 137–52.

Bannister, M. (1999) *Positively George Street: A Personal History of Sneaky Feelings and the Dunedin Sound*, Reed, Auckland.

Barber, K. and Waterman, C. (1995) Traversing the global and the local: Fuji music and praise poetry in the production of contemporary Yoruba popular culture, in D. Miller, ed., *Worlds Apart. Modernity through the Prism of the Local*, Routledge, London, 240–62.

Barr, T. (2000) *Techno: The Rough Guide*, Penguin, Ringwood.

Barrett, J. (1996) World music, nation and postcolonialism, *Cultural Studies*, 10, 237–47.

Barthes, R. (1972) *Mythologies*, Paladin, St Albans.

Baudrillard, J. (1981) *For a Critique of the Political Economy of the Sign*, Telos Press, St Louis.

Baulch, E. (1996) Punks, rastas and headbangers: Bali's Generation X, *Inside Indonesia*, October–December, 23–5.

Baumann, G. (1990) The re-invention of bhangra, *World of Music*, 32(2), 81–95.

Bayton, M. (1993) Feminist-musical practice: problems and contradictions, in T. Bennett, S. Frith, L. Grossberg, T. Shepherd and G. Turner, eds, *Rock and Popular Music*, Routledge, London, 177–92.

—— (1997) *Frock Rock: Women Performing Popular Music*, Oxford University Press, Oxford.

BBC (1996) *Dancing in the Street*, BBC, London.

Bell, D (1991) Insignificant others: lesbian and gay geographies, *Area*, 23, 323–9.

Bell, D. and Valentine, G., eds, (1995) *Mapping Desire. Geographies of Sexuality*, Routledge, London.

Bell, T.L. (1998) Why Seattle? An examination of an alternative rock culture hearth, *Journal of Cultural Geography*, 18, 35–48.

Bennett, A. (1997) Bhangra in Newcastle: music, ethnic identity and the role of local knowledge, *Innovation*, 10, 107–16.

—— (1999a) Rappin' on the Tyne: white hip hop culture in northeast England – an ethnographic study, *Sociological Review*, 47, 1–24.

—— (1999b) Hip hop am Main: the localization of rap music and hip hop culture, *Media, Culture and Society*, 21, 77–91.

—— (2000) *Popular Music and Youth Culture Music, Identity and Place*, Macmillan, Basingstoke.

Berger, H.M. (1999) *Metal, Rock and Jazz: Perception and the Phenomenology of Musical Experience*, Wesleyan University Press, Hanover.

Berland, J. and Straw, W. (1991) Getting down to business: cultural politics and policies in Canada, in B.D. Singer, ed., *Communications in Canadian Society*, Nelson, Vancouver, 276–94.

Berrian, B.F. (2000) *Awakening Spaces. French Caribbean Popular Songs, Music and Culture*, Chicago University Press, Chicago.

Bey, H. (1991) *T.A.Z. The Temporary Autonomous Zone, Ontological Anarchy, Poetic Terrorism*, Autonomedia, New York.

Bhabha, H. (1994) *The Location of Culture*, Routledge, London.

Bigot, Y. (1994) Liner notes, Manu Dibango, *Wakafrika*, TimeWarner.

Bilby, K. (1999) 'Roots explosion': indigenization and cosmopolitanism in contemporary Surinamese popular music, *Ethnomusicology*, 43, 256–96.

Binas, S. (1999) Sampling the world: some observations and questions, unpublished paper to International Association for the Study of Popular Music conference, *Changing Sounds*, University of Technology, Sydney.

Blair, M.E. and Hyatt, E.M. (1992) Home is where the heart is: an analysis of meanings of the home in country music, *Popular Music and Society*, 16, 69–82.

Blake, A. (1997) *The Land without Music*, Manchester University Press, Manchester.

Bloomfield, T. (1993) Resisting songs: negative dialectics in pop, *Popular Music*, 12, 13–31.

Blum, S. and Hassanpour, A. (1996) 'The morning of freedom rose up': Kurdish popular song and the exigencies of cultural survival, *Popular Music*, 15, 325–43.

Borja, J., Castells, M., Belil, M. and Brenner, C. (1997) *Local and Global: Management of Cities in the Information Age*, Earthscan, London.

Borneman, J. and Senders, S. (2000) Politics without a head: Is the 'Love Parade' a new form of political identification?, *Cultural Anthropology*, 15, 294–317.

Bourdieu, P. (1984) *Distinction: A Social Critique of the Judgement of Taste*, Routledge & Kegan Paul, London.

Bowen, D. (1997) Lookin' for Margaritaville: Place and Image in Jimmy Buffett's Songs, *Journal of Cultural Geography*, 16, 99–108.

Boyd, T. (1994) Check yo self before you wreck yo self: variations on a political theme in rap music and popular culture, *Public Culture*, 7, 289–312.

Boyes, G. (1993) *The Imagined Village. Culture, Ideology and the English Folk Revival*, Manchester University Press, Manchester.

Boyle, P., Halfacree, K. and Robinson, V. (1998) *Exploring Contemporary Migration*, Longman, London.

Brabazon, T. (1993) From Penny Lane to Dollar Drive: Liverpool tourism and Beatle-led recovery, *Public History Review*, 2, 108–24.

—— (2000) *Tracking the Jack. A Retracing of the Antipodes*, University of NSW Press, Sydney.

Brackett, D. (1995) *Interpreting Popular Music*, Cambridge University Press, Cambridge.

Bradby, B. (1993) Lesbians and popular music. Does it matter who is singing?, in G. Griffin, ed., *Outwrite Lesbians and Popular Culture*, Pluto, London, 148–71.

Bradby, B. and Torode, B. (1984) Pity Peggy Sue, *Popular Music*, 4, 183–206.

Bradley, L. (2000) *Bass Culture: When Reggae Was King*, Viking, London.

Breen, M. (1995) The end of the world as we know it: popular music's cultural mobility, *Cultural Studies*, 9, 486–504.

Brennan, T. (1997) *At Home in the World: Cosmopolitanism Now*, Harvard University Press, Cambridge.

Brenner, N. (1998) Between fixity and motion: accumulation, territorial organization and the historical geography of spatial scales, *Environment and Planning D*, 16, 459–81.

—— (1999) Globalisation as reterritorialisation: the re-scaling of urban governance in the European Union, *Urban Studies*, 3, 431–51.

British Tourism Authority (1998) *Rock and Pop Map: One Nation under a Groove*, London.

Brown, A., O'Connor, J. and Cohen, S. (2000) Local music policies within a global music industry: cultural quarters in Manchester and Sheffield, *Geoforum*, 31, 437–51.

Brown, M. (1993) *American Heartbeat: Travels from Woodstock to San Jose by Song Title*, Penguin, London.

Buchanan, M. (1998) Melancholy? Moi?, *Sydney Morning Herald Metro*, 17 April, 4–5.

Bull, A. (1993) *Coast to Coast: A Rock Fan's U.S. Tour*, Black Swan, London

Burnett, R. (1992) Dressed for success: Sweden from Abba to Roxette, *Popular Music*, 11, 141–50.

Caglar, A.S. (1998) Popular culture, marginality and institutional incorporation: German-Turkish rap and Turkish pop in Berlin, *Cultural Dynamics*, 10, 243–61.

Calmont, T. (2001) Paul Oscar: interview, *Tribal Move*, 27 (January), 42–3.

Cannon, S. (1997) Paname city rapping. B-boys in the banlieues and beyond, in A. Hargreaves and M. McKinney, eds, *Post-Colonial Cultures in France*, Routledge, London, 150–66.

Cantwell, R. (1984) *Bluegrass Breakdown. The Making of the Old Southern Sound*, University of Illinois Press, Urbana.

Carney, G.O. (1974) Bluegrass grows all around: the spatial dimensions of a country music style, *Journal of Geography*, 73(4), 34–55.

—— (1978) The Roots of American Music, in G. Carney, ed., *The Sounds of People and Places*, University Press of America, Washington, 286–321.

——, ed. (1978) *The Sounds of People and Places: Readings in the Geography of American Folk and Popular Music*, University Press of America, Washington.

—— (1992) Branson: the new mecca of country music, *Journal of Cultural Geography*, 14(2), 17–32.

—— (1996) Western North Carolina: culture hearth of bluegrass music, *Journal of Cultural Geography*, 16, 65–87.

—— (1998) Music geography, *Journal of Cultural Geography*, 18, 1–10.

Carroll, J. and Connell, J. (2000) "You gotta love this city". The Whitlams and inner Sydney, *Australian Geographer*, 31, 141–54.

Castells, M (1989) *The Informational City: Information Technology, Economic Restructuring and the Urban-regional Process*, Blackwell, Oxford.

—— (1994) European cities, the informational society and the global economy, *New Left Review*, 204, 18–32.

Catt, J. (1998) Indigo chemistry, *Sydney Star Observer*, 26 February, 31.

Caves, R.E. (2001) *Creative Industries: Contracts Between Art and Commerce*, Harvard University Press, Cambridge, MA.

Chamberlain, M. (1995) Family narratives and migration dynamics: Barbadians to Britain, *New West Indian Guide*, 69, 251–75.

Chambers, I. (1993) Cities without maps, in J. Bird, B. Curtis, T. Putman, G. Robertson, L. Tickner, eds, *Mapping the Futures: Local Cultures, Global Change*, Routledge, London, 188–98.

Chan, S. (1998) Music(ology) needs a context, *Perfect Beat*, 3(4), 93–7.

Charlesworth, B. (1994) George Michael: professional slavery or occupational hazard?, *Law Institute Journal*, 68(12), 1172–5.

Charlton, R. (1971) Liner notes, Billy Pigg, *The Border Minstrel*, Leader, London.

Chatwin, B. (1987) *Songlines*, Picador, London.

Chude-Sokei, L. (1997) Postnationalist geographies: rasta, ragga and reinventing America, in C. Potash, ed., *Reggae, Rasta, Revolution. Jamaican Music from Ska to Dub*, Schirmer, New York, 215–27.

Clarke, D. (1995) *The Rise and Fall of Popular Music*, Penguin, Harmondsworth.

Clarke, S. (1980) *Jah Music. The Evolution of the Popular Jamaican Song*, Heinemann, London.

Clawson, M. (1999) Where women play bass, *Gender and Society*, 13, 193–210.

Clay, B.J. (1986) *Mandak Realities. Person and Power in Central New Ireland*, Rutgers University Press, New Brunswick.

Clough, B. (1997) Version galore! Reggae translations in Sydney, *Sounds Australian*, 15(50), 12–13.

Cohen, S. (1991a) Popular music and urban regeneration: the music industries of Merseyside, *Cultural Studies*, 5, 332–46.

—— (1991b) *Rock Culture in Liverpool: Popular Music in the Making*, Clarendon, Oxford.

—— (1993) Ethnography and popular music studies, *Popular Music*, 12, 123–38.

—— (1994) Identity, place and the 'Liverpool sound', in M. Stokes, ed., *Ethnicity, Identity and Music: the Musical Construction of Place*, Berg, Oxford, 117–34.

—— (1995) Sounding out the city: music and the sensuous production of place, *Transactions of the Institute of British Geographers*, 20, 434–46.

—— (1997a) 'More than the Beatles: popular music, tourism and urban regeneration', in S. Abram, J.D. Waldren and D.V.L. MacLeod, eds, *Tourists and Tourism: Identifying People with Place*, Berg, Oxford, 71–90.

—— (1997b) Men making a scene. Rock music and the production of gender, in S. Whiteley, ed., *Sexing the Groove*, Routledge, London, 17–36.

—— (1999) Scenes, in B. Horner and T. Swiss, eds, *Key Terms in Popular Music and Culture*, Blackwell, Oxford, 239–50.

Cole, F. and Hannan, M. (1997) Goa trance, *Perfect Beat*, 3(3), 1–14.

Connell, J. (1999) 'My island home'. The politics and poetics of the Torres Strait, in R. King and J. Connell, eds, *Small Worlds, Global Lives. Islands and Migration*, Pinter, London, 195–212.

Cooper, C. (1998) 'Ragamuffin sounds': crossing over from reggae to rap and back, *Caribbean Quarterly*, 44, 153–68.

Coplan, D. (1995) *In the Time of Cannibals: The World Music of South African Basotho Migrants*, University of Chicago Press, Chicago.

Copper, B. (1971) *A Song for Every Season*, Heinemann, London.

Copping, A. (1997) Liner notes, *Siva Pacifica*, Aurora Music, Sydney.

Corbert, J. (1994) *Extended Play: Sounding off from John Cage to Dr. Funkenstein*, Duke University Press, Durham and London.

Cornish, V. (1928) Harmonies of scenery: an outline of aesthetic geography, *Geography*, 14, 275–82 and 383–94

—— (1934) The scenic amenity of Great Britain, *Geography*, 19, 195–202.

Cosgrove, S. (1988) Global style, *New Statesman and Society*, 1(14), 9 September, 50.

Coulthard, A. (1995) 'George Michael v Sony Music – a challenge to artistic freedom?', *Modern Law Review*, 58(5), 7731–44.

Coyle, M. and Dolan, J. (1999) Modelling authenticity, authenticating commercial models, in K. Dettmar and W. Richey, eds, *Reading Rock and Roll*, Columbia University Press, New York, 17–36.

Cox, C. (1996) Liner notes, *Two Paintings and a Drum,*, WWU, London.

Crang, M. (1998) *Cultural Geography*, Routledge, London.

Crang, P. (1996) Displacement, consumption, and identity, *Environment and Planning A*, 28, 47–67.

Creswell, T. (1991) The eighties, *Rolling Stone*, 459, July, 43–5.

Cross, B. (1993) *It's not about a salary....rap, race + resistance in Los Angeles*, Verso, New York.

Crush, J. (1995) Vulcan's brood. Spatial narratives of migration in Southern Africa, in R. King, J. Connell and P. White, eds, *Writing across Worlds, Literature and Migration*, Routledge, London, 229–47.

Curtis, J.R. and Rose, R.F. (1987) 'The Miami sound': A contemporary Latin form of place-specific music, in G.O. Carney, ed., *The Sounds of People and Places*, University Press of America, Washington, 285–99.

Cushman, T. (1991) Rich rastas and communist rockers: a comparative study of the origin, diffusion and defusion of revolutionary musical codes, *Journal of Popular Culture*, 25(3), 16–61.

Cusic, D. (1994) QWERTY, Nashville, and country music, *Popular Music and Society*, 18(4), 41–55.

Cvetkovich. R. and Kellner, D. (1997) Introduction: thinking global and local, in A. Cvetkovich and D. Kellner, eds, *Articulating the Global and the Local*, Westview, Boulder, 1–30.

Daniel, Y.P. (1996) Tourism dance performances: authenticity and creativity, *Annals of Tourism Research*, 23, 780–97.

Danielson, V. (1997) *The Voice of Egypt. Umm Kulthúm, Arabic Song and Egyptian Society in the Twentieth Century*, University of Chicago Press, Chicago.

Davis, S. (1983) *Bob Marley. Conquering Lion of Reggae*, Plexus, London.

de Certeau, M. (1984) *The Practice of Everyday Life*, University of California Press, Berkeley.

de Genova, N. (1995) Gangsta rap and nihilism in black America, *Social Text*, 43, 89–132.

DeNora, T. (2000) *Music in Everyday Life*, Cambridge University Press, Cambridge.

DeNora, T. and Belcher, S. (2000) When you're trying something on you picture yourself in a place where they are playing this kind of music – musically sponsored agency in the British clothing retail sector, *Sociological Review*, 48, 80–101.

Denselow, R. (1989) *When the Music's Over. The Story of Political Pop*, Faber, London.

Diallo, Y. and Hall, M. (1989) *The Healing Drum: African Wisdom Teachings*, Destiny Books, Rochester.

Diethrich, G. (1999) Desi music vibes: the performance of Indian youth culture in Chicago, *Asian Music*, 31, 35–61.

Dixon, R.M.W. and Koch, E. (1996) *Dyirbal Song Poetry. The Oral Literature of Australian Rainforest People*, University of Queensland Press, Brisbane.

DjeDje, J. and Meadows, E., eds (1998) *California Soul: Music of African Americans in the West*, University of California Press, Berkeley.

Dublin Tourism (1998) *Rock n Stroll: Pubs, Restaurants, Night Life and Music*, Dublin Tourism Centre, Dublin.

Duffy, M. (2000) Lines of drift: festival participation and performing a sense of place, *Popular Music*, 19, 51–64.

Dunbar-Hall, P. (1997) Site as song – song as site: constructions of meaning in an Aboriginal rock song, *Perfect Beat*, 3(3), 55–74.

Dunbar-Hall, P. and Gibson, C. (2000) Singing about nations within nations: geopolitics and identity in Australian indigenous rock music, *Popular Music and Society* 24(2), 45–74.

Duran, I. (1998) Liner notes, *Paranda*, Stonetree, Belize.

Durant, A. (1990) 'A new day for music? Digital technologies in contemporary music-making', in P. Haywood, ed., *Culture, Technology and Creativity in the Late 20th Century*, John Libbey, London, 175–96.

Dyer, R. (1990) In defence of disco, in S. Frith and A. Goodwin, eds, *On Record*, Routledge, London, 410–18.

—— (1997) *White: Essays on Race and Culture*, Routledge, London.

Dyson, M. (1996) *Between God and Gangsta*, Oxford University Press, Oxford.

Elflein, D. (1998) From Krauts with attitudes to Turks with attitudes: some aspects of hip-hop history in Germany, *Popular Music*, 17, 255–77.

El-Shawan Castelo-Branco, S. (1987) Some aspects of the cassette industry in Egypt, *The World of Music*, 29(2), 32–48.

Ellison, M. (1989) *Lyrical Protest: Black Music's Struggle against Discrimination*, Praeger, New York.

—— (1998) Thomas Mapfumo. The lion king, *The Guardian*, 26 June, 14–15.

Ennis, P. (1992) *The Seventh Stream. The Emergence of Rock 'n' roll in American Popular Music*, Wesleyan University Press, Hanover.

Erlmann, V. (1996) The aesthetics of the global imagination: reflections on world music in the 1990s, *Public Culture*, 8, 467–87.

—— (1998) How beautiful is small? Music, globalisation and the aesthetic of the local, *Yearbook for Traditional Music*, 30, 12–21.

Euba, A. (1970) Traditional elements as the basis of new African art music, *African Urban Notes*, 5(4), 52–6.

Evans, M. (1997) Princess Tabu. A Melbourne-Tongan synthesis, *Sounds Australian*, 15(50), 38–9.

Eyck, F.G. (1995) *The Voice of Nations: European National Anthems and Their Authors*, Greenwood Press, London.

Eyles, J. (1989) The Geography of Everyday Life, in D. Gregory and R. Walford, eds, *Horizons in Human Geography*, Macmillan, Basingstoke, 102–17.

Farrell, G. (1998) The early days of the gramophone record industry in India: historical, social and musical perspectives, in A. Leyshon, D. Matless and G. Revill, eds, *The Place of Music*, Guilford, New York, 57–82.

Feintuch, B. (1993) Musical revival as musical transformation, in N.V. Rosenberg, ed., *Transforming Traditions*, University of Illinois Press, Urbana, 183–93.

Feld, S. (1982) *Sound and Sentiment: Birds, Weeping, Poetics and Song in Kaluli Expression*, University of Philadelphia Press, Philadelphia.

—— (1984) Sound structures as social structure, *Ethnomusicology*, 28, 383–409.

—— (1991) Voices of the rainforest, *Public Culture*, 4, 131–40.

—— (1994) Notes on 'world beat', in C. Keil and S. Feld, *Music Grooves*, University of Chicago Press, Chicago, 238–46.

—— (1996a) Pygmy POP. A genealogy of schizophonic mimesis, *Yearbook for Traditional Music*, 28, 1–35.

—— (1996b) Waterfalls of song: an acoustemology of place resounding in Bosavi, Papua New Guinea, in S. Feld and K. Basso, eds, *Senses of Place*, School of American Research, Santa Fe, 91–135.

—— (2000a) A sweet lullaby for world music, *Public Culture*, 12, 145–71.

—— (2000b) Sound worlds, in P. Kruth and H. Stobart, eds, *Sound*, Cambridge University Press, Cambridge, 173–200.

Feld, S. and Fox, A. (1994) Music and language, *Annual Review of Anthropology*, 23, 25–53.

Fellman, J., Getis, A. and Getis, G. (1996) *Human Geography*, McGraw Hill, Boston.

Ferraro, B. (1997) Size ain't everything, *3D World*, 368, 29 September, 58.

Fine, B. and Leopold, E. (1993) *The World of Consumption*, Routledge, London.

Finnegan, R. (1989) *The Hidden Musicians: Music-Making in an English Town*, Cambridge University Press, Cambridge.

Flores, J. (1994) Puerto Rican and proud, boyee! Rap, roots and amnesia, in A. Ross, and T. Rose, eds, *Microphone Fiends: Youth Music and Youth Culture*, Routledge, London and New York, 89–98.

Ford, L. (1978) Geographic factors in the origin, evolution and diffusion of rock and roll music, in G. Carney, ed., *The Sounds of People and Places: Readings in the Geography of Music*, University Press of America, Washington, 212–32.

Ford, L. and Henderson, F. (1974) The image of place in American popular music: 1890–1970, *Places*, 1, 31–7.

Ford, L.R. (1971) Geographic factors in the origin, evolution and diffusion of rock and roll music, *Journal of Geography*, 70, 455–64.

Forman, M. (2000) 'Represent': race, space and place in rap music, *Popular Music*, 19, 65–90.

Fowler, P. (1972) Skins rule, in C. Gillett, ed., *Rock File*, New English Library, London. 10–24.

Friedson, S. (1996) *Dancing Prophets: Musical Experience in Tumbuka Healing*, University of Chicago Press, Chicago.

Frith, S. (1987) Towards an aesthetic of popular music, in R. Leppert and S. McClary, eds, *Music and Society*, Cambridge, Cambridge University Press, 133–49.

—— (1988) Why do songs have words?, in S. Frith, ed., *Music for Pleasure*, Polity, Cambridge, 105–28.

—— (1989) Introduction, in S. Frith, ed., *World Music, Politics and Social Change*, Manchester, Manchester University Press.

—— (1993) Popular music and the local state, in T. Bennett, S. Frith, L. Grossberg, T. Shepherd and G. Turner, eds, *Rock and Popular Music*, Routledge, London, 14–24.

—— (1997) The suburban sensibility in British pop and rock, in R. Silverstone, ed., *Visions of Suburbia*, Routledge, London, 269–79.

Frith, S. and Horne, H. (1987) *Art into Pop*, Routledge, London.

Frith, S. and McRobbie, A. (1990) Rock and Sexuality, in S. Frith and A. Goodwin, eds, *On Record*, Routledge, London, 371–89.

Fromartz, S. (1998) Anything but Quiet, *Natural History*, 107(3), 44–8.

Fryer, D.W. (1974) A geographer's inhumanity to man, *Annals of the Association of American Geographers*, 64, 479–82.

Gaar, G. (1992) *She's a Rebel*, Seal Press, Seattle.

Garofalo, R. (1992) *Rockin' the Boat: Mass Music for Mass Movements*, South End Press, Boston.

—— (1993a) Black popular music: crossing over or going under?, in T. Bennett, S. Frith, L. Grossberg, T. Shepherd and G. Turner, eds, *Rock and Popular Music*, Routledge, London, 231–48.

—— (1993b) Whose world, what beat: the transnational music industry, identity and cultural imperialism, *The World of Music*, 35, 16–32.

Gay, H.J. (1998) With all the charms of a woman, *New Art Examiner*, 25(9), June, 20–5.

Gee, M. (1999) Insider entertainment industry news, *Revolver*, 1 November, 60.

Geyrhalter, T. (1996) Effeminacy, camp and sexual subversion in rock: the Cure and Suede, *Popular Music*, 15, 217–24.

Gibson, C. (1998) 'We sing our home, we dance our land': indigenous self-determination and contemporary geopolitics in Australian popular music, *Environment and Planning D*, 16, 163–84.

—— (1999) Subversive sites: rave culture, spatial politics and the internet in Sydney, Australia, *Area*, 3, 19–33.

—— (2000) Systems of provision for popular music and a regional music industry, unpublished PhD thesis, University of Sydney.

Gibson, C. and Connell, J. (2000) 'Artistic dreamings: tinseltown, sin city and suburban wasteland', in J. Connell, ed., *Sydney: The Evolution of a World City*, Oxford University Press, Melbourne, 292–318.

—— (2003) Bongo fury: tourism, music and cultural economy at Byron Bay Australia, *Tijdschrift voor Economische en Sociale Geografie* [in press].

Gibson, C. and Dunbar-Hall, P. (2000) 'Nitmiluk': place and empowerment in Aboriginal popular music, *Ethnomusicology*, 44, 39–64.

Gibson, C. and Pagan, R. (2002) Mapping youth spaces in media discourse: rave cultures in Sydney, Australia, in M.A. Wright, ed., *Dance Culture: Party Politics and Beyond*, Verso, London [forthcoming].

Gilbert, J.(1997) Soundtrack to an uncivil society: rave culture, the Criminal Justice Act and the politics of modernity, *New Formations*, 31, 5–22.

Gilbert, J. and Pearson, E. (1999) *Discographies*, Routledge, London.

Giles, J. (1990) Pop's hair apparent, *Rolling Stone*, 442, March, 21.

Gill, A. (1999) Revolting middle age, *Independent*, 21 May, 15.

Gill, W. (1993) Region, agency, and popular music: the Northwest sound 1958–1966, *The Canadian Geographer*, 37, 120–31.

Gillett, C. (1970) *The Sound of the City*, London, Sphere.

Gilroy, P. (1987) *There Ain't No Black in the Union Jack*, Hutchinson, London.

—— (1991) Sounds authentic: black music, ethnicity and the challenge of a changing name, *Black Music Research Journal*, 11, 111–36.

—— (1993a) *Small Acts*, Serpent's Tail, London

—— (1993b) *The Black Atlantic. Modernity and Double-Consciousness*, Verso, London.

Goertzen, C. and Azzi, M.S. (1999) Globalization and the Tango, *Yearbook for Traditional Music*, 66, 67–76.

Goh, L. (1996) 'The hidden soul of harmony': gender and identity in the Sydney popular music scene, unpublished BA Honours thesis, Department of Geography, University of Sydney.

Goldberg, D.T. (1993) *Racist Culture: Philosophy and the Politics of Meaning*, Blackwell, Oxford.

Gondola, C.D. (1997) Popular music, urban society and changing gender relations in Kinshasa, Zaire (1950–1990), in M. Grosz-Nagaté and O.H. Kokole, eds, *Gendered Encounters*, Routledge, London, 65–84.

Goodwin, A. (1992) *Dancing in the Distraction Factory: Music, Television and Popular Music*, University of Minnesota Press, Minneapolis.

Goodwin, A. and Gore, J. (1990) World beat and the cultural imperialism debate, *Socialist Review*, 3, 63–80.

Gopinath, G. (1995) 'Bombay, UK, Yuba City'. Bhangra music and the engendering of diaspora, *Diaspora*, 4, 303–21.

Gordon, R. (1995) *It Came from Memphis*, Secker & Warburg, London

Gottlieb, J. and Wald, G. (1994) Smells like teen spirit: riot grrrls, revolution and women in independent rock, in A. Ross and T. Rose, eds, *Microphone Fiends*, Routledge, London, 250–74.

Gray, M. and Osborne, R. (1996) *The Elvis Atlas: A Journey through Elvis Presley's America*, Henry Holt, New York.

Green, T. (1987) James Lyons: singer and story-teller: his repertory and aesthetic, in M. Pickering and T. Green, eds, *Everyday Culture: Popular Song and the Vernacular Milieu*, Milton Keynes, Open University Press, 105–24.

Greene, P. (2001) Mixed message: unsettled cosmopolitanisms in Nepali pop, *Popular Music*, 20, 169–87.

Grenier, L. (1993) The aftermath of a crisis: Quebec music industries in the 1980s, *Popular Music*, 12, 209–27.

Grenier, L. and Guilbault, J. (1997) Créolité and Francophonie in music: socio-political positioning where it matters, *Cultural Studies*, 11, 207–34.

Gross, J., McMurray, D. and Swedenburg, T. (1994) Arab noise and Ramadan nights: rai, rap and Franco-Maghrebi identity, *Diaspora*, 3, 3–39.

Gross, J. and Mark, V. (2001) Regionalist accents of global music: the Occitan rap of Les Fabulous Trobadors, *French Cultural Studies*, 12, 77–94.

Grossberg, L. (1984) Another boring day in paradise: rock and roll and the empowerment of everyday life, *Popular Music*, 4, 225–58.

—— (1992) *We Gotta Get out of This Place: Popular Conservatism and Postmodern Culture*, Routledge, New York.

—— (1993) The Media Economy of Rock Culture: Cinema, Post-Modernity and Authenticity, in S. Frith, A. Goodwin and L. Grossberg, eds, *Sound and Vision*, London, Routledge, 185–209.

—— (1997) *Dancing in Spite of Myself: Essays on Popular Culture*, Duke University Press, Durham and London.

Guattari, F. (1984) *Molecular Revolution, Psychiatry and Politics*, Penguin, Harmondsworth.

Guilbault, J (1993a) *Zouk: World Music in the West Indies*, University of Chicago Press, Chicago and London.

—— (1993b) On redefining the 'local' through world music, *The World of Music*, 35, 33–47.

—— (1994) Créolité and the new cultural politics of difference in popular music in the French West Indies, *Black Music Research Journal*, 14, 161–78.

—— (1997) Interpreting world music: a challenge in theory and practice, *Popular Music*, 16, 31–44.

Gumprecht, B. (1998) Lubbock on everything: the evolution of place in popular music, a west Texas example, *Journal of Cultural Geography*, 18, 61–81.

Habell-Pallan, M. (1999) El Vez is 'taking care of business': the inter/national appeal of Chicago popular music, *Cultural Studies*, 13, 195–210.

Halfacree, K.H. and Kitchin, R.M. (1996) 'Madchester rave on': placing the fragments of popular music, *Area*, 28, 47–55.

Hall, P. (1998) *Cities and Civilization*, Weidenfeld & Nicholson, London.

Hall, S. (1995) New cultures for old, in D. Massey and P. Jess, eds, *A Place in the World*, Open University Press, Milton Keynes, 175–214.

Hall, S. and Jefferson, T. (1976) *Resistance through Rituals: Youth Subcultures in Post-War Britain*, Hutchinson, London.

Hampton, M. (1998) Backpacker tourism and economic development, *Annals of Tourism Research*, 23, 639–60.

Hannerz, U. (1987) The world in creolisation, *Africa*, 57, 546–59.

Hannigan, J. (1998) *Fantasy City*, Routledge, London and New York

Hansing, K. (2001) Rasta, race and revolution; transnational connections in socialist Cuba, *Journal of Ethnic and Migration Studies*, 27, 733–47.

Harker, D. (1980) *One for the Money. Politics and Popular Song*, Hutchinson, London.

—— (1985) *Fake Song*, Open University Press, Milton Keynes.

Harrell, J. (1994) The poetics of destruction: death metal rock, *Popular Music and Society*, 18, 91–104.

Harris, K. (2000) 'Roots'?: the relationship between the global and the local within the extreme metal scene, *Popular Music*, 19, 13–30.

Haro, C.M. and Loza, S. (1994) The evolution of banda music and the current banda movement in Los Angeles, *Selected Reports in Ethnomusicology*, 10, 59–71.

Harvey, D. (1985) The geopolitics of capitalism, in D. Gregory and J. Urry, eds, *Social Relations and Spatial Structures*, Macmillan, London, 128–63.

—— (1989) *The Condition of Postmodernity* Blackwell, Oxford.

Haslam, D. (2000) *Manchester, England: The Story of the Pop Cult City*, Fourth Estate, London.

Havens, D.F. (1987) Up the river from New Orleans: the jazz odyssey – myth or truth?, *Popular Music and Society*, 11(4), 61–74.

Hayward, P. (1998) *Music at the Borders. Not Drowning, Waving*, John Libbey, Sydney.

Heath, C. (1990) *Pet Shop Boys, Literally*, Viking, London.

Hebdige, D. (1979) *Subculture: The Meaning of Style*, Methuen, London and New York.

—— (1987) *Cut 'n' Mix: Culture, Identity and Caribbean Music*, Routledge/Comedia, London.

Heining, D. (1998) Cars and girls – the car, masculinity and pop music, in D. Thorns, L. Holden and T. Claydon, eds, *The Motor Car and Popular Culture in the 20th Century*, Ashgate, Aldershot, 96–119.

Henderson, F.M. (1974) The image of New York City in American popular music: 1890–1970, *New York Folklore Quarterly*, 30, 267–79.

Henderson, R. (2000) Pitching product, *Billboard*, 112 (45), 4.

Herman, E.S. and McChesney, R.W. (1997) *The Global Media: The New Missionaries of Corporate Capitalism*, Cassell, London.

Herman, T. (1990) Liner notes, Thomas Mapfumo, *Shumba. Vital Hits of Zimbabwe*, Virgin, London.

Hinton, B. (1995) *Message to Love: The Isle of Wight Festivals, 1968–1970*, Castle Communications, Chessington.

Hirshberg, J. (1992) In search of a model for transplanted music in migrant communities, *Musicology Australia*, 15, 26–46.

Hirshberg, J. and Seares, M. (1993) The displaced musician – transplantation and commercialization, *The World of Music*, 35(3), 3–34.

Hirshey, G. (1994) *The Story of Soul Music*, Times Books, New York.

Hisama, E. (1993) Postcolonialism on the make: the music of John Mellencamp, David Bowie and John Zorn, *Popular Music*, 12, 91–104.

Hobsbawm, E. (1983) Introduction: inventing traditions, in E. Hobsbawm and T. Ranger, eds, *The Invention of Tradition*, Cambridge University Press, Cambridge, 1–14.

Holden, S. (1994) How pop music lost the melody, *Sydney Morning Herald*, 16 July, 8A.

Hollows, J. and Milestone, K. (1998) Welcome to Dreamsville. A history and geography of northern soul, in A. Leyshon, D. Matless and G. Revill, eds, *The Place of Music*, Guilford, New York, 83–103.

Homan, S. (1998) After the law: Sydney's Phoenician Club and the death of Anna Wood, *Perfect Beat*, 4(1), 56–83.

—— (2000) Losing the local: Sydney and the Oz Rock tradition, *Popular Music*, 19, 31–49.

Horsley, A.D. (1978) Geographic distribution of American quartet gospel music, in G.O. Carney, ed., *The Sounds of People and Places*, University Press of America, Lanham, 213–35.

Hosokawa, S. (1994) East of Honolulu: Hawaiian music in Japan from the 1920s to the 1940s, *Perfect Beat*, 2(1), 51–67.

—— (1999) Strictly ballroom – the rumba in pre-World War Two Japan, *Perfect Beat*, 4(3), 3–23.

Howlett, S. (1990) Which one's Jack?, *Drum Media*, 5 December, 32–3

Hozic, A.A. (1999) Uncle Sam goes to Siliwood: of landscapes, Spielberg and hegemony, *International Review of Political Economy*, 6, 289–312.

Huck, P. (1999) Digital update for the copy cops, *Australian Financial Review Magazine*, July, 54–9.

Hudson, R. (1995) Making music work? Alternative regeneration strategies in a deindustrialised locality: the case of Derwentside, *Transactions of the Institute of British Geographers*, 20, 460–73.

Humphries, P. (1999) Never mind the ballads, *Sunday* (*Sunday Telegraph*), 17 January, 8–11.

Hutnyk, J. (2000) *Critique of Exotica. Music, Politics and the Culture Industry*, Pluto, London.

Hutson, C.K. (1993) Cotton pickin', hillbillies and rednecks: an analysis of Black Oak Arkansas and the perpetual stereotyping of the rural South, *Popular Music and Society*, 17(4), 47–62.

Hutson, S. (2000) The rave: spiritual healing in modern Western subcultures, *Anthropological Quarterly*, 73, 35–49.

Ingham, J. (1999) Listening back from Blackburn: virtual sound worlds and the creation of temporary autonomy, in A. Blake, ed., *Living through Pop*, Routledge, London, 112–28.

Ingham, J., Purvis, M. and Clarke, D. (1994) Hearing places, making spaces: sonorous geographies, ephemeral rhythms and the Blackburn warehouse parties, *Environment and Planning D*, 17, 283–305.

International Federation of Phonographic Industries (IFPI) (1999) 'Governments in Asia urged to step up fight against surging optical disc piracy', press release, Hong Kong, 4 November.

Jackson, P. (1999) Commodity cultures: the traffic in things, *Transactions of the Institute of British Geographers*, 24, 95–108

James, D. (1995) Forget the Beatles, this is the new revolution, *Business Review Weekly*, 4 December, 62–3.

Jameson, J. (1995) Heartland of country, *Daily Telegraph Mirror* (Sydney), 10 October, 45.

Jarvis, B. (1985) The truth is only known by guttersnipes, in J. Burgess and J.R. Gold, eds, *Geography, the Media and Popular Culture*, Croom Helm, London, 96–122.

Jensen, J. (1998) *The Nashville Sound: Authenticity, Commercialisation and Country Music*, Country Music Foundation Press and Vanderbilt University Press, Nashville.

Jinman, R. (1997) Pop is dead, long live pop, *The Australian Magazine*, February 15–16, 5.

Jipson, A. (1994) Why Athens? Investigations into the site of an American music revolution, *Popular Music and Society*, 16(3), 9–22.

Johnson, P. (1996) *Straight outa Bristol: Massive Attack, Portishead, Tricky and the Roots of Trip Hop*, Hodder & Stoughton, London

Johnston, R.J. (1986) *Bell-Ringing: The English Art of Change-ringing*, Viking, New York.

Johnston, R.J., Allsopp, G., Baldwin, J. and Turner, H. (1990) *An Atlas of Bells*, Blackwell, Oxford.

Jones, A. (1992) *Like A Knife. Ideology and Genre in Contemporary Chinese Popular Music*, Cornell East Asia Program, Ithaca.

Jones, Q. (1977) Liner notes, *Roots*, A and M, Los Angeles.

Kaemmer, J.E. (1993) *Music in Human Life. Anthropological Perspectives on Music*, University of Texas Press, Austin.

Kaplan, E.A. (1987) *Rocking around the Clock. Music Television, Postmodernism and Consumer Culture*, Methuen, New York.

Kassabian, A. (1999) Popular, in B. Horner and T. Swiss, eds, *Key Terms in Popular Music and Culture*, Blackwell, Oxford, 113–23.

Kaya, H. (2002) Aesthetics of diaspora: contemporary minstrels in Turkish Berlin, *Journal of Ethnic and Migration Studies*, 28, 43–62.

Kearney, M. (1995) The local and the global: the anthropology of globalisation and transnationalism, *Annual Review of Anthropology*, 24, 547–65.

Keil, C. (1994) 'Ethnic' music traditions in the USA (black music; country music; others; all), *Popular Music*, 13, 175–8.

Keyes, C.L. (1996) At the crossroads: rap music and its African nexus, *Ethnomusicology*, 40, 223–48.

Kibby, M.D. (2000) Home on the page: a virtual place of music community, *Popular Music*, 19, 91–100.

King, S. and Jensen, R. (1995) Bob Marley's 'Redemption Song': the rhetoric of reggae and Rastafari, *Journal of Popular Culture*, 29, 17–36.

Klein, C. (1991) A multi-level analysis of Billy Joel's 'Goodnight Saigon', *Popular Music and Society*, 15, 75–94.

Kong, L. (1995a) Music and cultural politics: ideology and resistance in Singapore, *Transactions, Institute of British Geographers*, 20, 447–59.

—— (1995b) Popular music in geographical analyses, *Progress in Human Geography*, 19, 183–98.

—— (1996a) Making 'music at the margins'? A social and cultural analysis of Xinyao in Singapore, *Asian Studies Review*, 19, 99–124.

—— (1996b) Popular music and a 'sense of place' in Singapore, *Crossroads*, 9(2), 51–77.

—— (1996c) Popular music in Singapore: exploring local cultures, global resources, and regional identities, *Environment and Planning D*, 14, 273–92.

—— (1997) Popular music in a transnational world: the construction of local identities in Singapore, *Asia Pacific Viewpoint*, 38(1), 19–36.

—— (1999) The invention of heritage: popular music in Singapore, *Asian Studies Review*, 23(1), 1–25.

Krim, A. (1992) Route 66: auto river of the American west, in D. Janelle, ed., *Geographical Snapshots of the American West*, Guilford Press, New York, 30–3.

—— (1998) 'Get Your Kicks on Route 66!' A song map of post-war migration, *Journal of Cultural Geography*, 18, 49–60.

Krims, A. (2000) *Rap Music and the Poetics of Identity*, Cambridge University Press, Cambridge.

Kruse, H. (1993) Subcultural identity in alternative music culture, *Popular Music*, 12, 33–41.

—— (1999) Gender, in B. Horner and T. Swiss, eds, *Key Terms in Popular Music and Culture*, Blackwell, Oxford, pp. 85–100.

Kun, J. (1998) Rap en Espanol: the red hot sound of Mexican hip hop, *Option*, 80, May–June, 54–61.

Kurosawa, S. (2000) Pump up the volume, *The Australian Magazine*, 17 June, 47.

Lahusen, C. (1993) The aesthetic of radicalism: the relationship between punk and the patriotic nationalist movement of the Basque country, *Popular Music*, 12, 263–80.

Laing, D. (1972) Roll over Lonnie (tell George Formby the news), in C. Gillett, ed., *Rock File*, New English Library, London, 45–51.

—— (1985) *One Chord Wonders. Power and Meaning in Punk Rock*, Open University Press, Milton Keynes.

Lanza, J. (1994) *Elevator Music: A Surreal History of Muzak, Easy-Listening and Other Moodsong*, Quartet Books, London.

Lawe Davies, C. (1993) Aboriginal rock music: space and place, in T. Bennett, S. Frith, L. Grossberg, T. Shepherd and G. Turner, eds, *Rock and Popular Music*, Routledge, London, 249–65.

Lazarus, N. (1993) 'Unsystematic fingers at the conditions of the times': 'Afropop' and the paradoxes of imperialism, in J. White, ed., *Recasting the World: Writing after Colonialism*, Johns Hopkins Press, Baltimore, 137–60.

Leach, E.V. (2001) Vicars of 'wannabe': authenticity and the Spice Girls, *Popular Music*, 20, 146–65.

Lebeau, V. (1997) The Worst of All Possible Worlds?, in R. Silverstone, ed., *Visions of Suburbia*, Routledge, London, 280–97.

Lee, J.C. (1992) Cantopop songs on emigration from Hong Kong, *Yearbook for Traditional Music*, 24, 14–23.

Lee, S. (1997) Wizards of Oz, *Sunday Times*, May 11, 27.

Lefebvre, H. (1991) *The Production of Space*, Blackwell, Oxford.

Lehr, J.C. (1983) 'Texas (when I die)': national identity and images of place in Canadian country music broadcasts, *The Canadian Geographer*, 27, 361–70.

Leming, J.S. (1987) Rock music and the socialisation of moral values in early adolescence, *Youth and Society*, 18, 363–83.

Lewin, S. (1997) Loft cause, in R. Benson, ed., *Nightfever*, Boxtree, London, 89–90.

Lewis, G.H. (1991) Tension, conflict and contradiction in country music, *Journal of Popular Culture*, 24, 103–17.

—— (1992) La Pistola y el corazon: protest and passion in Mexican-American popular music, *Journal of Popular Culture*, 26, 51–68.

—— (1997) Lap dancer or hillbilly deluxe? The cultural constructions of modern country music, *Journal of Popular Culture*, 31(3), 163–73.

Lewis, L. (1993) Being discovered: the emergence of female address on MTV, in S. Frith, A. Goodwin and L. Grossberg (eds) *Sound and Vision: The Music Video Reader*, Routledge, London and New York, 129–52.

Lewis, L.A. and Ross, M. (1995) The gay dance party culture in Sydney: a qualitative analysis, *Journal of Homosexuality*, 29, 41–70.

Leydon, R. (1999) Utopias of the Tropics: the exotic music of Les Baxter and Yma Sumac, in P. Hayward, ed., *Widening the Horizons. Exoticism in Post-War Popular Music*, John Libbey, Sydney, 45–71.

Leyshon, A (2001) Time-space (and digital) compression: software formats, music networks, and the reorganisation of the music industry, *Environment and Planning A*, 33, 49–77.

Leyshon, A., Matless, D. and Revill, G. (1995) The place of music, *Transactions, Institute of British Geographers*, 20, 423–33.

——, eds (1998) Introduction, in A. Leyshon, D. Matless and G. Revill, eds, *The Place of Music*, Guilford, New York, 1–30.

Lipsitz, G. (1994) *Dangerous Crossroads: Popular Music and the Poetics of Postmodernism*, Verso, London and New York.

—— (1999a) World cities and world beat: low-wage labor and transnational culture, *Pacific Historical Review*, 68, 213–31.

—— (1999b) 'Home is where the hatred is': work, music and the transnational economy, in H. Naficy, ed., *Home, Exile, Homeland: Film, Media and the Politics of Place*, Routledge, New York and London, 193–212.

Llewellyn, M. (2000) Popular music and the Welsh language and the affirmation of youth identities, *Popular Music*, 19, 319–39.

Lockard, C.A. (1998) *Dance of Life: Popular Music and Politics in Southeast Asia*, University of Hawaii Press, Honolulu.

Lomax, A. (1959) Folk Song Style, *American Anthropologist*, 61, 927–54.

—— (1962) Song structure and social structure, *Ethnology*, 1, 425–51.

—— (1976) *Cantometrics. An Approach to the Anthropology of Music*, University of California Press, Berkeley.

Lomax, A. and Erickson, E.E. (1971) The world song style map, in A. Lomax, ed., *Folk Song Style and Culture*, American Association for the Advancement of Science, Washington, 75–110.

Longhurst, B. (1995) *Popular Music and Society*, Polity Press, Cambridge

Lowenthal, D. (1989) Nostalgia tells it like it wasn't, in C. Shaw and M. Chase, eds, *The Imagined Past: History and Nostalgia*, Manchester University Press, Manchester, 18–32.

Luckman, S. (2001) What are they raving on about? Temporary Autonomous Zones and 'Reclaim the Streets', *Perfect Beat*, 5(2), 49–68.

Lury, C. (1993) *Cultural Rights: Technology, Legality and Personality*, Routledge, London and New York.

Lysloff, R.T.A. (1997) Mozart in mirrorshades: ethnomusicology, technology and the politics of representation, *Ethnomusicology*, 41, 206–19.

McClary, S. (1994) Same as it ever was. Youth culture and music, in A. Ross and T. Rose, eds, *Microphone Fiends*, Routledge, New York, 29–40.

McClure, S. (2000a) Utada, Dion among those to strike gold at Japan awards, *Billboard*, 112(14), 76.

—— (2000b) DIY indie music takes root in Japan, *Billboard*, 112(37), 65.

McCray Pattacini, M. (2000) Deadheads yesterday and today: an audience study, *Popular Music and Society*, 24(1), 1–14

McDonald, E. and Mayhew, E. (1999) Women in popular music and the construction of 'authenticity', *Journal of Interdisciplinary Gender Studies*, 4, 63–81.

Macdonald, I. (1994) *Revolution in the Head: The Beatles Records and the Sixties*, Fourth Estate, London.

McDonnell, E. and Powers, A. (1995) *Rock She Wrote*, Plexus, London.

McKay, G. (1996) *Senseless Acts of Beauty: Cultures of Resistance since the Sixties*, Verso, London.

McKee, M. and Chisenhall, F. (1993) *Beale Black and Blue: Life and Music on Black America's Main Street*, Louisiana State University Press, Baton Rouge.

McLaren, P. (1995) Gangsta pedagogy and ghettoethnicity: the hip-hop nation as counterpublic sphere, *Socialist Review*, 25(2), 9–56.

McLaughlin, N. and McLoone, M. (2000) Hybridity and national musics: the case of Irish rock music, *Popular Music*, 19, 181–99.

McLean, D. (1997) *Lone Star Swing*, Random House, London.

McLeay, C. (1994) The 'Dunedin sound' – New Zealand rock and cultural geography, *Perfect Beat*, 2(1), 38–50.

—— (1997) Popular music and expressions of national identity, *New Zealand Journal of Geography*, 103, 12–17.

—— (2001) Review of M. Bannister, *Positively George Street*, *Perfect Beat*, 5(2), 102–5.

McNeil, L. and McCain, G. (1997) *Please Kill Me: The Uncensored Oral History of Punk*, Viking Penguin, Ringwood.

McRobbie, A. (1994) *Postmodernism and Popular Culture*, Routledge, London.

Maffesoli, M. (1995) *The Time of the Tribes*, trans. 1998, Sage, London.

Maira, S. (1999) Identity dub: the paradoxes of an Indian American youth subculture (New York mix), *Cultural Anthropology*, 14, 29–60.

Malbon, B. (1999) *Clubbing: Dancing, Ecstasy, Vitality*, Routledge, London.

Malone, B. (1985) *Country Music USA: A Fifty Year History*, University of Texas Press, Austin.

Manuel, P. (1988) *Popular Music of the Non-Western World*, Oxford University Press, Oxford.

—— (1993) *Cassette Culture. Popular Music and Technology in North India*, University of Chicago Press, Chicago.

—— (1995) *Caribbean Currents*, Temple University Press, Philadelphia.

Marcus, G. (1975) *Mystery Train: Images of America in Rock 'n' Roll Music*, Dulton, New York.

—— (1989) *Lipstick Traces: The Secret History of the Twentieth Century*, Harvard University Press, Cambridge, MA.

—— (1993) *In the Fascist Bathroom*, Viking, London.

Mardesich, J. (1999) How the Internet hits big music, *Fortune*, 10 May, 139(9), 96–7.

Margetts, J. (1994) Pure caffeine, nicotine and music: Portishead, *3D World*, 28 November, 31.

Margolis, M. (1994) *Little Brazil: An Ethnography of Brazilian Immigrants in New York City*, Princeton University Press, Princeton.

Marling, K. (1996) *Graceland: Going Home with Elvis*, Harvard University Press, Cambridge.

Marsh, D. and Swenson, J., eds. (1993) *The New Rolling Stone Record Guide*, New York, Random House.

Martin, G. (1998) Generational differences amongst new age travellers, *The Sociological Review*, 46, 735–56.

Massey, D. (1994) *Space, Place and Gender*, Polity, Cambridge.

—— (1998) The spatial construction of youth cultures, in T. Skelton and G. Valentine, eds, *Cool Places: Geographies of Youth Cultures*, Routledge, London, 121–9.

Mathews, G. (2000) *Global Culture/Individual Identity*, Routledge, London.

Mathieson, C. (2000) *The Sell-In: How the Music Business Seduced Alternative Rock*, Allen & Unwin, Sydney.

Maxwell, I. (1997a) Hip hop aesthetics and the will to culture, *The Australian Journal of Anthropology*, 8, 50–70.

—— (1997b) On the flow – dancefloor grooves, rapping 'freestylee' and 'the Real Thing', *Perfect Beat*, 3(3) 15–27.

Mead, M. (1935) *Sex and Temperament in Three Primitive Societies*, Routledge, London.

Medhurst, A. (1997) Negotiating the gnome zone; versions of suburbia in British popular culture, in R. Silverstone, ed., *Visions of Suburbia*, Routledge, London, 240–68.

—— (1999) What did I get? Punk, memory and autobiography, in R. Sabin, ed., *Punk Rock. So What?*, Routledge, London, 199–218.

Meethan, K. (1996) Place, image and power: Brighton as a resort, in T. Selwyn, ed., *The Tourist Image: Myths and Myth Making in Tourism*, Wiley, Chichester, 179–95.

Meintjes, L. (1990) Paul Simon's Graceland, South Africa and the Mediation of Musical Meaning, *Ethnomusicology*, 34, 37–73.

Merriam, A.P. (1964) *The Anthropology of Music*, Northwestern University, Chicago.

Middleton, R. (1990) *Studying Popular Music*, Open University Press, Milton Keynes.

Milestone, K. (1997) The love factory: the sites, practices and media relationships of northern soul, in S. Redhead, ed., *The Club Cultures Reader*, Blackwell, Oxford, 152–67.

Mitchell, D. (2000) *Cultural Geography: A Critical Introduction*, Blackwell, Oxford.

Mitchell, T. (1996) *Popular Music and Local Identity*, Leicester University Press, London and New York.

—— (1999) Another root: Australian hip hop as a 'global' subculture – re-territorialising hip hop, in G. Bloustein, ed., *Musical Visions*, Wakefield Press, Adelaide, 85–94.

—— (2001) Dick Lee's Transit Lounge: Orientalism and pan-Asian pop, *Perfect Beat*, 5(3), 18–45.

Mockus, M. (1994) Queer thoughts on country music and k.d. lang, in P. Brett, E. Wood and G. Thomas, eds, *Queering the Pitch*, Routledge, London, 257–74.

Morley, D. (1995) Theories of consumption in media studies, in D. Miller, ed., *Acknowledging Consumption: A Review of New Studies*, Routledge, London and New York, 296–328.

Morton, T. (1991) *Going Home: The Runrig Story*, Mainstream, Edinburgh.

Moss, P. (1992) Where is the 'Promised Land'? Class and gender in Bruce Springsteen's rock lyrics, *Geografiska Annaler*, 74B, 167–86.

Moyle, R. (1986) *Alyawarra Music: Songs and Society in a Central Australian Community*, Australian Institute of Aboriginal Studies, Canberra.

Munro, A. (1996) *The Democratic Muse: Folk Music Revival in Scotland*, Aberdeen, Scottish Cultural Press.

Myers, P. (1997) Mixing it in a mad world, *Guardian*, 16 May, 16.

Nash, P. (1968) Music regions and regional music, *Deccan Geographer*, 6, 1–24.

—— (1978) Music and environment, in G. Carney, ed., *The Sounds of People and Places*, University Press of America, Washington, 1–53.

Nash, P. and Carney, G. (1996) The seven themes of music geography, *The Canadian Geographer*, 40, 69–74.

Negus, K. (1999) *Music Genres and Corporate Cultures*, Routledge, London and New York.

Nehring, N. (1993) *Flowers in the Dustbin: Culture, Anarchy, and Post-War England*, University of Michigan Press, Ann Arbor.

Nettl, B. (1956) *Music in Primitive Culture*, Harvard University Press, Cambridge.

Neuenfeldt, K. (1993) Yothu Yindi and Ganma: the cultural transposition of Aboriginal agenda through metaphor and music, *Journal of Australian Studies*, 38, 1–11.

Nowell, D. (1999) *Too Darn Soulful: The Story of Northern Soul*, Robson, London.

O'Brien, L. (1995) *She Bop. The Definitive History of Women in Rock, Pop and Soul*, Penguin, London.

O'Brien, R. (1992) *Global Financial Integration: the End of Geography*, Pinter, London.

O'Connor, J. (1998) Popular culture, cultural intermediaries and urban regeneration, in T. Hall, and P. Hubbard, eds, *The Entrepreneurial City: Geographies of Politics, Regime and Representation*, Wiley, Chichester, 225–39.

O'Duibhir, L. (1998) Homemade jam, *City Hub*, 10 December, 9–10.

O'Róchàin, M. (1975) Liner notes, Bernard O'Sullivan and Tommy McMahon, *Clare Concertinas*, Topic Records, London.

Ohmae, K. (1990) *The Borderless World: Power and Strategy in the Interlinked Economy*, Harper Collins, London.

—— (1995) *The End of the Nation State: The Rise of Regional Economies*, Harper Collins, London.

Oliver, D.L. (1955) *A Solomon Island Society*, Harvard University Press, Boston.

Oliver, P. (1990) *Blues Fell This Morning: Meanings in the Blues*, Cambridge University Press, Cambridge.

Olson, M. (1998) 'Everybody loves our town': scenes, spatiality, migrancy, in T. Swiss, J. Sloop and A. Herman, eds, *Mapping the Beat*, Blackwell, Oxford, 269–89.

Ortega, T. (1995) 'My name is Sue! How do you do?' Johnny Cash as lesbian icon, in C. Tichi, ed., *Readin' Country Music*, Duke University Press, Durham, 259–72.

Osgerby, B. (1999) 'Chewing out a rhythm on my bubble gum': the teenage aesthetic and geographies of American punk, in R. Sabin, ed., *Punk rock: so what?*, Routledge, London, 154–69.

Owens, D. (2000) *Cerys, Catatonia and the Rise of Welsh Pop*, Ebury Press, London.

Pacini-Hernandez, D.H. (1993) A view from the south: Spanish Caribbean perspectives on world beat, *The World of Music*, 35, 48–69.

Padilla, F. (1990) Salsa: Puerto Rican and Latino music, *Journal of Popular Culture*, 24, 87–104.

Palmer, G. (1997) Bruce Springsteen and masculinity, in S. Whiteley, ed., *Sexing the Groove*, Routledge, London, 100–17.

Palmer, R. (1974) *A Touch on the Times: Songs of Social Change 1770 to 1914*, Penguin, Harmondsworth.

Perry, T. and Glinert, E. (1996) *Rock & Roll Traveler USA*, Fodor's, New York.

Peterson, R.A. (1997) *Creating Country Music: Fabricating Authenticity*, University of Chicago Press, Chicago.

Peterson, R.A. and Berger, D.G. (1996) Measuring industry concentration, diversity and innovation in popular music, *American Sociological Review*, 61, 175–8.

Phua, S.C. and Kong, L. (1996) Ideology, social commentary and resistance in popular music: a case study of Singapore, *Journal of Popular Culture*, 30, 215–31.

Pickering, M. and Green, T. (1987) Towards a cartography of the vernacular milieu, in M. Pickering and T. Green, eds, *Everyday Culture, Popular Song and the Vernacular Milieu*, Open University Press, Milton Keynes, 1–38.

Pickering, M. and Shuker, R. (1994) Struggling to make ourselves heard: music, radio and the quota debate, in P. Hayward, T. Mitchell and R. Shuker, eds, *North Meets South: Popular Music in Aotearoa/New Zealand*, Perfect Beat Publications, Umina, NSW, 73–97.

Pini, M. (1997) Women and the early British rave scene, in A. McRobbie, ed., *Back to reality? Social experience and cultural studies*, Manchester University Press, Manchester, 152–69.

Police, D. (2000) Mauritian sega: the trace of the slave's emancipatory voice, *UTS Review*, 6(2), 57–69.

PolyGram N.V. (1995) 'PolyGram acquires majority stake in Indian record company', press release, Bombay, 16 January.

Potter, R.A. (1995) *Spectacular Vernaculars: Hip Hop and the Politics of Postmodernism*, SUNY Press, Albany.

Powell, D. (1993) *Out West: Perspectives of Sydney's Western Suburbs*, Allen & Unwin, Sydney.

Powers, A. (1993) *Rock She Wrote: Women Write about Rock, Pop and Rap*, Plexus, London.

Pratt, A. (2000) New media, the new economy and new spaces, *Geoforum*, 31, 425–36.

Pratt, A.C. (1994) *Uneven Re-production: Industry, Space and Society*, Pergamon, Oxford.

Pratt, R. (1990) *Rhythm and Resistance: Explorations in the Political Use of Popular Music*, Praeger, New York.

Quinn, B. (1996) The sounds of tourism: exploring music as a tourist resource with particular reference to music festivals, in M. Robinson, N. Evans and P. Callaghan, eds, *Tourism and Culture Towards the 21st Century*, Centre for Travel and Tourism and Business Education Publishers, Sunderland, 383–96.

Quinn, M. (1996) 'Never shoulda been let out the penitentiary': gangsta rap and the struggle over racial identity, *Cultural Critique*, 34, 65–89.

Qureshi, R.B. (1999) His master's voice? Exploring *qawwali* and 'gramophone culture' in South Asia, *Popular Music*, 18, 63–98.

Ramet, S. (1994) Rock: the music of revolution (and political conformity), in S. Ramet, ed., *Rocking the State*, Westview, Boulder, 1–14.

Ramet, S., Zamascikov, S. and Bird, R. (1994) The Soviet rock scene, in S. Ramet, ed., *Rocking the State*, Westview, Boulder, 181–218.

Raphael, A. (1994) *Never Mind the Bollocks: Women Rewrite Rock*, Virago, London.

Reck, D., Slobin, M. and Titon, J. (1992) Discovering and documenting a world of music, in J. Titon, ed., *Worlds of Music*, Schirmer, New York, 429–54.

Record Industry Association of America (RIAA) (1999) 'RIAA worldview', www.riaa.com.

Regev, M. (1997) Rock aesthetics and musics of the world, *Theory, Culture and Society*, 14, 125–42.

Reily, S.A. (1992) Musica sertaneja and migrant identity: the stylistic development of a Brazilian genre, *Popular Music*, 11, 337–58.

Remes, P. (1999) Global popular music and changing awareness of urban Tanzanian youth, *Yearbook for Traditional Music*, 31, 1–26.

Revill, G. (2000) English pastoral: music, landscape, history and politics, in I. Cook, D. Crouch, S. Naylor and J. Ryan, eds, *Cultural Turns/Geographical Turns*, Prentice Hall, Harlow, 141–58.

Reynolds, S. and Press, J. (1995) *The Sex Revolts: Gender, Rebellion and Rock 'n' Roll*, Harvard University Press, Cambridge.

Rheingold, H. (1993) The virtual community: homesteading on the electronic frontier, Addison-Wesley, Reading, MA.

Richard, B. and Kruger, H. (1998) Ravers' paradise?: German youth cultures in the 1990's, in T. Skelton and G. Valentine, eds, *Cool Places*, Routledge, London, 161–74.

Roberson, J.E. (2001) Ucinaa pop, place and identity in Okinawan popular music, *Critical Asian Studies*, 33, 211–42.

Roberts, M. (1993) 'World music' and the global cultural economy, *Diaspora*, 2, 229–42.

Robins, K. and Gillespie, A. (1992) 'Communication, organisation and territory', in K. Robins, ed., *Understanding Information: Business, Technology and Geography*, Belhaven Press, New York, 147–64.

Robins, K. and Morley, D. (1996) Almanci, Yabanci, *Cultural Studies*, 10, 248–54.

Robinson, D.C., Buck, E. and Cuthbert, J. (1991) *Music at the Margins: Popular Music and Global Cultural Diversity*, Sage, London.

Rochereau (1984) Liner notes, *Tabu Ley*, Shanachie, Dalebrook Park, NJ.

Rodman, G.B. (1996) *Elvis after Elvis: The Posthumous Career of a Living Legend*, Routledge, London.

Rolston, B. (2001) 'This is not a rebel song': the Irish conflict and popular music, *Race and Class*, 42(3), 49–67.

Romagnan, J.-M. (2000) La Musique: un nouveau terrain pour les géographes, *Géographie et cultures*, 36, 107–26.

Roman-Velazquez, P. (1999) The embodiment of salsa: musicians, instruments and the performance of a Latin style and identity, *Popular Music*, 18, 115–31.

Rommen, T. (1999) Home sweet home: junkanoo as national discourse in the Bahamas, *Black Music Research Journal*, 19(1), 71–92.

Rose, A. (1997) NW interview: Jessamine, *The Rocket*, March 26–April 9, 14.

Rose, G. (1994) The cultural politics of place: local representation and oppositional discourse in two films, *Transactions of the Institute of British Geographers*, 19, 46–60.

Rose, T. (1994) *Black Noise: Rap Music and Black Culture in Contemporary America*, Wesleyan University Press, Hanover.

Roseman, M. (1991) *Healing Sounds from the Malaysian Rainforest*, University of California Press, Berkeley.

—— (1998) Singers of the Landscape. Song, History and Property Rights in the Malaysian Rain Forest, *American Anthropologist*, 100, 106–21.

—— (2000) Shifting Landscapes: Musical Mediations of Modernity in the Malaysian Rainforest, *Yearbook of Traditional Music*, 32, 31–66.

Rosenberg, N. (1985) *Bluegrass: A History*, University of Illinois Press, Urbana.

—— (1993a) Introduction, in N.V. Rosenberg, ed., *Transforming Tradition*. University of Illinois Press, Urbana, 1–25.

—— (1993b) Starvation, serendipity and the ambivalence of bluegrass revivalism, in N.V. Rosenberg, ed., *Transforming Tradition: Folk Music Revivals Examined*, University of Illinois Press, Urbana, 194–202.

Ross, A. (1998) *Real Love: In Pursuit of Cultural Justice*, New York University Press, New York.

Russell, D. (1997) *Popular Music in England 1840–1914*, Manchester University Press, Manchester.

Ryback, T.W. (1990) *Rock around the Bloc*, New York, Oxford University Press.

Sabin, R., ed. (1999) *Punk Rock: So What? The Cultural Legacy of Punk*, Routledge, London.

Sadler, D. (1997) The global music business as an information industry: reinterpreting economies of culture, *Environment and Planning A*, 29, 1919–36.

Said, E. (1978) *Orientalism*, Routledge and Kegan Paul, London

—— (1990) Figures, confrontations, transfigurations, *Race and Class*, 32, 1–16.

—— (1995) *The Politics of Dispossession: The Struggle for Palestinian Self-Determination 1969–1994*, Vintage, London.

Saldanha, A. (2000) Fear and loathing in Goa, *Unesco Courier*, July/August, 51–2.

—— (2001) Music-bodies-politics: geographies of psychedelic rave culture in Goa, paper presented at the 93rd Annual Meeting of the Association of American Geographers, 27 February–2 March.

Sampath, N. (1997) 'Mas' identity: tourism and global and local aspects of Trinidad carnival, in S. Abram, J.D. Waldren and D.V.L. MacLeod, eds, *Tourists and Tourism: Identifying People with Place*, Berg, Oxford, 149–71.

Sanchez-Gonzalez, L. (1999) Reclaiming salsa, *Cultural Studies*, 13, 237–50.

Sandford, C. (1999) *Springsteen: Point Blank*, Little Brown, London.

Sant Cassia, P. (2000) Exoticising discources and extraordinary experiences: 'traditional' music, modernity and nostalgia in Malta and other Mediterranean societies, *Ethnomusicology*, 44, 281–301.

Sapoznik, H. (1997) Klezmer music: the first one thousand years, in K. Lornell and A. Rasmussen, eds, *Musics of Multicultural America*, Schirmer, New York, 49–71.

Sarkissian, M. (1995) 'Sinhalese Girl' meets 'Aunty Annie': competing expressions of ethnic identity in the Portuguese settlement, Melaka, Malaysia, *Asian Music*, 27, 37–62.

Sassen, S. (1999) Servicing the global economy: reconfigured states and private agents, in K. Olds, P. Dicken, L. Kong and H. Wai-chung Yeung (eds) *Globalisation and the Asia-Pacific*, Routledge, London, 149–63.

Savage, J. (1991) *England's Dreaming: Sex Pistols and Punk Rock*, Faber, London.

Savan, L. (1993) Commercials go rock, in S. Frith, A. Goodwin and L. Grossberg, eds, *Sound and Vision*, London, Routledge, 85–90.

Sawhney, N. (1999) Liner notes, *Beyond Skin,* Outcaste Records, London.

Scanlon, A. (1997) *Those Tourists Are Money: The Rock n Roll Guide to Camden*, Tristia, London.

Scatena, D. (1995) Hi-fi way: King Loser, *On the Street*, 16 May, 19.

Schade-Poulsen, M. (1995) The power of love. Rai music and youth in Algeria, in V. Amit-Talai and H. Wulff, eds, *Youth Culture: A Cross-Cultural Perspective*, Routledge, London, 81–113.

—— (1997) Which world? On the diffusion of Algerian rai to the West, in K. Olwig, ed., *Siting Cultures: The Shifting Anthropological Object*, Routledge, London, 59–85.

—— (1999) *Men and Popular Music in Algeria*, University of Texas Press, Austin.

Schieffelin, E. (1978) *The End of Traditional Music, Dance and Body Decoration in Bosavi, Papua New Guinea*, Port Moresby, Institute of Papua New Guinea Studies Discussion Paper Nos, 30–2.

Schulz, D.E. (2001) Music videos and the effeminate vices of urban cultures in Mali, *Africa*, 71, 346–72.

Scott, A.J. (1986) High technology industry and territorial development: the rise of the Orange County Complex, *Urban Geography*, 7, 3–45.

—— (1999) The US recorded music industry: on the relations between organization, location, and creativity in the cultural economy, *Environment and Planning A*, 31, 1965–84.

Scruggs, T.M. (1999) 'Lets enjoy as Nicaraguans': the use of music in the construction of a Nicaraguan national consciousness, *Ethnomusicology*, 43, 297–321.

Seeger, A. (1987) *Why Suyá Sing: A Musical Anthropology of an Amazonian People*, Cambridge University Press, Cambridge.

—— (1994) Music and Dance, in T. Ingold, ed., *Companion Encyclopedia of Anthropology*, Routledge, London, 686–705.

Sellars, A. (1998) *The Influence of Dance Music on the UK Youth Tourism Market*, Travel and Tourism Working Paper 3, University of Hertfordshire Business School, Hertford

Shaar Murray, C (1989) *Crosstown Traffic: Jimi Hendrix and Post-War Pop*, Faber & Faber, London

Shankar, R. (1969) *My Music, My Life*, Cape, London.

Sharma, A. (1996) Sounds Oriental: the (Im)possibility of theorizing Asian musical cultures, in S. Sharma, J. Hutnyk and A. Sharma, eds, *Disorienting Rhythms: The Politics of the New Asian Dance Music*, Zed, London, 15–31.

Shedden, I. (2001) Echoes of the past, *The Australian*, July 21, R16–17.

Shelemay, K.K. (1998) *Let Jasmine Rain Down: Song and Remembrance among Syrian Jews*, University of Chicago Press, Chicago.

Sherman, C (1999a) 'Protecting music rights in the digital era: the U.S. experience', paper presented to the MIDEM annual conference, Cannes, France.

—— (1999b) Presentation to the SDMI Organizing Plenary, Los Angeles.

Shuker, R. (1994) *Understanding Popular Music*, Routledge, London.
—— (1998) The New Zealand sound recording industry: cultural policy and New Zealand On Air, *Australian Journal of Communication*, 25, 129–39.
Sibley, D. (1994) The sin of transgression, *Area*, 26, 300–3.
Simonett, H. (2001) Narcocorridos: an emerging micromusic of Nuevo L.A., *Ethnomusicology*, 45, 315–37.
Siriyuvakak, U. (1998) Thai pop music and cultural negotiation in everyday politics, in K.-H. Chen, ed., *Trajectories. Inter-Asia Cultural Studies*, Routledge, London, 206–27.
Skelton, T. (1995) 'Boom, Bye Bye'. Jamaican ragga and gay resistance, in D. Bell and G. Valentine, eds, *Mapping Desire. Geographies of Sexualities*, Routledge, London, 264–83.
Slater, D. (1997) Spatialities of power and postmodern ethics – rethinking geopolitical encounters, *Environment and Planning D*, 15, 55–72.
Slobin, M. (1976) *Music in the Culture of Northern Afghanistan*, University of Arizona Press, Tuscon.
—— (1993) *Subcultural Sounds: Micromusics of the West*, Wesleyan University Press and the University Press of New England, Hanover and London.
—— (1994) Music in diaspora: the view from America, *Diaspora*, 3, 243–52.
Smith, A. (1997) Northern delights, *Sunday Times*, 9 February, 24.
Smith, M. (1992) Sexual mobilities in Bruce Springsteen: performance as commentary, in A. de Curtis, ed., *Present Tense: Rock & Roll and Culture*, Duke University Press, Durham, 197–218.
Smith, P.J. (1999) 'Ask any girl'. Compulsory heterosexuality and girl group culture, in K. Dettmar and W. Richey, eds, *Reading Rock and Roll: Authenticity, Appropriation, Aesthetics*, Columbia University Press, New York, 93–124.
Smith, S.J. (1994) Soundscape, *Area*, 26, 232–40.
—— (1997) Beyond geography's visible worlds: a cultural politics of music, *Progress in Human Geography*, 21, 502–29.
Solomon, T. (2000) Dueling landscapes: singing places and identities in Highland Bolivia, *Ethnomusicology*, 44(2), 257–80.
Sorce Keller, M. (1994) Reflections of continental and Mediterranean traditions in Italian folk music, in M. Kartomi and S. Blum, eds, *Music Cultures in Contact: Convergences and Collisions*, Currency Press, Sydney, 40–7.
Stapleton, C. (1990) African connections: London's hidden music scene, in P. Oliver, ed., *Black Music in Britain*, Open University Press, Milton Keynes, 79–86.
Stapleton, C. and May, C. (1987) *African All Stars: The Pop Music of a Continent*, Quartet, London.
Stebbins, R.A. (1996) Cultural tourism as serious leisure, *Annals of Tourism Research*, 23, 948–50.
Stephens, G. (1999) 'You can sample anything': *Zebrahead*, 'black' music and multiracial audiences, *New Formations*, 39, 113–29.
Stephens, M. (1998) Babylon's 'natural mystic': the North American music industry, the legend of Bob Marley, and the incorporation of transnationalism, *Cultural Studies*, 12, 139–67.
Sterne, J. (1997) Sounds like the Mall of America: programmed music and the architectonics of commercial space, *Ethnomusicology*, 41, 22–50.
Steward, S. (1999) *Salsa: Musical Heartbeat of Latin America*, Thames & Hudson, London.
Steward, S. and Garratt, S. (1984) *Signed, sealed and delivered: The Life Stories of Women in Pop*, South End Press, Boston.
Stockbridge, S. (1992) Rock'n'roll and television, in A. Moran, ed., *Stay Tuned*, Allen & Unwin, Sydney, 135–42.
Stockmann, D. (1994) Synthesis in the culture of scholarship: problems in investigating and documenting the archaic and modern styles of yoiking by the Sami in Scandinavia, in M. Kartomi and S. Blum, eds, *Music Cultures in Contact: Convergences and Collisions*, Currency Press, Sydney, 1–12.

Stokes, D. (1997) F... karaoke, we want rock!?, *Asian Studies Review*, 20, 54–61.

Stokes, M. (1994) Place, exchange and meaning: Black Sea musicians in the west of Ireland, in M. Stokes, ed., *Ethnicity, Identity and Music*, Berg, Oxford, 97–116.

——, ed., (1994) *Ethnicity, Identity and Music: The Musical Construction of Place*, Berg, Oxford.

Straw, W. (1991) Systems of articulation, logics of change: communities and scenes in popular music, *Cultural Studies*, 5, 368–88.

—— (1993) Popular music and postmodernism in the 1980s, in S. Frith, A. Goodwin and L. Grossberg, eds, *Sound and Vision*, Routledge, London, 3–21.

Street, J. (1986) *Rebel Rock: The Politics of Popular Music*, Blackwell, Oxford.

Strong, M.C. (1999) *The Great Alternative and Indie Discography*, Canongate, Edinburgh.

Symon, P. (1997) Music and national identity in Scotland: a study of Jock Tamson's Bairns, *Popular Music*, 16, 203–16.

Szwed, J.F. (1997) *Space is the Place: The Life and Times of Sun Ra*, Payback Press, New York.

Taffet, J.F. (1997) 'My guitar is not for the rich'. The New Chilean Song Movement and the politics of culture, *Journal of American Culture*, 20, 91–103.

Tanenbaum, S. (1995) *Underground Harmonies: Music and Politics in the Subways of New York*, Cornell University Press, Ithaca.

Tannenbaum, R. (1992) Are you ready for the country?, *Rolling Stone*, 471, June, 20–1.

Taylor, T.D. (1997) *Global Pop: World Music, World Markets*, Routledge, New York and London.

Textor, A.R. (1994) A close listening of the Pet Shop Boys' 'Go West', *Popular Music and Society*, 18, 91–6.

Thomas, A. (1992) Songs as history. A preliminary analysis of two songs of the recruiting era recently recorded in west Futuna, Vanuatu, *Journal of Pacific History*, 27, 229–36.

Thomas, L. (1990) A sense of community: blues music as primer for urbanisation, *Popular Music and Society*, 14(2), 77–86.

Thomas, S. (1999) Angélique Kidjo. Manifest power, *City Hub*, 4(30), 25 March, 15.

Thornton, S. (1995) *Club Cultures: Music, Media and Subcultural Capital*, Polity Press, Cambridge.

Thrills, A. (1999) *You're Not Singing Anymore*, Ebony Press, London.

Tichi, C. (1994) *High Lonesome: The American Culture of Country Music*, University of North Carolina Press, Chapel Hill.

—— (ed.)(1998) *Reading Country Music*, Duke University Press, Durham, NC.

Tillman, R.H. (1980) Punk rock and the consumption of pseudo-political movements, *Popular Music and Society*, 7, 165–75.

Toop, D (1984) *The Rap Attack*, Pluto Press, London.

—— (1992) *Rap Attack 2*, Consortium Press, Boston.

—— (1995) *Ocean of sound: aether talk, ambient sound and imaginary worlds*, Serpent's Tail, London and New York.

—— (1999) *Exotica: Fabricated soundscapes in a Real World*, Serpent's Tail, London.

Trager, D.S. (2001) L.A.'s 'White Minority'. Punk and the contradictions of self-marginalization, *Culture Critique*, 48, 30–64.

Train, M. (1999) Ringing in the changes, *Geographical*, 71(1), 23–8.

Triantafillou, N. (1996) Grunge: why it won't go away, *Sun Herald Tempo*, 17 November, 20.

Trimboli, A. (1999) Detroit delight, *Sydney Morning Herald Metro*, 1 October, 21.

Troitsky, A. (1987) *Back in the USSR. The True Story of Rock in Russia*, Omnibus, London.

Turino, T. (1989) The coherence of social style and musical creation among the Aymara in southern Peru, *Ethnomusicology*, 33, 1–30.

—— (1993) *Moving away from Silence: Music of the Peruvian Altiplano and the Experience of Urban Migration*, University of Chicago Press, Chicago.

—— (2000) *Nationalists, Cosmopolitans, and Popular Music in Zimbabwe*, University of Chicago Press, Chicago.

Urry, J. (1990) *The Tourist Gaze: Leisure and Travel in Contemporary Society*, London, Sage.

—— (1996) Is the global a new space of analysis?, *Environment and Planning A*, 28, 1977–82.

Usinger, S. (1997) Presidents still stupidly exhilarating, *The Georgia Straight*, April 10–17, 63.

Valentine, G. (1995) Creating transgressive space: the music of kd lang, *Transactions of the Institute of British Geographers*, 20, 474–85.

van der Lee, P. (1998) Sitars and bossas: world music influences, *Popular Music*, 17, 45–68.

Van Elteren, M. (1994) Populist rock in postmodern society: John Cougar Mellencamp in perspective, *Journal of Popular Culture*, 28(3), 95–123.

Velez, M.T. (1994) Eya Arnala: overlapping perspectives on Santeria group, *Diaspora*, 3, 289–304.

Venanda Lovely Boys (1989) Liner notes, *Bo-Tata*, Rounder Records, Cambridge, MA.

Verhagen, S., van Wel, F, ter Bogt, T. and Hibbel, B. (2000) Fast on 200 beats per minute. The youth culture of gabbers in the Netherlands, *Youth and Society*, 32, 147–64.

Virolle, M. (1995) *La chanson rai*, Karthala, Paris.

Virolle-Souibès, M. (1989) Le Rai entre résistance et récupération, *Revue d'études du monde musulman et méditerranéen*, 51, 47–62.

Wachsmann, K. (1953) Musicology in Uganda, *Journal of the Royal Anthropological Institute*, 83, 50–7.

Wade, L. (1994) New Orleans' Bourbon Street: the evolution of an entertainment district, in Browne, R.B. and Marsden, M.T., eds, *The Cultures of Celebrations*, Bowling Green State University Popular Press, Bowling Green, 181–201.

Wade, P. (2000) *Music, Race, and Nation: Música Tropical in Colombia*, University of Chicago Press, Chicago.

Walker, C. (1996) *Stranded: The Secret History of Australian Independent Music 1977–1991*, Macmillan, Sydney.

—— (2000) *Buried Country: The Story of Aboriginal Country Music*, Pluto, Sydney.

Wall, M. (2000) The popular and geography. Music and racialized identities in Aotearoa, New Zealand, in I. Cook, D. Crouch, S. Naylor and J. Ryan, *Cultural Turns/Geographical Turns*, Prentice Hall, Harlow, 75–87.

Wallis, R. and Malm, K. (1984) *Big Sounds from Small Peoples*, Pendragon Press, New York.

Walser, R. (1993) *Running with the Devil: Power, Gender and Madness in Heavy Metal Music*, Wesleyan University Press, Hanover.

—— (1994) Prince as queer poststructuralist, *Popular Music and Society*, 18, 79–89.

—— (1995) Rhythm, rhyme and rhetoric in the music of Public Enemy, *Ethnomusicology*, 39, 193–217.

Wang, N. (1999) Rethinking authenticity in tourism experience, *Annals of Tourism Research*, 26(2), 349–70.

Ward, C. (1992) Anarchy in Milton Keynes, *The Raven*, 18, 116–31.

Wark, M. (1994) *Virtual Geography: Living with Global Media Events*, Indiana University Press, Bloomington.

Warne, C. (1997) The impact of world music in France, in A. Hargreaves and M. McKinney, eds, *Post-Colonial Cultures in France*, Routledge, London, 133–49.

Waterman, C. (1990a) Our tradition is a very modern tradition: popular music and the construction of a pan-Yoruba identity, *Ethnomusicology*, 34, 367–79.

—— (1990b) *Juju: A Social History and Ethnography of an African Popular Music*, University of Chicago Press, Chicago.

Waterman, S. (1998) Place, culture and identity: summer music in Upper Galilee, *Transactions of the Institute of British Geographers*, 23, 253–67.

Watkins, L. (2001) 'Simunye, we are not one': ethnicity, difference and the hip-hoppers of Cape Town, *Race and Class*, 43, 29–44.

Watson, I. (1983) *Song and Democratic Culture in Britain*, Croom Helm, London.

Waxer, L. (2001) Record grooves and salsa dance moves: the Viejoteca phenomenon in Cali, Colombia, *Popular Music*, 20, 61–81.

Webb, M. (1993) *Lokal Musik: Lingua Franca Song and Identity in Papua New Guinea*, National Research Institute, Port Moresby.

Weber, T. (1999) Raving in Toronto: peace, love, unity and respect in transition, *Journal of Youth Studies*, 2, 317–36.

Webster, D. (1988) *Looka Yonder! The Imaginary America of Populist Culture*, Routledge, London.

Wee, C.J.W.-L. (1996) Staging the New Asia: Singapore's Dick Lee, pop music and a counter-modernity, *Public Culture*, 8, 489–510.

Weiner, J.F. (1991) *The Empty Place, Poetry, Space and Being amongst the Foi of Papua New Guinea*, Indiana University Press, Bloomington.

Weintraub, A.N. (1993) Jawaiian and local cultural identity in Hawai'i, *Perfect Beat*, 1(2), 78–89.

Westerhausen, K. (2002) Western travellers in Asia: always one step ahead?, in *Current Issues in Tourism Research* [in press].

Wheeler, M. (1994) Cruising USA, *Planet Talk* (Lonely Planet), 19, 3.

Wheeller, B. (1996) No particular place to go: travel, tourism and popular music, a mid-life crisis perspective, in M. Robinson, N. Evans and P. Callaghan, eds, *Tourism and Culture towards the 21st Century*, Centre for Travel and Tourism and Business Education Publishers, Sunderland, 333–40.

White, B. and Day, F. (1997) Country Music Radio and American Culture Regions, *Journal of Cultural Geography*, 16, 21–35.

Wightman, R. (1971) Foreword, in B. Copper, *A Song for Every Season*, Heinemann, London, xi–xiii.

Wilson, H. (1996) Papua New Guinea and the South-West Pacific, in S. Cunningham and E. Jacka, *Australian Television and International Mediascapes*, Cambridge University Press, Cambridge.

Wilson, M. (1986) *Dreamgirl: My Life as a Supreme*, New York, St Martin's Press.

Wilson, P. (1995) Mountains of contradiction: gender, class and region in the star image of Dolly Parton, *South Atlantic Quarterly*, 94, 109–34.

Winders, J.A. (1983) Reggae, Rastafarians and revolution: rock music in the Third World, *Journal of Popular Culture*, 17, 61–73.

Wise, S. (1990) Sexing Elvis, in S. Frith and A. Goodwin, eds, *On Record*, Routledge, London, 390–8.

Wolf, E. (1982) *Europe and the People without History*, University of California Press, Berkeley.

Wong, K. (1993) Asia's beaming new faces, *Rolling Stone*, 483, May, 38–9, 94.

Woods, L. and Gritzner, C.F. (1990) A million miles to the city: country music's sacred and profane images of place, in L. Zonn, ed., *Place Images in Media*, Rowman & Littlefield, Savage, MD, 231–54.

Wright, M.A. (1993) The rave scene in Britain: a metaphor for metanoia, unpublished dissertation, Centre for Human Ecology, University of Edinburgh.

Yano, C. (1997) Inventing selves: images and image-making in a Japanese popular music genre, *Journal of Popular Culture*, 31(2), 113–27.

Yates, M. (1975) Liner notes, *Packie Manus Byrne, Songs of a Donegal Man*, Topic Records, London.

Yeung, H. (1998) Capital, state and space: contesting the borderless world, *Transactions of the Institute of British Geographers*, 23, 291–309.

Young, R. (2001) Desolation angels, *The Wire*, 203 (January), 28–33.

York, F.A. (1995) Island song and musical growth: toward culturally based school music in the Torres Strait islands, *Research Studies in Music Education*, 4, 28–38.

Zanes, R. (1999) Too much Mead? Under the influence (of participation observation), in K. Dettmar and W. Richey, eds, *Reading Rock and Roll*, Columbia University Press, NewYork, 37–71.

Zemke-White, K. (2001) Rap music and Pacific identity in Aotearoa: popular music and the politics of opposition, in C. Macpherson, P. Spoonley and M. Anae, eds, *Tangata O Te Moana Nui: The Evolving Identities of Pacific Peoples in Aotearoa/New Zealand*, Dunmore Press, Palmerston North, 228–42.

Zhang, J., Harbottle, G., Wang, C. and Kong, Z. (1999) Oldest playable musical instruments found in Jiahu early Neolithic site in China, *Nature*, 401, 366–8.

Zilberg, J. (1995) Yes, it's true: Zimbabweans love Dolly Parton, *Journal of Popular Culture*, 29, 111–25.

Zukin, S. (1991) *Landscapes of Power: From Detroit to DisneyWorld*, University of California Press, Berkeley.

INDEX